God and the Nature of Time

GARRETT J. DEWEESE
*Talbot School of Theology, Biola University,
La Mirada, California, USA*

ASHGATE

© Garrett J. DeWeese 2004

All rights reserved. No part of this publication may be reproduced, stored in a retrieval system or transmitted in any form or by any means, electronic, mechanical, photocopying, recording or otherwise without the prior permission of the publisher.

Garrett J. DeWeese has asserted his moral right under the Copyright, Designs and Patents Act, 1988, to be identified as the author of this work.

Published by
Ashgate Publishing Limited
Gower House
Croft Road
Aldershot
Hampshire GU11 3HR
England

Ashgate Publishing Company
Suite 420
101 Cherry Street
Burlington, VT 05401-4405
USA

Ashgate website: http://www.ashgate.com

British Library Cataloguing in Publication Data
DeWeese, Garrett J.
  God and the nature of time. – (Ashgate philsophy of religion series)
  1. God – Immutability 2. Time – Religious aspects 3. Time – Philosophy
  I. Title
  212.7

Library of Congress Cataloging-in-Publication Data
DeWeese, Garrett J., 1947-
  God and the nature of time / Garrett J. DeWeese.
    p. cm. – (Ashgate philosophy of religion series)
  Includes bibliographical references (p.) and index.
  ISBN 0-7546-3518-X (alk. paper) – ISBN 0-7546-3519-8 (pbk. : alk. paper)
  1. God–Immutability. 2. Time. I. Title. II. Series.

BT153.I47D49 2003
231'.4–dc21
                                                                                2003050034

ISBN 0 7546 3518 X (HBK)
ISBN 0 7546 3519 8 (PBK)

Printed and bound in Great Britain by MPG Books Ltd, Bodmin, Cornwall

To Barbara
whose loving companionship is my joy.

*He who finds a wife finds what is good,
and receives favor from the Lord.*
Proverbs 18:22

Contents

Preface	ix
Acknowledgements	xi

1 Introduction	1
The Nature of the Problem	1
A Survey of Proposed Solutions	2
Constraints on a Solution	5
Summary of the Book	7
Kinds of Time	8

PART I: METAPHYSICS AND PHYSICS

2 The Metaphysics of Time	15
Static versus Dynamic Time	15
Defense of a Dynamic Theory of Time	18
A Causal Theory of Dynamic Time	36
The Topology of Causal Dynamic Time	53
Conclusion	63
3 Physics and Time	65
The Special Theory of Relativity	65
The General Theory of Relativity	75
Quantum Mechanics	84
Conclusion	88

PART II: SCRIPTURE AND TRADITION

4 The Evidence from Scripture	93
Philology: Hebrew and Greek Words for Time and Eternity	93
Hebrew Words	95
Greek Words	101
Exegesis: Central Texts that Speak of Time and Eternity	104
Conclusion	109

5 The Medieval Consensus: God is Atemporally Eternal — 111
    St Augustine — 111
    Boethius — 134
    St Anselm — 145
    St Thomas Aquinas — 151
    Conclusion — 157

6 Atemporality: Contemporary Statements — 159
    Eternal-Temporal Simultaneity: Eleonore Stump and Norman Kretzmann — 160
    Eternity as a Reference Frame: Brian Leftow — 167
    The 'Classical View' Revisited: Paul Helm — 175
    Atemporality and Static Time — 179
    Conclusion — 184

7 A Medieval Dissent: God is Temporally Everlasting — 185
    John Duns Scotus — 186
    William of Ockham — 194
    Later Medieval Developments — 200
    Luis de Molina — 201
    Conclusion — 207

8 Temporality: Contemporary Statements — 209
    The Argument from Personality — 209
    The Argument from Dynamic Time — 217
    The Argument from the Incarnation — 232
    Conclusion — 235

PART III: OMNITEMPORAL GOD

9 Omnitemporality — 239
    Omnitemporality: An Informal Account — 239
    Analysis of Key Concepts — 243
    Conclusion — 255

10 Implications of Omnitemporality — 257
    What Cannot Be Affirmed — 258
    What is Problematic: God's Knowledge of Future Contingents — 262
    What Can Be Asserted — 264
    Conclusion — 276

Bibliography — 277
Index — 291

Preface

Anyone who thinks deeply about the fundamental nature of reality should be moved to awe and humility; anyone who thinks deeply about the nature of God should be moved to worship. But awe, humility and worship are not attitudes normally perceived in analytic philosophy.

Still, even as a boy watching the hands creep around the clock, or the digital hundredths of a second blurring on the stopwatch, I was in awe that something was slipping away forever, being replaced by something that had never been before. I wondered what time would look like to God. Dimly aware of what I later learned to call a past light cone, I supposed that God would see time *sub specie aeternitatis* as a three-dimensional cartoon of everything I had ever done. If that affected my behavior, I don't recall. Only later did the mathematical formalisms of a four-dimensional Minkowski manifold bring thought that God might also see my future light cone, and I felt a fatalistic chill.

In church we sang the words of Isaac Watts' hymn, 'O God, Our Help In Ages Past':

> Time, like an ever rolling stream
> Bears all her sons away;
> They fly, forgotten, as a dream
> Dies at the op'ning day.

Sometime later I read T. S. Eliot's very different version of time in 'The Four Quartets':

> Time present and time past
> Are both perhaps present in time future,
> And time future contained in time past.
> If all time is eternally present
> All time is unredeemable.[1]

My growing theological understanding brought further tensions. My moderate Calvinism easily accommodated a God who was 'outside' of time, the Great Cartoonist who had already sketched my life in full, and now was (as I lived it) coloring it in panel by panel. On the other hand, the emphasis on petitionary prayer in my evangelical tradition, and several years teaching the Old Testament with its rich stories of God's immanent involvement with the lives of his people, comported better with libertarian freedom and a God who was himself temporal.

---

[1] T. S. Eliot, 'The Four Quartets: Burnt Norton,' in *The Complete Poems and Plays 1909-1950* (New York: Harcourt, Brace and World, 1971), p. 117.

Now, not every conceptual difficulty need be solved before a person accepts a worldview such as the Christian faith, but (*pace* Tertullian, the Red Queen, and Zen Buddhism with its koans) I find it doubtful that a rational person can knowingly believe a contradiction. And it was becoming clear that I faced a contradiction in beliefs I held about the nature of time, and also about God's relation to time.

This book is the result of my attempts to understand and resolve those conflicts. A number of technical works on the nature of time have appeared in the past several years; I trust there is enough technical argument here to satisfy philosophers of time that I'm operating with a coherent and defensible view of the nature of time, but I make no claim to great originality. Similarly, several books on the topic of God's relation to time have appeared recently, some of which (the works of Alan Padgett and most notably William Lane Craig, references to which appear throughout this book) offer a theory of time as well as of God's temporal nature. Again, in offering my theory, I make no claim to great originality.

Why then add another book to the already long shelf? Because I have become convinced that nothing else in print attempts the synoptic view that I offer here. Metaphysics and physics, Scripture and the philosophical/theological tradition, contemporary philosophers and theologians—all are addressed, some briefly but all (I trust) fairly.

Finally, if the attitudes of awe, humility and worship appropriate to these topics seem not to radiate through the terse style of analytic philosophy, then let me urge you to stop every so often and search for them in yourself, lest my heap of words detract from the very motives which gave them birth.

Acknowledgements

I am acutely aware of debts, intellectual and personal, which I have accumulated in the progress of this study. In addition to intellectual debts apparent from the footnotes, I'm happy to acknowledge the following.

To many teachers too numerous to mention at the United States Air Force Academy, Dallas Theological Seminary and the University of Colorado, who sharpened my scholarship; to colleagues who prodded and probed my concepts; and to students who invariably asked the hard questions, I owe an aversion to facile and lazy thought.

To Ed. L. Miller (my *Doktorvater*), Michael Tooley, Wes Morriston, Luc Bovens, Bill Craig, JP Moreland, Dave Horner and Greg Ganssle, I owe thanks for many helpful comments on different parts of this book, and for holding me to commitment to first-rate thinking. I could wish they had caught all my errors, but those that remain are mine; they are absolved.

To Talbot School of Theology, Biola University, and especially Dennis Dirks, Dean, Michael Wilkins, Dean of the Faculty and R. Dougls Geivett, Chair of the Department of Philosophy of Religion and Ethics, I owe hearty gratitude for a generous policy which made possible a study leave, Fall Semester 2002, during which this work was completed.

To Joe Gorra, research assistant without peer, I owe appreciation for time freed up to work on content by his meticulous work on references and indexing.

To the editors at Oxford University Press and Philosophia Christi, I owe thanks for permission to adapt, respectively, in Chapter 9, 'Atemporal, Sempiternal, or Omnitemporal: God's Temporal Mode of Being,' in *God and Time: Essays on the Divine Nature*, ed. Gregory E. Ganssle and David M. Woodruff (New York: Oxford, 2002), and in Chapter 6, 'Timeless God, Tenseless Time,' (*Philosophia Christi* 2:2:1 [2000], pp. 53-9).

On the personal level: to my parents, I owe a love of learning and the desire to pursue truth, and gratitude for showing me that the Christian faith is relevant to daily life. They didn't realize that they were raising a philosopher and theologian!

To the congregation of Bethany Baptist Church, Boulder, Colorado, who first freed me from certain pastoral duties so I could pursue doctoral studies, and to a number of faithful friends whose financial contributions to Dayspring Center for Christian Studies, Boulder, partially underwrote my studies, I owe a return on dollars invested (wisely, I trust).

To my children (and their spouses), Geoff (and Melody), Gini, and Greg (and Laile), I owe enduring gratitude for many delightful conversations, warm family moments, and mature patience. This father's greatest joy is the friendship of his adult children.

And finally to my wife, Barbara, whose love and encouragement never failed, whose belief that I really could do this might not have been irrational after all, and whose loving companionship in faith and life is a priceless treasure, I owe the most. To her this is dedicated.

Chapter 1

Introduction

What then is time? If no one asks me, I know. If someone asks me to explain, I'm baffled.
                                                                St Augustine[1]

If the nature of time is perplexing in itself, as Augustine's famous lament attests, the perplexity increases for the theist who must explain not only the nature of time but also God's relation to time, or, more fundamentally, God's temporal mode of being. Historically, it has been the case theologians—both exegetical and philosophical—systematized their treatments of God, time and eternity with more concern for concepts such as divine simplicity and divine immutability than for the nature of time. However in the last several decades, paralleling a surge of interest in the metaphysics of time, significant studies in philosophical theology have dealt with the question of God's relation to time. Characterized by respect for, and interaction with, the philosophical theological tradition, these recent studies also attempt to come to grips with the very difficult issues that distinguish contemporary philosophy of time as well as issues arising from the treatment of time in the physical sciences. We are about to join this conversation.

The Nature of the Problem

On the one hand, it would seem that if God created time, he himself must somehow transcend time. On the other hand, it would seem impossible for a God outside of time to interact with his creation at moments of time. This, in a nutshell, is the problem. But many other issues in philosophical theology turn on the view one takes on God's temporal mode of being and his relation to temporal creation. A satisfactory solution to the dilemma of divine foreknowledge and human freedom might well turn on this relation. But so might adequate treatments of topics such as creation, providence, omniscience, immutability, and divine simplicity. And God's relation to time certainly figures in discussions of petitionary prayer as well as attempts to explicate the Incarnation.

---

[1] *Quid est ergo tempus? Se nemo ex me quaerat, scio; si quaerenti explicare velim, nescio. Confessions*, XI.14.

A Survey of Proposed Solutions

Generally, solutions to the question of God's relation to the temporal order must choose among two pairs of concepts. God may exist either atemporally or temporally, and time may be either static or dynamic. While the details of these solutions will be presented in Chapters 5 through 8, a brief survey at this point will be helpful.

God is Atemporal

From the fourth through the fourteenth centuries, with few exceptions, God was regarded as being outside of time.[2] Some philosophers and theologians understood God's temporal mode of being as timeless duration (i.e., duration without temporal succession); others took it as simply atemporal (i.e., something to which temporal concepts cannot be applied). I shall refer to both as divine atemporality. For most of the medieval thinkers, as we shall see, divine was grounded in two theses rooted in an acceptance of a certain metaphysical point of view. The metaphysics that informed early medieval philosophy and theology was that of Neoplatonism.[3] In the Neoplatonic tradition the realm of Being, the realm of the Forms, was eternal and immutable, while the realm of Becoming was temporal and mutable. The first-century Jewish theologian Philo of Alexandria (13 BC-AD 50?) saw in Plato a philosophical ally, while the mystical view of the One, developed by Plotinus (205-270), was virtually identified with the God of the Bible by early Christian theologians such as Origen (185-254).[4] With Augustine (354-430) the influence of Plotinus is felt directly in his view of time, and Neoplatonist metaphysics becomes incorporated into the orthodox philosophical theological tradition.[5]

The acceptance of Neoplatonism brought with it certain metaphysical commitments that, for the next millennium, largely dominated philosophical

---

[2] For the most part I will avoid using the locutions 'in time' or 'outside of time.' The reason is that such locutions seem to spatialize (or at least reify) time, leading to the inference that time is *necessarily* something distinct from God's being. But that is to beg the question. It could turn out that the analysis of time shows that any concrete entity 'experiences' successive states in its being; or an analysis of life shows that any living being (divine, human, angelic, Martian, . . .) experiences life as at least a succession of internal mental events. In either case, temporality would be an essential property of God, and in neither case would it be proper to conceive of time as a realm 'within which' God was imprisoned. So I prefer to say that God is either temporal or atemporal in his being, rather than that he is in or outside of time.

[3] William Kneale, 'Time and Eternity in Theology,' *Proceedings of the Aristotelian Society* 61 (1960-61): 87-108; Richard Swinburne, *The Coherence of Theism*, 2nd ed. (Oxford: Oxford University Press, 1993), pp. 223-7.

[4] More precisely, Plotinus's One is 'above' anything eternal, while his *Noûs*, 'below' the One, is eternal. Augustine and later Christian writers conflated the One and the *Noûs* in their concepts of God. See further discussion of Neoplatonism in Chapter 4.

[5] Richard Sorabji, *Time, Creation and the Continuum: Theories in Antiquity and the Early Middle Ages* (Ithaca: Cornell University Press, 1983), pp. 165-72.

thinking about God and time. Two theses which flowed from Neoplatonism and were accepted virtually as axiomatic were (i) the simplicity of God's being, and (ii) compatibilism regarding God's foreknowledge and human freedom. The unity and pure, unmixed nature of Plato's Forms and Plotinus' One translated into the doctrine of God's simplicity. A metaphysically simple being is one of which its intrinsic attributes are identical with that entity's individual essence, so it has no parts. So if God is simple, then for every intrinsic attribute F, if God is F, then God is identical to F-ness. Since simplicity entails immutability, and immutability entails atemporality, a simple being such as God is atemporal.[6] And beginning with Boethius (480-524) the doctrine of God's eternity served a crucial role in efforts to explain the compatibility of divine foreknowledge and human freedom.

Both the doctrine of simplicity and the problem of foreknowledge and freedom are significant for many contemporary supporters of divine atemporality as well. It seems clear enough that divine simplicity leads to divine timelessness, and timelessness seems to be an attractive way to resolve the foreknowledge/freedom dilemma by placing God's foreknowledge outside of time altogether, rather than temporally prior to putatively free actions.

Thus, for both the majority of medieval philosophical theologians, as well as many contemporary thinkers, divine atemporality is the preferred solution.

God is Temporal

In the late medieval philosophers, John Duns Scotus (1266?-1308) and William of Ockham (1285-1347), we see the first possibility of conceiving God's being as temporal. In Duns Scotus' doctrine of contingency we have the first arguments for dynamic time, and in Ockham's proposed resolution to the foreknowledge/freedom dilemma we may discern the first medieval suggestion that God is temporal—that is, while unchanging, God endures through time. And although he himself was a proponent of divine atemporality, the sixteenth-century Spanish Jesuit Luis de Molina makes a significant contribution to contemporary accounts with his doctrine of Middle Knowledge.

Many contemporary philosophers have concluded that proposed explanations of how a timeless God could stand in real relations with temporal things are either inadequate or incoherent. Further, the doctrine of divine simplicity seems to have lost its appeal as the Neoplatonic metaphysical background has faded, and divine immutability has been interpreted in a weak rather than a strong sense. These results, together with metaphysical arguments supporting dynamic time, have spawned several variations of the view that, rather

---

[6] Both St Anselm in *Monologion* 21 (in *Saint Anselm: Basic Writings*, tr. S. W. Deane, 2nd ed. [La Salle, IL: Open Court, 1962]), and St Thomas Aquinas in *Summa Theologica* Ia (tr. The Fathers of the English Dominican Province [New York: Benzinger Brothers, 1948] hereafter *ST*), work diligently on the logic. For example, Aquinas offers three arguments for immutability, two of which are based on simplicity (*ST* Ia, 9, 1). He then offers an argument that 'the notion of eternity follows from immutability' (*ST* Ia, 10, 2). More will be said about this later.

than being timeless, God does in fact experience temporal succession, both internally (in his own mental states) and externally (in his relation to creation) and thus is properly termed temporal. Indeed, the majority of contemporary philosophers and theologians seem to be on the side of divine temporality.

Time is Static

In spite of some vague similarities with Parmedian stasis and Heraclitean flux, the distinction between the static and the dynamic views of time is a relatively recent one. To claim that time is static is to claim that all moments of time coexist, are equally real; that there is no ontological difference between past, present and future. According to this view, the common notion of temporal passage is a myth. Where we think pre-theoretically of ourselves as approaching possible future events that become actual in the present and then recede ever further into the past, we should realize upon reflection that such a view is reflective only of the phenomenology of our psychology. In the language of the Theory of Relativity, all points in four-dimensional space-time coexist.

We will see that the views of many of the medieval atemporalists entail a static view of time. Static time was not clearly understood as such by many of the medievals, I believe; nevertheless, it is the only view of time that comports with their views concerning God's nature and foreknowledge. The medieval philosophical theologians worked from doctrinal considerations, and their view of time came along as part of the bargain, relatively unexamined.

On the other hand, a static view of time has become quite popular among metaphysicians in this century, largely due to (i) difficulties arising from McTaggart's Paradox and from linguistic puzzles about time, and (ii) the successes of modern physical theories, notably the Special Theory of Relativity, and certain philosophical interpretations of these theories. Consequently some contemporary philosophical theologians begin with a static theory of time, and then proceed to develop philosophical theological views that cohere with such a theory.

Time is Dynamic

According to the dynamic theory of time, the present (and the past) have a quite different ontological status than the future. The present (and the past) are (or has been) real, while the future is not yet real. It follows that temporal becoming is a real feature of the world, not merely a psychological one.

A dynamic theory of time is not without difficulties, however; in addition to certain philosophical puzzles, the standard interpretations of the Special and General Theories of Relativity, as well as certain aspects of Quantum Mechanics, seem to present serious problems to any theory of dynamic time. These problems must be faced, since a metaphysical theory that entails the denial of well-confirmed experimental results predicted by contemporary scientific theories would be highly suspect.

Constraints on a Solution

I believe that the search for a solution to God's temporal mode of being and his relation to temporal creation should be shaped by certain constraints. I shall not offer arguments for the following constraints, but only indicate the role they play in guiding the dialectic of this project.

Biblical Exegesis

This is a study in the Judeo-Christian tradition of philosophical theology. My perspective is that of St Anselm who articulated the principle of *fides quarens intellectum*, or 'faith seeking understanding.'[7] In this tradition, the doctrines derived from God's special revelation in Scripture are examined, clarified, evaluated, and explained using the tools of philosophical analysis. The result of this approach might indeed be, in the end, the rejection of a traditional formulation of a doctrine as incoherent or implausible, and a consequent re-evaluation of the proper interpretation of the biblical texts.[8] But it is Scripture that constitutes the core of the tradition, so the exegesis of scriptural material regarding God and time must be examined.

Philosophical Analysis

As the medieval philosopher theologians maintained, 'Theology is the queen of the sciences, and philosophy is her handmaid.' Although the queen did not even get out of bed without the help of the handmaid, there was no question as to who held the authority. Much has been written about the relationship between faith and reason,[9] and this is not the place to add to that literature. Still, I should explain briefly how I see philosophical analysis serving as a constraint. First, philosophy plays a crucial role in clarifying theological concepts. For example, while theologians maintain that God is omnipotent, it remains for philosophers to clarify the notion by exploring whether omnipotence entails the ability to do the logically impossible, and the relation between the essential moral nature of God and his ability to do moral evil. Second, philosophical analysis applies the rigor of logic to theological claims. Contra the tradition associated with Tertullian and

---

[7] Anselm, Preface to *Proslogion*, in *Saint Anselm: Basic Writings*.
[8] See, for example, Ed. L. Miller, *God and Reason*, 2nd ed. (Englewood Cliffs, NJ: Prentice Hall, 1995), pp. 134-7.
[9] For a recent survey, see Paul Helm, *Faith and Understanding* (Grand Rapids, MI: Eerdmans, 1997). The Jewish and Muslim traditions also struggled with the issue of faith and reason. Tamar Rudavsky introduces her study of medieval Jewish concepts of time and space by noting, 'These issues comprise the intellectual agenda of medieval Jewish philosophers, as well as their Christian and Muslim peers, who must construct a philosophical theology that is sensitive to religious constraints while at the same time incorporating compelling elements of (the current) science and philosophy.' T. M. Rudavsky, *Time Matters: Time, Creation, and Cosmology in Medieval Jewish Philosophy* (Albany, NY: State University of New York Press, 2000), p. xi.

Kierkegaard, I hold that it is not rational to (attempt to) believe a logical contradiction,[10] and a rational God would not desire irrational beliefs in his creatures. Third, philosophy is the second-order discipline that does the integrative work of assembling and adjudicating conflicts between biblical exegesis and other disciplines. So any solution to the problem of God and time must be philosophically satisfying in the end.

The Weight of the Historical Theological Tradition

A tradition of philosophical theology provides a certain kind of constraint on developments as the tradition evolves. In the sense in which I understand this constraint, a theological tradition plays a role analogous to that of a scientific research program as understood in contemporary philosophy of science.[11] In addition, the medieval philosophers and theologians exhibit a depth and sophistication in dealing with the issues at hand that we ignore to our own detriment.[12]

Pre-Philosophical Intuitions about the Dynamic Nature of Time

Pre-philosophical intuitions can, of course, be mistaken; but any theory which concludes that such intuitions, especially ones highly resistant to modification, are nevertheless mistaken should at least attempt to account for the resilience of the false intuitions. If a theory can be developed which coheres with deep intuitions, that theory, other things being equal, is to be preferred. In this study, common intuitions concerning the reality of temporal passage and the unreality of the future are regarded as highly resistant to change, and in need of explanation.

Theories of Modern Physics

Contrary to some philosophers who conceive of metaphysics as secondary to the empirical sciences, or perhaps as rendered obsolete by them, I believe that metaphysics is an independent enterprise. But independent does not mean isolated, and the deliverances of the empirical science constitute some of the evidence that

---

[10] Some might believe this is unfair to either Tertullian or Kierkegaard. Be that as it may, there is a tradition in Christianity (and in Judaism and Islam, as well as in Eastern religions) that holds that logic cannot be applied to God or to religious claims. But this is hardly a coherent claim. How could God be 'beyond' the Law of Identity? If the Law of Non-Contradiction doesn't apply to him, then—surprise—it does at the same time! This is not to say that logic somehow is 'sovereign' over God, but that logic itself is a feature of being.
[11] This understanding of the constraints of an historical tradition is explored by Nancey Murphy, *Theology in the Age of Scientific Reasoning* (Ithaca, NY: Cornell University Press, 1990), and by Michael C. Banner, *The Justification of Science and the Rationality of Religious Belief* (Oxford: Oxford University Press, 1990); see also F. F. Bruce, *Tradition: Old and New* (Grand Rapids, MI: Zondervan, 1970).
[12] An argument strongly made by Alfred J. Freddoso, 'The "Openness" of God: A Reply to William Hasker,' *Christian Scholars Review*, 28 (1998), pp. 124-33.

metaphysical theories must explain. Conversely, since data will always underdetermine theory, purely metaphysical considerations should be brought to bear in the process of theory adjudication. A successful theory will result from respectful interaction between the scientist and the philosopher. Thus, a constraint on the search for a solution to the nature of time must be consistency with acceptable interpretations of the theories of modern physics, notably the Special and General Theories of Relativity and Quantum Mechanics.

Greater Explanatory Power

Greater explanatory power is one of the virtues by which competing theories have generally been judged. So a successful theory of the nature of time will explain more with respect to our daily experience of the world, including empirical science. And a successful account of God's temporal mode of being should point towards solutions for other associated problems in philosophical theology, and the best explanation will offer more and/or better answers than its competitors.

Summary of the Book

Part I deals with the nature of time from the perspectives of metaphysics and physics. I shall begin in Chapter 2 with contemporary metaphysical theories of time and argue that the best available theory is one in which time is seen as dynamic, where the past and present are real but the future is not. This contrasts with the static theory of time, and I will argue that the static theory gives an inferior account of common linguistic practice, of our common experience of time, and of causation. This leads to a causal theory of time, one that explains both the flow and the direction of time in terms of causation.

Next, in Chapter 3 we will take up issues arising from the treatment of time in modern physics. I must show that the theory of time defended in Chapter 2 is in fact immune from problems arising from the three most significant theories of modern physics, Special and General Relativity (STR and GTR respectively) and Quantum Mechanics (QM). The chapter's immersion in contemporary physics is necessary in order to show that a causal theory of dynamic time fares at least as well as—indeed, I shall argue that it fares better than—the static theory in giving acceptable interpretations of physical theory. This chapter completes Part I.

Part II turns to matters explicitly concerned with philosophical theology. Chapter 4 begins an historical survey of the theological tradition by investigating the conceptual framework of the documents formative of the Judeo-Christian tradition, the Old and New Testaments. I shall argue that biblical exegesis cannot decide the issue, because the Hebrew and Greek Scriptures allow either for a temporalist or for an atemporalist view of God's being.

Chapters 5 and 6 continue the historical survey with an eye on the development of the theological tradition. Chapter 5 examines the traditional understanding that God is timelessly eternal by summarizing the views of philosophers and theologians from St Augustine through St Thomas Aquinas. The survey will demonstrate that while the atemporal view of God's being predominates the tradition, it is likely that the atemporalist view would not have gained traditional status had these thinkers not adopted a Neoplatonist metaphysics which entailed the doctrine of divine simplicity, and with it a particular doctrine of strong immutability which seems to have led to a tacit commitment to a static theory of time. In Chapter 6 we will consider the arguments of contemporary defenders of divine atemporality and find that telling objections may be brought against all. I will close Chapter 6 with a general argument to the effect that a timeless view of God entails a tenseless view of time. If, however, we have good independent reasons to reject tenseless theories of time, then we must also reject timeless theories of God's being.

While the atemporalist view was predominant in the medieval period, it never reached the status of orthodoxy, any denial of which would constitute heresy. In Chapter 7 we will look at John Duns Scotus, William Ockham, and Luis de Molina, medieval philosophers in whose work the beginnings of a doctrine of divine temporality can be discerned. Arguments of contemporary exponents of the temporalist view will then come into view in Chapter 8. I shall conclude that there are indeed very good reasons to prefer the view that God is temporal.

Part III builds on the metaphysical theory of time defended in Part I and the conclusions reached in Part II. Chapter 9 develops in much greater detail the concepts of temporality, atemporality, sempiternity, and omnitemporality. While this chapter is somewhat heavy going philosophically, I hope to show that the concept of omnitemporality provides the most plausible explanation of God's temporal mode of being.

In conclusion, Chapter 10 explores the explanatory power of an omnitemporal concept of God. We will ask what features of traditional theology can no longer be affirmed on the proposed theory, and I will claim that the two clear losses—the doctrines of divine simplicity and strong immutability—are only losses of certain consequences of Neoplatonist metaphysics. The question of God's foreknowledge of future contingents is also an important one, but since a thorough consideration of the question is beyond the scope of this book, I will only sketch possible solutions. We will then briefly explore the implications of the theory of omnitemporality in three relevant areas of philosophical theology: creation, petitionary prayer, and providence. I conclude that omnitemporality offers better explanations in these areas than any other theory of God's temporal mode of being, and so is the preferable theory of God and time.

Kinds of Time

Before plunging headlong into the study, it would be wise to pause for a moment and consider what we refer to when we use the word 'time.' We should not assume that every occurrence of the word refers to the same concept. Consequently we must at the outset recognize four distinct conceptions of time that, while common in the technical philosophical literature on the subject, are not generally made in either theological or popular discussions.

Physical (Clock) Time

Physical time (sometimes called 'measured' or 'clock' time) refers to the time most familiar to us as we go about our daily lives, and is the time that figures in physical theories. At the very least, what we understand by time is the succession of states of affairs, both in the physical world around us, and in our own interior mental lives.

'Physical' or 'clock time' refers to time in any temporal world where laws of nature allow for the measurement of time with some physical clock. This is possible only if the laws of nature in that world allow for regular physical processes that form the basis of the operation of the clock. As Whitrow says,

> Greater accuracy in the measurement of time can be obtained by means of atomic and molecular clocks. Implicit in these developments is the hypothesis that all atoms of a given element behave in exactly the same way, irrespective of place and epoch. The ultimate scale of time is therefore a theoretical concomitant of *our concept of universal laws of nature*.[13]

The metric of time in this sense would depend on a clock appropriate to the laws of nature of that particular temporal world.[14] Since it is demonstrable that the measurement of time by a particular clock in the actual universe is affected by acceleration and gravity,[15] Physical time thus is relative to a local reference frame. So it would be proper to speak of 'local time' to underscore the fact that any measurement of time will necessarily depend upon local reference frames, and that observations of the same temporal processes will not necessarily be equivalent from one local reference frame to another.

But this raises a controversial question. How does the measurement of time relate to the nature of time? Stated differently, is time an operationalist concept? The answer to this question will play a significant role in the discussion of STR and GTR in Chapter 3. For the most part, however, in the theoretical development to follow, these complications can be overlooked, and I will generally speak of physical time as if it were in fact universal in the sense described in the next section.

---

[13] G. J. Whitrow, *The Natural Philosophy of Time*, 2nd ed. (Oxford: Clarendon Press, 1980), p. 43, my emphasis.
[14] While the numbers assigned to temporal measurements depend upon the physical clocks used in the measurement, the existence of quantitative temporal intervals is necessary as truth-makers for statements of the laws of nature in that particular world. See Michael Tooley, *Time, Tense and Causation* (Oxford: Oxford University Press, 1997), pp. 274-82.
[15] The time dilation predicted in Einstein's theory of relativity was first experimentally confirmed in 1941, and has been reconfirmed numerous times. Paul Davies, *About Time: Einstein's Unfinished Revolution* (New York: Simon and Schuster, 1995), pp. 55-8, 81-3.

In this study, references to time should be understood to mean physical time; where this is not the case, or where it is especially important to avoid possible confusion, I will be explicit as to which kind of time is under consideration.

Cosmic (Universal) Time

Cosmic time, if it exists, would be the standard of time by which all events in the universe could be located and measured. On a standard Einsteinian interpretation of STR the relation of simultaneity is relative to a reference frame, and since there is no privileged reference frame in the universe, there is no cosmic time. However, it can be argued that there is indeed a privileged reference frame, that one in which the universe is expanding isotropically from its initial singularity. Alternatively, in the theoretical framework of GTR, cosmic time would be proper time as measured along a time-like geodesic normal to the space-like hypersurface that represents the present. I will suggest in Chapter 3 that there is in fact some observational support for such a view, and that this feature of our universe may serve to provide the frame of reference for defining absolute simultaneity and, with it, cosmic time.

Personal (Psychological) Time

Personal time refers to the individual conscious experience of the passage of time. A metric of personal time cannot be global since different people, and indeed the same person at different times, seem to experience the passage of time at different rates. While it is of considerable phenomenological interest that the rate of passage of personal time varies from one person to another, the simple fact remains that however fast or slow the lecture passed, all listeners 'arrive' at the end at the same time. Hence personal time cannot be the basis of universal metaphysical theorizing, and will enter into the discussion only in relation to the question of whether temporal passage is merely psychological or if it is an objective ontological feature of the universe.

Metaphysical Time

Metaphysical time refers to the succession of moments (events) through which concrete objects persist. Since there are possibly concrete objects that are not physical, metaphysical time is not identical to physical time. Thus, metaphysical time would be basic to all other kinds of time. (Spelling out just what these claims mean will be part of the task of Chapter 9.) If it turns out that God experiences succession in his being, then metaphysical time would be equivalent to God's time.

Whether or not such time has a metric is a question that I will leave open, but it is not necessary that the metric of time derived from any temporal world be applicable to metaphysical time. I will argue that the topology of metaphysical time is linear and unidirectional. Further, I will claim that if two temporal series are related, then the same topology applies to both. Hence, since metaphysical time is related to physical time in this world, the topology of physical time is also linear and unidirectional. This is so even if physical time is wholly contained in (or is a proper subset of) metaphysical time.

Having completed the necessary preliminaries, then, let us begin the study.

PART I
METAPHYSICS AND PHYSICS

Chapter 2

The Metaphysics of Time

This century has seen a high level of interest in the metaphysics of time, largely due to two factors. The first factor was certain puzzles about time that followed from Einstein's Special and General Theories of Relativity and from the development of the theory of Quantum Mechanics (STR, GTR, and QM, respectively). As a result, discussions of the metaphysics of time have become highly technical and sophisticated, to the point where the philosophy of time has become a sub-discipline in its own right, a hybrid of philosophy and physics. Any proposed solution to the problem of God's temporal mode of being and his relation to time must take into account contemporary metaphysical and scientific discussions of time, something much theological writing fails to do. If a theory of God and time leads to metaphysical inconsistency or confusion, it cannot commend itself for our acceptance. Hence, we shall seek consistency with metaphysical constraints of a theory of time that is realist with respect to our deep intuitions about such things as the dynamic nature of time, temporal becoming, and the ontological status of the past, present, and future. Further, a successful theory must comport well with modern physics or it will not be taken seriously. Thus, we require consistency with empirical observations and acceptable interpretations of the theories of physics, especially notably STR, GTR and QM. The purpose of this chapter and the next is to summarize the metaphysics and physics of time and to understand the constraints which each imposes on a solution to the problem.

Static versus Dynamic Time

There is little doubt that the general, pre-philosophical view is that time is dynamic. That is, the future does not yet exist, but events, which we may have regarded as (possible, non-existent) future events, become real in the present and then flow away from us, receding into our past. Temporal becoming is a genuine feature of the world and not merely of our psychology. The past and the present are real, but the future is not. And there can be little doubt that, from the days of Aristotle until the present century, we would be hard-pressed to find a well-articulated competing view of time.

With the development of relativity theory, however, a rather new concept of time has emerged, one in which time is joined with the three dimensions of space to form a four-dimensional space-time continuum in which past, present and future events all exist on an ontological par. Time, then, is static, and what had been taken as temporal becoming is simply a psychological phenomenon resulting from displacement along the temporal axis. And there can be little doubt that, given the generally impressive experimental evidence supporting STR, and the more common interpretations of STR, the majority of serious philosophers hold to a static view of time.

We must now articulate more carefully the distinction between dynamic and static time. The clearest way to distinguish the two views is in terms of how they describe change. Roughly, an entity E undergoes change if it has (or lacks) property P at time t_1 and lacks (has) P at t_2, where $t_1 < t_2$.[1] Now a dynamic view of time holds that possibly E has certain accidental properties at t_1 which E lacks at t_2. If at t_2 E has different accidental properties than E had at t_1, then E has undergone change. On a static view of time, the analysis is different. E is an object extended in both the three spatial (x, y, z) dimensions and the one temporal (t) dimension. Thus E can have (or lack) property P at (x, y, z, t_1) of its four-dimensional extension, and lack (have) P at (x, y, z, t_2) of that same extension. For the adherent of the static view of time, this is no more mysterious than claiming that an object can have different properties at different points of its spatial extension. For example, a fire-poker might be hot at one end and cold at another—that is, E at (x_1, y_1, z_1, t_1) has different properties than E at (x_2, y_2, z_2, t_1), exhibiting a 'change' in its spatial extension. Similarly, the same end of the poker can be hot at one time (x_1, y_1, z_1, t_1) and cold at another time (x_1, y_1, z_1, t_2), thus exhibiting a change in its temporal extension. Change, thus, is not something that 'comes to pass,' but merely a matter of change in locations in the space-time continuum.[2] Because the static view of time envisions a four-dimensional continuum within which all events (or states of affairs) exist, it is sometimes referred to as the block model of space-time.

The distinction between the static and the dynamic views of time is a clear difference in ontology. But there is a second kind of distinction that is often made, a linguistic distinction between sentences that describe the world in tensed terms, and sentences that describe the world in tenseless terms. Following J. M. E. McTaggart,[3] the tensed and tenseless views of time are generally designated the A-theory and the B-theory respectively. More precisely, the A-theory of time claims that time is defined by the properties of pastness, presentness, and futurity; while the B-theory claims that time is defined by the relations of earlier than,

---

[1] The properties must be monadic so as to exclude 'Cambridge' change.
[2] The literature on four-dimensionalism versus three-dimensionalism is growing. On the side of a four-dimensional ontology, see Theodore Sider, *Four Dimensionalism: An Ontology of Persistence and Time* (Oxford: Clarendon Press, 2001); or Hud Hudson, *A Materialist Metaphysics of the Human Person* (Ithaca, NY: Cornell University Press, 2001). On the side of a three-dimensional ontology, see Trenton Merricks, *Objects and Persons* (Oxford: Clarendon Press, 2001); Peter van Inwagen, 'Four-Dimensional Objects,' *Noûs* 24 (1990), pp. 245-55.
[3] J. M. E. McTaggart, *The Nature of Existence*, (Cambridge: Cambridge University Press, 1927), vol. II, chapter 33. This chapter has been widely reprinted, e.g. as 'The Unreality of Time,' in *The Philosophy of Time*, ed. Robin Le Poidevin and Murray MacBeath (Oxford: Oxford University Press, 1993), pp. 23-34.

simultaneous with, and later than. For the A-theory, the crucial fact is the relation of a temporal entity to 'now,' whereas for the B-theory, the crucial fact is the date (i.e. individual temporal instant or interval) of the temporal entity.[4] But this raises the question, what is the relation between the static/dynamic distinction and the tenseless/tensed distinction?

It is generally thought that A-theoretic descriptions correspond to a theory of dynamic time, and B-theoretic descriptions correspond to a theory of static time. Proponents of a dynamic view of time often argue that A-determinations are analytically more basic than B-determinations. They claim that there is something about an A-determination that cannot be analyzed in terms of a B-determination without significant loss of meaning. B-theorists, of course, claim the opposite. It is assumed that if it could be proved that either A- or B-determinations were indeed analytically basic, then that fact would argue decisively that either a dynamic or static theory of time was correct. But there are problems with this assumption. It is not at all clear that there is an entailment relation between the ontology of a particular theory of time and the form that sentences of that theory must take. Thus, even if it could be shown that one class of determinations could be reduced to the other, it is not clear that anything of ontological significance regarding a theory of time would follow. For such a reduction would be merely a semantic reduction, and it is quite problematic to assert that semantic reduction effects an envisioned ontological reduction.[5] Indeed, Michael Tooley, who holds firmly to a dynamic theory of time, claims that A-determinations can be analyzed in terms of B-determinations plus the idea of being actual at a time, or the idea of truth at a time.[6]

My own view here is that while there are obvious correlations, no entailment relation holds between the semantic and the ontological theories. Our language—at least, the English we speak at the end of the twentieth century—is so rich in vocabulary and offers such syntactical flexibility that it might well turn out that any statement in one theory can be expressed without loss of significant meaning in terms of the other theory.[7] I believe, however, that this is so not primarily because of the flexibility or the imprecision inherent in language, but because ordinary language is at best a flawed guide to metaphysical reality. Alternatively, perhaps better, language shows the rough shape of the territory while

---

[4] Thus, statements made about temporal entities that use A-theoretic notions are called 'A-determinations' (e.g. 'Bonds hit a home run two days ago'—that is, two days before the present) while statements which use B-theoretic notions are called 'B-determinations' (e.g. 'Bonds hit a home run on 9 August 2003').

[5] Terrence E. Horgan, 'Reduction, Reductionism,' in *A Companion to Metaphysics*, ed. Jaegwon Kim and Ernest Sosa (Cambridge, MA: Basil Blackwell, 1995), pp. 438-9.

[6] Michael Tooley, *Time, Tense and Causation* (Oxford: Oxford University Press, 1997), pp. 16-20; also his chapters 6 and 7, pp. 155-212, hereafter *TT&C*.

[7] The receptor theory must admit theoretical statements conveying the idea of truth at a time or the idea of being actual as of a time, or such re-expression will fail.

metaphysics shows the detailed contours. Metaphysics precedes linguistics, not vice versa.[8] (I will say more on this below.)

This survey of the two competing theories of the nature of time provides the necessary background for my defense of the dynamic theory of time.

Defense of a Dynamic Theory of Time

The defense of a philosophical thesis may proceed either positively or negatively; that is, one may offer positive arguments for a thesis, or one may consider arguments against the thesis and show them to be unsuccessful, thus leaving acceptance or rejection of the thesis up to intuitive plausibility, or coherence with other accepted views, or some other criterion. I shall take the first approach, offering positive arguments for a dynamic view of time. But I shall not ignore the negative approach completely. It turns out that arguments for static time cluster in two classes: linguistic arguments (for example, the eliminability of tensed language) and scientific arguments (for example, the relativity of simultaneity under standard interpretations of STR). I shall deal with the first class as responses to my positive arguments in this chapter, and with the second class, by far the more popular, in the following chapter.

Outline of a Dynamic Theory of Time

I cannot offer here a full-blown account of dynamic time.[9] Different dynamic theories have their own idiosyncrasies, of course, and it is beyond the scope of this study to enter into the details of different accounts, let alone offer an alternative. I will, however, trace the outlines of a causal theory of dynamic time. I shall adopt and defend a 'generic' theory that has certain features, subject to qualifications that will emerge in the discussion.

In defending a dynamic theory of time, I shall offer three independent arguments, beginning with what I take to be the weakest of the three. I shall suggest first that common linguistic practice favors adopting a dynamic view of time. Second, I shall argue that we form beliefs about time based on common experience which are A-theoretic in nature, and that the variety and pervasiveness of these beliefs warrants the conclusion that time is indeed dynamic. Third, I shall briefly argue that only a dynamic theory of time can sustain a successful theory of causation.

---

[8] See, for example, the argument by Eli Hirsch, *Dividing Reality* (New York: Oxford University Press, 1993).

[9] Three recent book-length defenses of dynamic time are Quentin Smith, *Language and Time* (Oxford: Oxford University Press, 1993); Tooley, *TT&C*; William Lane Craig, *The Tensed Theory of Time: A Critical Examination* (Dordrecht: Kluwer Academic Publishers, 2000).

> **Features of a Dynamic Causal Theory of Time**
>
> - *Time is constituted by the causal succession of states of affairs.* It is perhaps best, in this context, to think of a state of affairs as a total description of the world at any single time. As that state of affairs causally brings about another, a temporal series is created.
> - *The direction of time is explained by the direction of causation.* The implications of this feature are that simultaneous causation (with the exception of agent causation and backward causation are not possible; I return to this in Chapters 9 and 10).
> - *The flow of time is explained by the operation of causation.* The explanation for one state of affairs succeeding another is causal, and so causation is the explanation for the succession.
> - *The causal relation is analytically prior to the temporal relation.* Causation explains time, not vice versa.

One might wonder why we do not begin with an argument from the asymmetry of the past and the future. Intuitively, the past and the future differ in ontological status. Generally, this difference is cashed out in terms of preventability: I can now prevent certain states of affairs from obtaining in the future, but I cannot now prevent any past state of affairs from obtaining. Events and states of affairs in the past seem to have a property of fixity that events and states of affairs in the future seem not to have. Classically, this fixity has been termed the necessity of the past, or necessity *per accidens*, or accidental necessity. The idea is that while I can bring it about that I do not eat cold cereal for breakfast tomorrow, I cannot bring it about that I do not eat cold cereal for breakfast today, given that I did so four hours ago. Indeed, no one can now bring it about that I did not eat cereal for breakfast four hours ago. Hence, my eating cereal for breakfast four hours ago is not preventable; it is necessary *per accidens* (not logically or metaphysically necessary, for it certainly was possible in both these senses that I did not eat cereal, but once I did eat cereal it became fixed by the necessity of the past).[10]

Technically, the argument from preventability claims that if R is a fact or a state of affairs that obtains at time t, then it is logically impossible for there to exist anyone who could have, at time t, prevented R from obtaining. From this it follows that what facts or states of affairs do obtain is time-dependent, that the past and the future differ in ontological status, and that time is dynamic. Formally,

> R is accidentally necessary at t $=_{df}$ R obtains at t and it is not possible both that R obtains at t and there exists an agent S and an action A such that (i) S has the power to perform A at t or later, and (ii) necessarily, if S were to perform A at t or later, then R would have failed to obtain.[11]

---

[10] The classical position is summarized succinctly in Linda Trinkaus Zagzebski, *The Dilemma of Freedom and Foreknowledge* (New York: Oxford University Press, 1991), pp. 15-23.
[11] This definition could just as easily be formulated in terms of propositions and truth.

Many defenses of dynamic time rely on some form of the argument from preventability, but it seems that no form of the argument is successful.[12] In particular, any argument from preventability will have as a premise the impossibility of backward causation. Most arguments against backward causation rely on showing the logical impossibility of causal loops.[13] Now, it may be possible to show that causal loops are impossible,[14] but for the argument from preventability to go through, the impossibility of causal loops must entail that backward causation is impossible. But Tooley shows that this premise 'cannot be established without appealing to a tensed view of time, thus rendering the argument circular.'[15] So I cannot appeal to the argument from preventability to establish the ontological asymmetry of the past and the future. It will turn out that the intuitive asymmetry in question is indeed real, but it is the result of, and not a premise for, an argument for a causal theory of dynamic time.

Summary of the Arguments

For Static Time:
- Linguistic arguments
  - the translatability of tensed to tenseless sentences ('old tenseless theory of time')
  - the possibility of giving tenseless truth conditions for tensed sentences ('new tenseless theory of time')
- McTaggart's Paradox
- Einsteinian interpretations of Relativity Theory

For Dynamic Time:
- Common linguistic practice
- The experience of time
- The nature of causation

The Argument from Common Linguistic Practice

It is common—indeed, so common as to appear to be required—for philosophers of time to devote a significant portion of their discussion to the nature of language. In particular, discussion focuses on the phenomenon of tensed and tenseless

---

[12] Tooley, *TT&C*, chapter 3, 'Temporally Relative Facts and the Argument from Preventability,' pp. 43-71.
[13] For example, Richard Swinburne, *The Christian God* (Oxford: Clarendon Press, 1994), pp. 82-4.
[14] The standard argument runs something like this: Take the case of a simple causal loop where *A* causes *B*, and *B* in turn causes *A*, and assume, as is generally done, that causation is transitive. In this case we have the state of affairs in which *A* is both the cause of, and the effect of, *B*. Hence, *A* possesses contradictory properties, and so by the Law of Non-Contradiction this state of affairs is logically impossible.
[15] Tooley, *TT&C*, pp. 43, 63-8. The reason is that the logical impossibility of causal loops does not entail the impossibility of causally isolated, oppositely directed causal chains. Further argument is needed to rule out this possibility.

statements, on attempts to reduce tensed to tenseless statements (or vice versa), on attempts to show the ineliminability (or reducibility) of temporal indexicals, and the like.[16] Oaklander and Smith write:

> One way in which philosophers of time have approached these questions [concerning the reality of temporal passage, the nature of temporal relations, etc.] is through a consideration of the language of time. Many of them have argued that if we can determine what we mean or intend to express by the use of temporal language, then we will have an accurate picture of what reality must be like if our words and thoughts about time are to be true. . . . It used to be thought that the question as to whether tensed discourse could be translated into tenseless discourse determined which theory of time is true.[17]

The Argument from Common Linguistic Practice

Although the relationship between language and reality is not straightforward, it is reasonable to think that the way reality is has to some degree determined how our language describing reality operates. So there is a weak argument from common linguistic practice:
1. Common linguistic practice provides prima facie evidence of the nature of objective reality.
2. Common linguistic practice is irreducibly tensed.
   a. Attempts to translate tensed into tenseless language are unsuccessful.
   b. Attempts to give tenseless truth conditions for tensed statements fail to convey significant information contained in the tensed statement.
3. Therefore, irreducibly tensed common linguistic practice provides prima facie evidence that objective reality itself is tensed (i.e. that time is dynamic).

There are two dimensions to this debate. First is the meta-question of whether reality determines language or vice versa, and second is the specific question of whether the translatability (or nontranslatability) of A-sentences into B-sentences, or vice versa, is relevant to determining the truth of a theory of time. I shall take up these questions one at a time.

The meta-question asks about what Willard van Orman Quine called 'semantic ascent,' the shift of attention by philosophers from the things themselves to the words used to describe the things. Is the shift involved in semantic ascent necessary, or justified, or even interesting? While it is clear enough that a language-using community can construct certain realities,[18] I would argue that the

---

[16] The significant—and sometimes dominant—role of language in the philosophy of time may be seen in D. H. Mellor, *Real Time* (Cambridge: Cambridge University Press, 1981); Smith, *Language and Time*; L. Nathan Oaklander and Quentin Smith, eds. *The New Theory of Time* (New Haven, CT: Yale University Press, 1994), in which the 'new theory' of time is directly related to theories of reference in the philosophy of language; Tooley, *TT&C*, especially Part II, 'Semantical Issues,' and Part III, 'Tensed Facts,'; Craig, *Tensed Theory*, pp. 3-130.
[17] Oaklander and Smith, *The New Theory of Time*, p. xiii.
[18] John Searle, *The Construction of Social Reality* (New York: Free Press, 1995), pp. 59-78.

extreme position, which holds that such a shift is necessary, is incorrect. To maintain, as is common in the Continental tradition of postmodernism or deconstructionism, that metaphysical concepts are reducible to forms of linguistic expression, that different narrative structures offer incommensurable ways of understanding reality, is, it seems to me, both self-defeating and unnecessary. It is self-defeating in that the very claim of incommensurability itself is derived in a certain 'narrative' and thus either is not binding upon other narratives, or else can only undercut its own claims.[19] And it is unnecessary, for we can easily admit to the inadequacies of our linguistic structures for giving us a ready-made representation of reality, and still argue that, by the careful and critical use of logical and analytical tools, we can achieve a significant degree of verisimilitude in the description of objective reality. The postmodern view of the nature of language only leads to a radical relativism that denies any (significant) role for metaphysics.[20]

Certainly there is much more that could be said about the Continental tradition in philosophy of time marked by the trajectory from Nietzsche through Husserl and Heidegger to Derrida, but I shall leave it aside. It is not that phenomenology has no important insights into our experience of time,[21] or that deconstructionism has nothing interesting to say about our reading of the history of the philosophy of time. It is, rather, a judgment that the Anglo-American tradition of analytical philosophy is better equipped to develop an understanding of time that is consistent with the best theories of the natural sciences. This analytical approach will be metaphysical in nature, in marked contrast to the decidedly anti-metaphysical nature of the Continental tradition.

Still, the shift involved in 'semantic ascent' might be justified. Language might be the only guide we have to the nature of the world. On the issue of the relation of language and reality, Bertrand Russell argued that we should locate ourselves among that class of philosophers 'who infer properties of the world from properties of language.'[22] The intuition here seems to be that modern logic provides us with the true forms of ordinary language sentences, and that the logical forms are guides to the structure of reality. This view has become known as the 'syntactic priority thesis.'

---

[19] In this regard, note how Wood uses Jacques Derrida's methods to deconstruct Derrida's own claims about time: David Wood, *The Deconstruction of Time* (Atlantic Heights, NJ: Humanities Press International, 1989), pp. 3-8, 269-77.
[20] Steven Connor, *Postmodernist Culture: An Introduction to Theories of the Contemporary* (Cambridge, MA: Blackwell, 1989), pp. 124-34, 224-44; James F. Harris, *Against Relativism: A Philosophical Defense of Method* (La Salle, IL: Open Court, 1992), pp. 95-122.
[21] For a concise introduction to a phenomenological approach to time, see Robert Sokolowski, *Introduction to Phenomenology* (New York: Cambridge University Press, 2000), chapter 9, 'Temporality.'
[22] Bertrand Russell, *An Inquiry into Meaning and Truth* (London: George Allen and Unwin, 1940), chapter 25.

But the syntactic priority thesis cannot be correct. Ordinary language admits of different, mutually exclusive interpretations, and the decision among the interpretations will have nothing to do with language and everything to do with metaphysics. Andrew Newman shows that the logical forms of predicate calculus into which ordinary language sentences may be analyzed (the syntax) are capable of different interpretations, ranging from Platonist to Aristotelian immanent realist to Fregean to Quinean:

> This argument shows why we quantify over particulars if we believe the world is a world of universals and particulars, and . . . why we quantify over objects if we believe the world is a world of concepts and objects. It shows that those reasons are metaphysical in nature. . . . It is metaphysics that explains the syntax.[23]

So the syntactic priority thesis should be rejected. And if it goes, so goes also any attempt to draw firm conclusions about the world from the putative analytic basicality of either A- or B-determinations. But might there not be a 'semantic priority thesis'? Quine, after all, spoke of 'semantic ascent,' and perhaps it is the words we use rather than the sentence structures in which they are used that are metaphysically significant.

But this suggestion also is incorrect. For words themselves, isolated from their context, do not, strictly speaking, 'mean' anything. If we have learned anything from Frege and Carnap, from Grice and Evans and Searle, and from many other philosophers of language, then we will not be at all tempted to find metaphysical insights in an ordinary language's vocabulary stock. Thus, I would maintain that Quine's 'semantic ascent' is not only unnecessary, but that it is also unjustified. That is, while it is crucially important to consider carefully how we use the words we do, and while a fine-grained intensional logic can indeed aid us in understanding why in fact we speak of reality in the way we do,[24] we are not justified in assuming that investigations of language alone can yield firm metaphysical conclusions.

Still, however, looking at common linguistic practices might lead us to something interesting. It is possible that by investigating the ways in which temporal considerations are expressed in ordinary language, we might uncover some interesting basic intuitions about time. For when we search for basic intuitions about a phenomenon that is as fundamental to our existence as time seems to be, we should expect to find that those intuitions are deeply embedded in linguistic practice and widely shared across linguistic communities.

But we need to tread carefully here here. Two examples will serve to show that the intuitions are weak. Western students of ancient Hebrew (or cognate Ancient Near Eastern languages) often find the tense structure of the language puzzling, for it is the 'aspect' (*Aktionsart*) of a verb rather than the time indicated

---

[23] Andrew Newman, *The Physical Basis of Predication* (Cambridge: Cambridge University Press, 1992), chapter 2, 'What Can Logic and Language Tell Us About Reality?,' p. 36.
[24] Hirsch, *Dividing Reality*; cf. George Bealer, *Quality and Concept* (New York: Oxford University Press, 1982).

by its tense that is fundamental. In Hebrew, aspect looks at the degree of completion of an action. For example, the Perfect tense refers to completed action, while the Imperfect tense refers to incomplete action. (Both 'Perfect' and 'Imperfect' are, of course, terms imported by grammarians from grammars of Indo-European languages.) And while it is true that completed action generally lies in the past, and incomplete action in the future, no translation can adopt the past or future tense as exact equivalents of the Hebrew Perfect or Imperfect. For instance, from the perspective of the ancient Hebrew people, any action attributed to God was 'as good as done' and thus the Perfect tense was used, even if the action lay in the future at the time of the utterance. The point here is that no 'basic intuition' about time can be discerned in the aspect of the Hebrew tense.[25]

Closer to home, there is an interesting anomaly between English and French, both living members of the Indo-European family of languages. French lacks distinct terms for 'time' and 'tense,' using the word *temps* for both (and *temps* may be used for 'weather' as well). Of course, since French has the linguistic apparatus for conveying the distinct concepts denoted by the English 'time' and 'tense,' it would be illegitimate to see in the absence of a distinct French word for the two concepts some 'basic French intuition' about time or tense.

Thus, in answer to the meta-question about the relationship between language and reality, I would express skepticism that any investigation of language, *in the absence of independent metaphysical arguments*, could yield firm evidence about the fundamental structure of reality.[26] Nevertheless, common linguistic practice can offer prima facie evidence as to the nature of reality— evidence which is admissible along with independent metaphysical argument in establishing a conclusion about the nature of reality.

What then should be said about the specific question of the translatability (or nontranslatability) of A- and B-sentences? I think that if it could be shown that, say, A-sentences could not be translated *salva veritate* into B-sentences, and if the same propositions expressed by an English A-sentence could be expressed by a sentence of a particular class in several other (preferably disparate) languages but could not be translated *salva veritate* into sentences of a class in those languages corresponding to English B-sentences, then we might have solid ground for affirming we had discovered something of fundamental metaphysical interest. But I know of no undisputed examples showing the success or failure of translation between English A- and B-sentences, let alone the corresponding success or failure in other languages. And this is despite many attempts to do so on both sides.

Recently, some B-theorists have agreed that A-sentences cannot be translated *salva veritate* into B-sentences, because they claim that tensed

---

[25] G. L. Klein, 'The "Prophetic Perfect",' *Journal of Northwest Semitic Languages* 16 (1990), pp. 45-60. More generally, Lyons shows that tense categories are not found in all languages. John Lyons, *Introduction to Theoretical Linguistics* (Cambridge: Cambridge University Press, 1971), p. 305. For further discussion of Hebrew tense, see Chapter 4.

[26] See E. J. Lowe, *The Possibility of Metaphysics: Substance, Identity, and Time* (Oxford: Oxford University Press, 1998), pp. 22-7.

statements are implicitly indexical. Rather than translatability, they wish to speak of truth conditions and analysis.[27] D. H. Mellor, for example, rejects the translatability thesis, but argues that the truth conditions of tensed statements may be analyzed in tenseless terms. While tensed language may be essential in expressing certain aspects of our experience, tense in those sentences is not real, since the truth conditions for those sentences can be given in tenseless terms.[28] However, I do not see this move as of any particular importance to the question of the relation of the language of time to the metaphysics of time. For once the move has been made from translatability to a truth-conditional analysis, one has moved from semantic to ontological issues.[29]

Nevertheless, and with all the appropriate caveats noted above, I do wish to make a modest claim about linguistic usage. Aside from those who would embrace the radical relativism that follows from a deconstructionist view of language, most philosophers would agree that one role of language is to make it possible to refer to objective reality. Consequently we should expect language to provide some sort of indication as to metaphysical realities. And this, I suggest, is salient in considerations of temporal indexicals and attitudes about the past, present or future.

It has long been commonplace for advocates of tensed theories of time to assert that statements like, 'Today is my birthday,' or 'A bomb will explode in the station five minutes from now,' are not analyzable in tenseless terms.[30] Tenseless advocates regularly offer token-reflexive analyses of these statements, so that the first becomes something like, 'My birthday is simultaneous with this utterance,' while the second becomes, 'A bomb explodes in the station five minutes later than the time of this utterance.' The tensed, temporally indexed statements in the first pair, which use A-properties such as presentness and futurity, have been restated in tenseless terms by the tenseless token-reflexive statements in the second pair which use B-relations of 'simultaneous with' and 'later than.'

A-theorists have responded to the token-reflexive analysis by saying that, while the truth conditions for the A-sentence and the token-reflexive B-sentence might in the end be the same, nevertheless there is something missing in the analysis.[31] Just what that missing thing is, is given classical expression by Arthur

---

[27] The view that tensed sentences could be translated into tenseless sentences has become known as the 'old tenseless theory of time,' while the view that, even though such translation is not possible, nevertheless tenseless truth conditions can be stated for all tensed sentences, thus rendering tense otiose, is known as the 'new tenseless theory of time.'
[28] Mellor, *Real Time*, pp. 40-46.
[29] Assuming, of course, that we hold a classical correspondence theory of truth rather than adopting the suggestion that truth is an epistemic concept.
[30] See Richard Swinburne, 'Tensed Facts,' *American Philosophical Quarterly* 27 (1990), pp. 117-30.
[31] Quentin Smith, however, argues that the two different sentences in question have different truth conditions as well as different confirmation conditions, and are not logically equivalent. While he acknowledges 'that they are intersubstitutable *salva veritate* in extensional contexts,' he claims that this intersubstitutivity is insufficient for the two

Prior.³² Suppose you undergo a very painful experience—a root canal procedure, say—and after leaving the dentist's office you exclaim, 'Thank goodness that's over!' On a token-reflexive analysis of this sentence, it seems that what you are thankful for is that the token you uttered is later than the dental procedure. But, says the A-theorist, this is preposterous. Surely what your exclamation expresses is your thankfulness that the procedure is in the past. Tooley responds to the A-theory criticism by suggesting that what the temporal indexical in 'That's over' picks out is not the token itself, as is generally assumed, but rather the time of the token; hence, the analysis, strictly speaking, is not a 'token-reflexive' analysis:

> When this view is adopted, what is happening in the sentence 'thank goodness that's over' is that a certain time is being picked out by its being simultaneous with the utterance, and one is saying, of that time, that one is glad that it does not involve a continuation of an earlier pain. What one is glad about, therefore, does not itself involve any relation between an event and a token.³³

This seems right to me. It seems that all objections based on the alleged failure of so-called token-reflexive analyses of tensed sentences containing indexicals are misguided. This is because what the indexicals pick out are not tokens of the utterance, but times.

Nevertheless, a significant problem still remains. For the time picked out by the indexical is not itself contained in the utterance, and yet is necessary for comprehending its meaning. As Frege noted,

> The mere wording, as it is given in writing [or in any tokening of the sentence], is not the complete expression of the thought, but the knowledge of certain accompanying conditions of utterance, which are used as means of expressing the thought, are [sic] needed for its correct apprehension .... If a time indication is needed by the present tense one must know when the sentence was uttered to apprehend the thought correctly. Therefore the time of utterance is part of the expression of the thought.³⁴

Frege held that the sentential content of an expression could exceed the propositional content of the expression, the difference being the extra-grammatical information conveyed by indexicals, which gives the propositional content under a different mode of presentation. Frege held that human behavior is motivated by the propositional content that we grasp, not the sentential content. The mode of presentation is essential to the cognitive significance of a proposition.

---

sentences in question to express identical propositions. Smith, *Language and Time*, p. 33. See also the debate represented by the articles in Part I of Oaklander and Smith, *The New Theory of Time*.
³² Arthur N. Prior, 'Thank Goodness That's Over,' *Philosophy* 34 (1959), pp. 12-17.
³³ Tooley, *TT&C*, p. 247.
³⁴ Gottlob Frege, 'The Thought,' quoted by Smith, *Language and Time*, p. 29.

More recently, David Kaplan and John Perry have given somewhat different analyses of how sentences containing indexicals work.[35] Kaplan, for example, distinguishes the propositional content of an expression from the linguistic rules governing the use of the expression, its character. Its character determines the cognitive significance of an expression. Kaplan and Perry seem to assert that the character of an expression is tensed, even if the propositional content is tenseless (or analyzable into a tenseless form).

The issues involved here are deep ones in the philosophy of language, specifically in the theory of direct reference. Fortunately, it is not necessary for our purposes here to go any deeper into these issues. Whatever analysis ultimately is adopted, it is clear that Frege, Kaplan and Perry would in the end agree that tensed sentences and sentences containing temporal indexicals are informative in ways that tenseless or token-reflexive translations are not. (Of course, this point applies to sentences containing indexicals in general, not just temporal indexicals. However, it is the latter that are of interest here.) The missing element in tenseless translations of tensed discourse is the apprehension by the speaker and/or the listener (or writer and reader) of what time it is now. Expressions conveying certain attitudes towards events depend crucially upon knowing whether those events are past, present or future—that is, upon knowing A-determinations. This additional information is, in very many cases, precisely what motivates human beliefs and actions.

Consider the following sentences:

A. A bomb will explode in the station at noon, March 1.
B. A bomb will explode in the station five minutes from now.
C. A bomb explodes in the station five minutes later than this utterance.

Assume that the truth conditions for (A), (B) and (C) are identical. The extra-grammatical information conveyed by (B) is just the information that will lead to the evacuation of the station. Unless the tenseless and token-reflexive sentences are augmented with additional information they will not motivate the same action, or at least not in the same way. It is also true that spatial indexicals will change the action-motivating character of a sentence, for example,

D. A bomb will explode here five minutes from now

may motivate action differently than (B). So while there may be nothing special about temporal indexicals as opposed to indexicals in general, it is clear that indexicals contribute extra information that is crucial to knowing what to do.

A strong objection to this line of argument comes from Hugh Mellor, who agrees that tensed sentences cannot be translated into tenseless ones, but asks what

---

[35] David Kaplan, 'Thoughts on Demonstratives,' in *Demonstratives*, ed. Palle Yourgrau (Oxford: Oxford University Press, 1990), pp. 34-49; John Perry, 'Frege on Demonstratives,' in ibid., pp. 50-70.

the implications are for advocates of tenseless time. 'It is tempting to infer from it that tensed sentences mean more than their token-reflexive truth conditions reveal; and that tensed facts may after all be needed to say what.'[36] But he argues that the only thing that tokens of tensed sentences supply is the truth conditions for the token, since truth conditions vary with temporal position. 'That is why no tenseless sentence can mean the same as a tensed one does: tenseless token-reflexive and not-token-reflexive truth conditions are bound to differ.'[37] Nevertheless, says Mellor, this implies only that tensed and tenseless sentences are not intertranslatable, not that there are tensed facts.

But this raises an obvious question, which Mellor anticipates: 'If there are no tensed facts, why must we think and speak as if there were?' The explanation, he argues, is that we need tensed language to express tensed beliefs (that is, beliefs whose truth conditions are token-reflexive) and we cannot help having tensed beliefs. Actions need, and are taken because of, tensed beliefs (e.g. a belief that [B] above is true). But—and here is Mellor's key move—'What action needs, in short, is the belief, not the fact. . . . My beliefs are what make me act, not the facts that make them true or false. . . . In particular, action will be timely if it satisfies the token-reflexive truth conditions of the tensed belief it depends on.'[38] And since the token-reflexive truth conditions serve to locate the tensed belief in a B-series, there are no tensed facts that are the objects of the belief. He concludes:

> In short, tensed beliefs are indispensable, not only for timely action in general and timely conversation in particular, but also for the conscious communication, sincere or insincere, of anything at all, tensed or tenseless. And the reason in both cases is that no agent, and above all no user of language, can do without token-reflexive beliefs. So far from the tenseless view of time, with its token-reflexive analysis of tensed belief, implying that tensed beliefs are dispensable, it alone explains exactly why they are not.[39]

Is Mellor's objection successful? Clearly it depends upon the success of his analysis of tense in token-reflexive terms. But this analysis relies on the assumption that any tensed sentence must specify the relevant time by means of a temporal indexical.[40] While common, this assumption is open to strong rejoinders. One rejoinder would be Tooley's view that truth conditions for common tensed sentences with indexicals can be formulated in terms of non-indexical tensed sentences of the form 'Event E lies (tenselessly) in the present (past, future) at time t.'[41] A second would follow Craig's careful distinction between a use of 'now' as

---

[36] Mellor, *Real Time*, p. 75.
[37] Ibid., p. 78.
[38] Ibid., p. 83.
[39] Ibid., p. 88.
[40] Mellor argues for this in ibid., chapter 2.
[41] Tooley, *TT&C*, pp. 192-7.

an indexical and a use of 'now' as an irreducibly tensed expression.[42] Either rejoinder is sufficient to defeat Mellor's objection.

In the end, the point is this. Whether or not it turns out that tenseless language is analytically prior to tensed language, it is clear that human thought and action depend inescapably upon tensed determinations. The ineliminability of tense in guiding our beliefs and actions explains why tense is so pervasive in common linguistic practice. I suggest that this serves as prima facie evidence that reality itself is tensed; that is, that time is dynamic.

The Argument from the Experience of Time

If the A-theory is correct, then human beings experience the world dynamically. It is an obvious fact that we automatically form beliefs about our experience in terms of the past, the present and the future. Call those A-beliefs. My argument in this section will be that common experiences elicit beliefs about the dynamic nature of those experiences, and the variety and pervasiveness of the A-beliefs we form strongly warrants the conclusion that our experience itself is dynamic, and therefore time is A-theoretic.

The Argument from the Experience of Time

1. Humans pervasively form A-beliefs about a wide variety of experiences.
2. B-theoretic explanations of such A-beliefs fail because:
    a. B-theoretic accounts of the presentness of experience incorrectly invoke indexicals or token-reflexivity;
    b. B-theoretic accounts of asymmetric attitudes about time are inadequate;
    c. B-theoretic accounts of the A-theoretic nature of mental experience lead to a vicious regress.
3. The best explanation of the variety and pervasiveness of A-beliefs about experience is that the experiences are in fact dynamic.
4. Therefore, belief in the dynamic nature of experience is warranted.

Tensed beliefs, A-beliefs, are universal. Psychologist William Friedman claims, on the basis of extensive studies of the consciousness of time, that 'the division between past, present, and future so deeply permeates our experience that it is hard to imagine its absence. . . . Most of us find quite startling the claim of some physicists and philosophers that the present has no special status in the physical world, that there is only a sequence of times, that the past, present, and future are only distinguishable in human consciousness.'[43] Even J. J. C. Smart, a staunch B-theorist, agrees that 'it is very common for people to doubt the reality of the future, and it is very hard to convince them that the future is real.'[44]

---

[42] William Lane Craig, *The Tenseless Theory of Time: A Critical Examination* (Dordrecht: Kluwer Academic Publishers, 2000), pp. 128-31.
[43] William Friedman, *About Time* (Cambridge, MA: MIT Press, 1990), p. 92.
[44] J. J. C. Smart, 'The Reality of the Future,' in *Essays Metaphysical and Moral* (New York: Basil Blackwell, 1987), p. 91.

Mellor argued that belief in tensed facts about the world is so deeply ingrained that only by showing that such belief entailed a logical contradiction would one be justified in abandoning such belief. But Mellor took McTaggart's Paradox as showing just that, and so concluded that reality was static. I am not persuaded by McTaggart's Paradox (see box), and suggest that it will be worthwhile to consider some reasons why our belief in dynamic time is so deeply ingrained.[45]

McTaggart's Paradox

J. M. E. McTaggart offered his eponymous paradox in 1908, believing thereby to have shown that time was unreal. It is generally used today by B-theorists to show that tense is unreal—that is, not essential to time.

Recall McTaggart's distinction between the A-theory, ordered by A-determinations (past, present, future) and the B-theory, ordered by B-determinations (earlier, simultaneous, later). McTaggart thought that both theories were true in some sense. He held that it was essential to time that the sequence of events form an A-series, since if change were to be real, then A-determinations would be necessary. On the other hand, he believed, B-determinations apply to events as they really are. Now, the same event could have both A-properties and B-properties. So a particular event could be future (A-determination) on a particular date (B-determination), present on another, and past on yet a third, meaning, McTaggart claimed, that it could have contradictory properties. But this was absurd. If, however, one argued, that an event had one property, future-at-t_1, and a second property present-at-t_2, and yet a third, past-at-t_3, then the puzzle was merely pushed back to the next level, for the opponent could ask when the event had those properties. This resulted in a vicious infinite regress. Hence, McTaggart concluded, time itself is unreal.

There's a quick route to block the paradox, once we see clearly a crucial assumption of McTaggart's reasoning. He held that both the A-series and the B-series existed side by side, and that the phenomenon of the flow of time was merely the sliding of the A-series along the B-series (or vice versa). Of course, for this view to be correct, the events comprising the B-series must exist. B-series events all exist tenselessly and change merely in the acquisition or loss of some A-property. On a dynamic theory of time, of course, this ontology is simply false. So the use of McTaggart's Paradox to argue for static time simply begs the question.

First, stating the obvious, we all live in the present. Our experiences are of the present. For example, I am conscious that the sun is shining. My experience of sunshine is present tense. I do not experience that the sun's shining is simultaneous with this token, or that the sun is shining at 3:00 p.m. PDT on 3 September 2003. I do not infer that this is the present from my 'being appeared to presently,' nor do I need to include an indexical 'now' in reporting my experience

---

[45] Mellor, *Real Time*, chapter 6. McTaggart's Paradox is an important and controversial argument in the philosophy of time. Defenders have offered quite different interpretations of the paradox, and opponents have offered various rebuttals. I have nothing original to add to the debate, so I shall simply refer the reader to two recent discussions and the references cited therein: Michael Tooley, *TT&C*, pp. 323-9; and Craig's very thorough examination of the paradox in *Tensed Theory,* chapter 6, pp. 169-217.

that the sun is shining. To say it awkwardly, I experience the presentness of the sun's shining.

Mellor agrees: 'The presence [presentness] of experience is the crux of the matter. Without a tenseless account of it, tenseless truth conditions on their own will never dispose of tensed facts.'[46] What Mellor and other B-theorists must deny is that the presentness of my experience is present-tensed. Mellor's strategy is to claim that what I am conscious of is not a present-tense experience, but a present-tense judgment about my experience. 'I emphasize this distinction,' he says, 'because there is a temptation to identify our experiences with our present tense judgments about them, a temptation which it is essential to my argument to resist. . . . Since this judgment is about the experiences I am having now, it will have token-reflexive truth conditions characteristic of the present tense.'[47] No one actually infers the presentness of an experience; presentness is something in experience that we are directly aware of. So to say that an experience that I am having now has the A-property of being present is a trivial truth. But it is this truth that makes the token-reflexive judgment 'I am having this experience now' true, and infallibly true. Mellor concludes his tenseless account of the presentness of experience confidently:

> For once it is not merely an alternative to a tensed explanation of the same thing. There is no tensed explanation of this phenomenon. If events can in reality have a range of tenses, I see no good reason for experience to be confined as it is to present events. In tensed terms, this is just a brute fact about experience. . . . Only our experiences, including our judgments (i.e. our thoughts), and our intentions, decisions and actions appear to be restricted to the present. Of that contrast, the token-reflexive account I have just given alone provides a serious explanation.[48]

Before critiquing Mellor's account, let's try to set it out clearly. It runs like this:

1. Events are tenseless.
2. Experiences are events.
3. Therefore, experiences are tenseless.
4. Presentness is essential to experience.
5. Judgments of the presentness of experiences are present-tense.
6. Therefore, judgments about experiences are indexed to the time of the judgment.
7. Indexed judgments have token-reflexive (i.e. tenseless) truth conditions.
8. Therefore, judgments about the presentness of experiences do not entail that experience is present-tense.

---

[46] Mellor, *Real Time*, p. 50.
[47] Ibid., p. 52.
[48] Ibid., p. 54.

One notices immediately that (4), the essential presentness of experiences, is just as much an unexplained brute fact about experience for Mellor as is the dynamic theorists' claim that experience is present-tense. But this is not simply a *tu quoque* response. For the causal theory of time to be proposed shortly can ground the claim that experience has the property of presentness in the causal nature of perception.

But Mellor's account fails for another reason. The essential presentness of experience leads to a present-tense judgment about the experience. Mellor believes that any present-tense judgment (an A-belief) contains an indexical and so is capable of a token-reflexive analysis. But this is not correct, as was shown above. My present-tense judgment that 'The sun is shining' can equally well be stated in terms of a non-indexical present-tense sentence plus a reference to the time at which the sentence is true: 'My experience that the sun is shining lies (tenselessly) in the present at time t.' All Mellor's objection shows is why my inference from present experience to what is presently occurring is true, but it does not de-tense the present-tense nature of my belief about the presentness of my experience.

Given the failure of Mellor's objection, the present-tense A-belief that I am presently experiencing x seems to be sufficient grounds to warrant the conclusion that my experience of x itself, and not just my beliefs about the experience, has the property of presentness.

A second aspect of our experience that warrants belief in dynamic time is the asymmetric character of intentional attitudes towards the past and the future. In addition to Prior's example of 'Thank goodness that's over,' discussed above, we may consider George Schlesinger's example of dread as a painful surgery approaches contrasted to the experience of relief when it is past.[49]

C. S. Lewis also notes the asymmetry of our attitudes about time:

> The Future is, of all things, the thing least like eternity. It is the most completely temporal part of time—for the Past is frozen and no longer flows, and the Present is all lit up with eternal rays .... Hence nearly all vices are rooted in the Future. Gratitude looks to the Past and love to the Present; fear, avarice, lust and ambition all look ahead.[50]

Certainly I experience (present-tense) relief, nostalgia, or regret when I remember a past event, but I experience (present-tense) dread, anxiety or anticipation when I think about a future event. Further, the strength of those experiences increases or decreases as the event in question 'comes closer' to the present or 'passes further' into the past. How can my experience of those attitudes be accounted for on the B-theory?

Mellor argues that I feel relief *simpliciter*, and not relief about something. So since the attitude has no object, it must be tenseless. But as to just why it is

---

[49] George N. Schlesinger, *Aspects of Time* (Indianapolis: Hackett, 1980), pp. 34-5.
[50] C. S. Lewis, *The Screwtape Letters* (New York: Macmillan, 1982), pp. 68-70.

rational to experience (tenselessly) relief later than some event, Mellor admits he simply does not know.

Paul Horwich suggests that the intentional attitude is due to a value asymmetry between past and future, and he grounds the value asymmetry in the evolutionary survival advantage conferred by a desire to have future desires satisfied.[51] I find this explanation of Horwich's value asymmetry to be less than compelling, but even if it succeeds, the evolutionary origin of attitudes about the past and the future is irrelevant to the question of the truth of the tensed propositions that are the objects of those attitudes.

Tooley thinks that argument along these lines is misguided, since, given that causes precede their effects,

> concern about future events is useful, since it increases our chances of avoiding those states that are bad, and realizing those that are good. A comparable concern about past events would not have that function. Accordingly, the adoption of a tenseless view of time does not preclude perfectly satisfactory explanations both of the fact that we have different attitudes towards past events and future events, and of the rationality of those attitudes.[52]

Tooley has apparently changed his mind about the validity of the argument from the asymmetry of intentional attitudes towards the past and the future. J. J. C. Smart cites an example Tooley put to him several years before he (Tooley) wrote the above. 'You have amnesia and someone tells you that there are two possibilities: (a) you have had twenty years of pain and will have one year of pleasure; (b) you have had twenty years of pleasure and will have one year of pain. Would you not prefer (a) to (b)?'[53] This type of example seems compelling to me. Unless there is indeed an asymmetry between past and future experience, and not merely between temporally indexed judgments about experience, such an attitude is inexplicable. Smart himself says the differing attitudes towards past or future pains and pleasures are explained in part 'by reference to the asymmetry of the causal grain of the universe: there are good reasons why our preferences and choices should be forward looking.'[54] However, if time is static, some type of reductionist analysis of causation must be given, and, if that analysis is correct, it is hard to see what is to be made of the statement that concern about the future could increase our chances of avoiding bad states and realizing good ones, or why our preferences should be forward looking.

A third class of experiences which warrants the belief in dynamic time is that of mentally tokening a sequence, such as counting to 50 when playing hide-and-seek, or counting sheep to fall asleep, or reciting the phonetic alphabet ('alpha, bravo, charlie, delta, . . .'), or mentally playing the score of a piano sonata, and so forth. Now whatever might be said about events in the external world, it does not

---

[51] Paul Horwich, *Asymmetries in Time* (Cambridge, MA: MIT Press, 1987).
[52] Tooley, *TT&C*, pp. 247-8.
[53] Smart, 'The Reality of the Future,' p. 99.
[54] Ibid.

seem possible to avoid the conclusion that I experience each successive member of the sequence as first future, in that I am at least tacitly aware of it as 'coming,' then as present, and then as past ('I got past the diminished-7th chord this time!'). My experience is of the successive passage of mental states, and surely I am justified in forming A-beliefs based on this experience.

Isn't it open to the B-theorist to reply that an experience of the 'passage' of mental states is itself merely psychological and not reflective of any ontological reality? Yes, but only at the pain of an infinite regress, which does seem to be vicious. If the B-theorist claims that my experience of the succession of my mental states is psychological, then I can counter-claim that if my experience of the succession of mental states while driving home is psychological, and my experience of mental states while typing this chapter is psychological, still it is an A-determination that the psychological experience of driving is past and the experience of typing is present. So once again the B-theorist will invoke a higher-order psychological experience, and again I can counter with another A-determination with respect to those higher-order psychological states. Since the *explanans* in each case is the same, and can be rephrased in terms of the original *explanandum*, the regress is vicious.

To avoid the force of this argument, the B-theorist must offer an analysis of the presentness of experience, and it is not at all clear that such analysis can be given in B-theoretic terms. (Mellor, it will be remembered, took presentness as essential to experience and offered no further account.) Consequently, I take it that the experience of temporal passage of mental states strongly warrants belief in the reality of dynamic time.

If belief in the objective dynamic nature of time is as strongly warranted as I have claimed, why would anyone want to maintain that temporal becoming is mind-dependent? Adolf Grünbaum argues for the thesis, 'Becoming is mind-dependent because it is not an attribute of physical events per se but requires the occurrence of states of conceptualized awareness.'[55] His theory 'makes nowness (and thereby pastness and futurity) depend on the existence of conceptualized awareness that an experience is being had.'[56] Thus temporal becoming is mind-dependent.

In support of his thesis Grünbaum offers three arguments. The first is to the effect that the temporal indexical 'now' is trivial, adding nothing to the assertion that a certain event occurs at a certain clock time. The second is that 'now' does not figure into the theories of physics. And the third is the absence of

---

[55] Adolf Grünbaum, 'The Status of Temporal Becoming,' in *The Philosophy of Time*, ed. Richard M. Gale (New Jersey: Humanities Press, 1978), p. 324; see also Grünbaum's *Philosophical Problems of Space and Time*, 2nd ed. (Dordrecht: Reidel, 1973). See Chapter 5 below for a comparison of Grünbaum's psychological theory of time with that of St Augustine.

[56] Grünbaum, 'The Status of Temporal Becoming,' p. 335.

explanation, on the A-theory, of why events become present when they do, rather than at some other time.[57]

In all three of these arguments, Grünbaum assumes the universe is a four-dimensional Minkowskian space-time, in which events are ordered by B-relations. In such space-time there is no past or future, but A-properties come to be applied to events in the B-series when a conscious mind experiences one of the events. This leads Grünbaum to misconstrue the meaning of 'now' in the dynamic theory of time.[58] And it surely begs the question to assume that the Minkowski model of space-time is correct.

I conclude that arguments for the mind-dependence of temporal becoming fail to offer plausible reasons for rejecting the common-sense view that becoming is real, and so do not constitute defeaters of the warranted beliefs about temporal becoming that we form on the basis of our experience. But I would go further and claim that arguments for the B-theory, which resort to ascribing temporal becoming solely to the realm of the mental, suffer one and all from the even more egregious failure to explain just what it means to say that minds traverse the time-dimension of space-time. Why is it that only consciousness 'moves'? What makes a mind's awareness 'move'? Why does consciousness only move in one direction? Why is it, in a classroom full of students who have very different phenomenal experiences of psychological time (for some the lecture drags on interminably, for others the hour flies by), that all arrive at the end of the class at the same time? Mustn't a psychological theory invoke some sort of mysterious occasionalism? Advocates of psychological theories of time are notoriously silent on these points.[59] It is hard to escape the conclusion that advocates of the mind-dependence of temporal becoming fail to take the mental phenomena seriously enough.

I believe it is unnecessary to multiply arguments, for the point is clear enough. We form A-beliefs based on our direct experiences. Further, I think it would be easy to show that all humans do in fact form such A-beliefs, regardless of theoretical commitments to a particular metaphysical theory of time. And the variety and pervasiveness of these beliefs call for explanation. Since the accounts offered by B-theorists are weak at best, the conclusion is that we are warranted in accepting an A-theoretic account of those beliefs—namely, an account that sees in the dynamic nature of experience a strong argument for the objectivity of dynamic time.

---

[57] Ibid., pp. 336-9.
[58] Craig, *Tenseless Theory*, pp. 127-45.
[59] Another salient point: most of those who argue for the mind-dependence of temporal becoming are naturalists who would give some sort of physicalist analysis of 'mind.' But if 'mind' is simply an epiphenomenon of the neurophysiology of the brain, how is it that the physical and chemical properties of brain matter give rise to something which 'experiences' non-existent temporal passage?

The Argument from Causation

A very strong case can be made that in our world, a vast number of events, which occur with great frequency, are prima facie highly improbable. But given certain assumptions about causation, that prima facie improbability can be explained. Those assumptions include a set of postulates that analyze causation in terms of a transfer of probability from a prior state to a posterior state. But the causal laws explaining the prima facie improbability can satisfy those postulates only in a dynamic world.

The Argument from Causation

1. The best account of causation is realist.
2. A realist account of causation is only possible in a world in which time is dynamic.
3. Therefore, the actual world is one in which time is dynamic.

There is, however, an extended aspect to this argument, which sees in the final causal theory of dynamic time an explanation of time's flow, direction and metrical relations, thus giving the theory a significant explanatory advantage over its competitors.

Since I will have much more to say about causation in what follows, and since the actual details of this argument are rather involved, I shall not pursue the argument further here.[60] For now I shall simply assume that if such an argument can be successfully mounted, then it offers independent line of support for dynamic time.

Summary

The three independent arguments in this section are sufficient, I believe, to establish the thesis that time is dynamic and that temporal becoming is real. Now an account of dynamic time must be offered. The theory of time that I shall defend argues that the dynamic nature of time is analyzable in causal terms. It is causation which ultimately explains why time flows, and in which direction. If such a theory can provide an account of time's dynamic nature, its flow and direction, and if (as was shown in this section) dynamic time is the best explanation of common linguistic practice, of beliefs about the tensed nature of experience, and of causal phenomena, then in virtue of its coherence and explanatory power, this theory may rightfully claim to be the best available theory of the nature of time itself.

A Causal Theory of Dynamic Time

Any theory of time must account for (at least) these three features of time: flow, direction and distance (or metrical relations). Addressing these three issues in turn will allow me to show the power of a causal theory of time.

---

[60] See the following sections; also Tooley, *TT&C*, pp. 103-11.

The Flow Question: 'Time's River'

The previous section established the conclusion that time is dynamic, that temporal becoming is real. Now, having argued that time flows, I must also explain *why* time flows. Some advocates of dynamic time have not recognized this burden. But the question is a very real one: given that the past and the present are real, but the future is not, what is it in virtue of which the present 'advances' into the future and what was present 'recedes' into the past?[61] Whitrow suggests that 'any theory which endeavours to account for time ought at least to throw some light on why everything does not happen at once.'[62] What explains the flow of time?

Talk of 'time flowing' is, of course, just a *façon de parler*. Most A-theorists do not think of time itself as a moving river. They understand the expression as referring to the dynamic nature of the temporal series. The most common explanation of that dynamic nature—the fact that time flows—is causal. But even those who appeal to a causal theory of time often fail to distinguish between two different questions. First, what accounts for the direction of time? And second, what accounts for the flow of time?[63] Hans Reichenbach, for example, writes, 'How can we find a suitable explication of time? It is clear that it can be sought only by a study of the relationships of causality.'[64] But Reichenbach only answers the direction question in the remainder of his treatise, and ignores the flow question. Richard Swinburne also adopts a causal theory of time, but applies the notion of causation solely to the direction question.[65] Tooley's thorough defense of a causal theory of time does not explicitly address the flow question, but in the argument for dynamic time the question is indirectly answered.[66]

I believe that a causal theory of time is better able than competing theories to answer the direction question, and that it is also in a unique position to offer an answer to the flow question. Thus I shall introduce my defense of a causal theory of time with a consideration of possible answers to the flow question.

---

[61] Advocates of static time, of course, simply deny that the 'flow' of time is a real feature of the world. Among the most significant of such denials is Donald C. Williams, 'The Myth of Passage,' *Journal of Philosophy* 48 (1951); reprinted in *The Philosophy of Time*, ed. Richard M. Gale (New Jersey: Humanities Press, 1978), pp. 98-116. See also Keith Seddon, *Time* (London: Methuen, 1987).

[62] G. J. Whitrow, *The Natural Philosophy of Time*, 2nd ed. (Oxford: Clarendon Press, 1980), p. 327. In a personal letter to astrophysicist Igor Novikov, the eminent physicist John Archibald Wheeler characterized time as 'what keeps everything from happening at once.' Cited in Igor D. Novikov, *The River of Time* (Cambridge: Cambridge University Press, 1998), p. 199.

[63] Few authors draw this distinction. An exception is P. C. W. Davies, 'Time and Reality,' in *Reduction, Time and Reality: Studies in the Philosophy of the Natural Sciences*, Richard Healey, ed. (Cambridge: Cambridge University Press, 1981), p. 63.

[64] Hans Reichenbach, *The Direction of Time*, ed. Maria Reichenbach (Berkeley: University of California Press, 1956), p. 24.

[65] Richard Swinburne, *The Christian God*, pp. 81-90.

[66] Tooley, *TT&C*, especially chapter 9, 'Causation and Temporal Relations.'

Some attempts to give an account of time's passing are somewhat less than clear. Irwin Lieb, for example, strongly defends a dynamic view of time, often speaking of time's passage in terms of action:

> Individuals act in every present, even when all they do is persist. There are no moments when they are paused and time is either stopped or the only thing that moves. Individuals and time suffuse one another, and neither can change without affecting the other . . . . As the future becomes present, time also acts. Its activity is called 'passing.' Time is made to pass partly because of what individuals do, and individuals are made to act partly because of passing time. The action of each calls for the action of the other; neither can act alone.[67]

There are a number of puzzles here. What does it mean to say that individual and time 'suffuse one another'? What does it mean to say that 'time acts'? Could there be time in the absence of individuals? And most crucially, what is meant by 'action'? In the absence of a theory of action (and Lieb does not offer one) I do not find much illumination in this account.

George Schlesinger also holds to a dynamic view of time.[68] He offers 'a coherent picture of the moving "now",' an ingenious account which attempts to avoid the force of some of the tenseless arguments, but which is, in my view, ultimately unsuccessful. Schlesinger asks us to imagine a whole family of worlds $\Sigma = \ldots, W_{n-1}, W_n, W_{n+1}, \ldots$. Each world W in Σ is identical with every other world in Σ, in that every moment (of small but finite duration) in one world exists also in every other world in the family, and the B-determinations are also equivalent from one world to the next. What makes the members of Σ numerically distinct is that, in each successive world, a successive moment is actual. Thus in W_{n-1}, moment m_{n-1} is actual; in W_n, m_n is actual; and in W_{n+1}, m_{n+1} is actual. Consequently, the successive worlds of Σ are ordered by an A-relation. (Schlesinger allows that presentness is a dyadic relation between a particular time and a particular world W_n.) Figure 2.1 represents Schlesinger's model.

The consequence of Schlesinger's view is that while he gives a realist interpretation of the A-series, he does so by having actuality move from world to world. Thus, at the actual moment when I type this, I am, strictly speaking, in a different world than I was in when I typed the previous sentence. If personal identity through time is a difficult issue, it is significantly more so when identity through time involves identity across worlds as well. Further, Schlesinger suggests no mechanism at all by which any continuent moves or is transferred from one world to the next, as actuality moves from one world to the next. It seems, then, that Schlesinger's view is implausible in the extreme.

---

[67] Irwin C. Lieb, *Past, Present, and Future: A Philosophical Essay About Time* (Urbana, IL: University of Illinois Press, 1991), p. 43.

[68] George N. Schlesinger, *Timely Topics* (New York: St Martin's Press, 1994), pp. 70-77. This model depicts a dynamic world in which the past, the present and the future all exist on equal ontological footing. This is significantly different from the dynamic causal view I am advocating, in which the past and the present are real but the future is not.

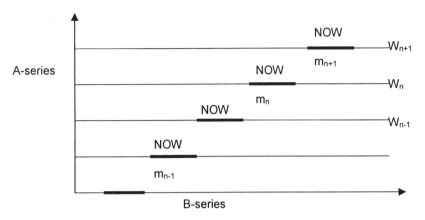

Figure 2.1 Schlesinger's dynamic time

Source: Adapted from Schlesinger, *Timely Topics*, p. 71.

This brief consideration of Lieb's and Schlesinger's attempts to answer the flow question should, at least, tip us off as to its difficulty. But if no easy explanation of why time flows is forthcoming, might we not simply take time as a metaphysical primitive? We might simply claim that it is a brute, unanalyzable fact about time that it flows, or that passage is an intrinsic essential property of time. We could take comfort in Newton's famous assertion that 'Absolute, true, and mathematical time, of itself, and from its own nature, flows equably without relation to anything external, and by another name is called duration.'[69] Time, then, is a substance that possesses the property of flowing as a brute, primitive feature.[70] We would then seek to analyze other concepts in terms of analytically basic temporal concepts.

Nevertheless, there are reasons against taking the flow of time as a brute, unanalyzable primitive. Metaphysical investigation has traditionally been guided by the principle of parsimony, a principle that has commended itself to many philosophers as a rational methodological principle, 'Minimize bruteness!'[71] If temporal flow can be analyzed in terms of some other, more basic, concept, then we would have a more parsimonious ontology. Until it becomes clear that such

---

[69] From Isaac Newton, 'Scholium on Absolute Space and Time,' appended in 1729 to *Mathematical Principles of Natural Philosophy* (1686), reproduced in full in John Earman, *World Enough and Space-Time: Absolute versus Relational Theories of Space and Time* (Cambridge, MA: MIT Press, 1989), pp. 20-26.

[70] This is the approach of Craig, who adopts a substantivalist view of time. See his *Tensed Theory*, chapter 7, and *God, Time and Eternity: The Coherence of Theism II: Eternity*, (Dordrecht: Kluwer Academic Publishers, 2000), chapters 5 and 6. See also idem, *Time and Eternity: Exploring God's Relationship to Time* (Wheaton, IL: Crossway Books, 2001), p. 13; hereafter *T&E*.

[71] Hudson, *A Materialist Metaphysics*, p. 23.

analysis is not possible, then, we should not simply opt to take time's flow as a brute given.

In looking for an answer to the flow question, it would seem that the concept of causation might be successfully employed. It is our experience that, in certain contexts, we are able to perform certain actions that bring about changes in those contexts. Ordinary language descriptions of such events use the vocabulary of cause and effect.[72] Causes bring about their effects; or, in the usual counterfactual reporting of such an event, if the cause had not occurred, the effect (other things being equal) would not have occurred either. Strongly implicated in this common notion of causation is the idea that causes are, or in some sense contain, power. So unless some other power intervenes, the power of (contained in) a cause will bring about a change in a state of affairs.

Arthur Danto has claimed, 'There exists no satisfactory analysis of the concept of causality in the philosophical literature, and it is possible to argue that it really is among the primitive concepts.'[73] But even if causality is a primitive concept, there still is much to be said about it. In my opinion, the correct account of causation will closely approximate the accounts found in the work of Armstrong and Tooley, where causation is taken to be a second-order relation between universals.[74] This is not the place to develop and defend such an account in detail. Such an account, however, will exhibit certain features that are germane to the causal theory of time.[75]

First, the account is realist. The facts of causation cannot be determined merely by examination of a complete history of the world, or even by such a history supplemented by a set of causal laws. Second, causation as described in this account possesses certain formal properties: causation is transitive, asymmetrical and irreflexive. Third, this account allows for causal laws that are probabilistic or indeterministic as well as deterministic. Fourth, this account also allows for singular instances of causation, instances where the causal relation is not the instantiation of a causal law.[76] The last two features say that causal laws are

---

[72] For the sake of simplicity here I am ignoring the possibility of causal chains. Nothing of importance hangs on this simplification.

[73] Arthur C. Danto, *Connections to the World: The Basic Concepts of Philosophy* (Berkeley: University of California Press, 1989), p. 257.

[74] David M. Armstrong, *What is a Law of Nature?* (Cambridge: Cambridge University Press, 1983); Michael Tooley, 'Causation: Reductionism versus Realism,' *Philosophy and Phenomenological Research*, suppl. 50 (1990), pp. 215-36; 'The Nature of Causation: A Singularist Account,' *Canadian Journal of Philosophy*, suppl. 16 (1990), pp. 271-322.; *Causation: A Realist Approach* (Oxford: Clarendon Press, 1987). See also Newman, *The Physical Basis of Predication*, chapters 4, 6 and 8.

[75] For a much more thorough development of this account, see Tooley, *TT&C*, chapters, 3, 4 and 9.

[76] A singularist conception of causation seems much more satisfactory than other realist accounts which take causal laws as basic. Tooley has offered a thorough defense of the concept in 'The Nature of Causation: A Singularist Account,' and in *TT&C*, pp. 93-101.

what they are in virtue of the causal relation, and not vice versa. The causal relation is the more basic concept.

Incorporation of these features allows realist accounts of causation to offer a compelling answer to the flow question. They do so by offering an explanation for the fact that entity A gives rise to entity B[77] in terms of dispositions possessed by A and B.[78] Generally, on a realist account, the categorical base for such dispositions are universals instantiated in the entities, and the conjunction of certain entities in a complex state of affairs is the activating condition for the realization of the disposition.[79] Put a bit differently, the causal relations between properties possessed by various entities in an actual state of affairs suffice to determine what other state(s) of affairs will become actual

In defending a causal theory of time, I shall deploy certain causal postulates developed by Michael Tooley. Tooley begins with the intuitive notion that a certain cause increases the probability that a certain effect will occur, or, metaphorically speaking, a cause 'transfers probability' to an effect.[80] I shall use the following symbolization in the statement of the postulates: P, Q are properties; x is a variable ranging over concrete entities; k is a value such that $0 \leq k \leq 1$; E represents a body of evidence; and C refers to information that is either tautological or concerns what causal laws there are. '$P \rightarrow Q$' means that it is a causal law that possession of P is causally sufficient to give rise to possession of Q. '$Pr(Px/E) = k$' says that the probability that x has property P, on evidence E, is k. With this symbolism, the postulates may be expressed as follows:

See also D. H. Mellor, *The Facts of Causation* (New York: Routledge, 1995), pp. 6-7; chapter 11.

[77] I use 'entity' here to indicate neutrality with respect to the question of whether events are the terms of causal relations. It does seem, however, that it is the 'constituent objects' of events, which I take to be states of affairs (particulars instantiating [dispositional] properties at a time), rather than the events themselves, which are the relata of causal relations. See Newman, *The Physical Basis of Predication*, pp. 168-78; Sydney Shoemaker, 'Causality and Properties,' in *Time and Cause*, ed. P. van Inwagen (Dordrecht: D. Reidel, 1980); reprinted in *Properties*, ed. D. H. Mellor and Alex Oliver (New York: Oxford University Press, 1997), pp. 228-9; Mellor, *The Facts of Causation*, pp. 130-39. On states of affairs as the basic constituents of reality, rather than events—successions of states of affairs—see D. A. Armstrong, *A World of States of Affairs* (Cambridge: Cambridge University Press, 1997), pp. 37, 103-105.

[78] Shoemaker argues that 'disposition' should be used as a 'predicate of predicates,' and 'power' as that in virtue of which a certain property has a certain effect. Shoemaker, 'Causality and Properties,' pp. 231-5. Nothing of significance for my argument hangs on this terminological debate.

[79] Characterizing a disposition in terms of its categorical base is less popular among metaphysicians today than is giving a functionalist account of a disposition. But as far as I am aware, functionalist accounts appeal to the concept of causation, and so cannot enter into the explanation at this point, on pain of circularity.

[80] Tooley, *TT&C*, pp. 59-63. Mellor's characterization of causation in terms of an 'increase in the chances' is related, although his account differs in significant ways from Tooley's. Mellor, *The Facts of Causation, passim*, especially pp. 21-8.

C1. $Pr(Px/P \rightarrow Q \& C) = Pr(Px/C)$
C2. $Pr(Qx/P \rightarrow Q \& C) = Pr(Px/C) + Pr(\sim Px/C) \times Pr(Qx/\sim Px \& P \rightarrow Q \& C)$
C3. $Pr(Qx/\sim Px \& P \rightarrow Q \& C) = Pr(Qx/\sim Px \& C)$
C4. $Pr(Qx/P \rightarrow Q \& C) = Pr(Px/C) + Pr(\sim Px/C) \times Pr(Qx/\sim Px \& C)$

According to postulate (C1), the posterior probability that entity x possesses P (i.e., Px), given information concerning the relevant causal laws, is the same as the prior probability of that entity's possession of P. That is, (C1) expresses the very intuitive notion that the posterior probability that Px is not a function of any state of affairs such as Qx to which Px gives rise.

Postulate (C2) says that the posterior probability that Qx, given that P causally gives rise to Q together with specific information about causation, C, is a function of the prior probability that Px.

Postulate (C3) expresses the claim that the posterior probability that Qx, given that it is not the case that Px and that P causally gives rise to Q, together with C, is just the prior probability that Qx given that it is not the case that Px, together with C. In other words, if P causally gives rise to Q and we do not have a state of affairs containing P, the posterior probability of a state of affairs containing Q is just the same as the prior probability of a state of affairs containing Q.

Postulates (C2) and (C3) together entail (C4), which asserts that the posterior probability that Qx, given that P causally gives rise to Q, together with C, is equal to the prior probability that Px plus the prior probability that Qx when it is not the case that Px.

If these postulates are correct and complete, then they capture the basic intuition that if we know something about the causal powers of a certain property, then there is a very precise way in which the presence of that property 'transfers probabilities' to its effect. Tooley concludes:

> Causal laws, rather than being merely regularities in the history of the world, *control* the course of history; they underlie, and account for any patterns that the world may exhibit over time. But how is this to be understood? One way—and, I think, the only satisfactory way—is if causal laws, in conjunction with what is actual as of a given time, determine what states of affairs will be added to what is already actual, and, thereby, what will ultimately exist, in a tenseless sense. And, if laws together with what is actual as of a given time do determine how the world will be, then causal laws will necessarily conform to postulates (C_1) through (C_4), and the asymmetry that is expressed, in particular, by postulates (C_1) and (C_4) will be captured and explained.[81]

Commenting on Tooley's argument, Robin Le Poidevin notes that:

> Causal laws must obey certain probabilistic constraints, among them the principle that the posterior probability of the effect is a function of the prior probability of the cause, but not vice versa. If the world were 'static' (i.e. tenseless) it would be

---

[81] Tooley, *TT&C*, p. 111.

impossible to explain why this principle obtains, whereas the dynamic world view offers us a very natural explanation.[82]

Only a dynamic world in which temporal becoming is real will be able to satisfy postulates (C1) through (C4). And while more would need to be said on the nature of causation if I were attempting a thorough account of the concept, enough has been said to give the basis for the assertion that a causal account can answer the question of the flow of time. I turn now to a defense of the claim that a causal theory of time will give a convincing answer to the direction question.

The Direction Question: 'Time's Arrow'

Explanations of the direction of time, or 'time's arrow,' as it is often called,[83] fall into one of three general classes: those that appeal to some physical law or process; those that appeal to observable traces in the world, and those that appeal to causation. An examination of the first two will show that the only viable explanation of the direction of time will be causal.[84]

Entropy and 'time's arrow' First, and probably the most widely known due to popularizations of science, are those explanations which equate the direction of time with some physical process or law.[85] Now, it is widely understood that the mathematical equations describing physical processes or laws are time-symmetric, and so, while such physical processes can give an order to time, they cannot specify its direction. Even Maxwell's equations, describing the propagation of electromagnetic radiation, are time-symmetric, and thus have the consequence that, say, a television signal, precisely 'obeying' Maxwell's equations, could be at my TV set before it left the studio. Of course, such a phenomenon goes against all our experience and intuition, and solutions to Maxwell's equations representing such a phenomenon are generally disregarded by physicists as 'non-physical.' We can easily envision an expanding sphere of electromagnetic energy propagating from a point source and being absorbed at quite different times by physical objects in space. But we cannot as easily envision—and once envisioned, we cannot overcome our intuitions of the extraordinary improbability of—a situation in which numerous physical objects, scattered randomly in space, somehow 'conspire

---

[82] Robin Le Poidevin, review of *Time, Tense and Causation*, by Michael Tooley, *The British Journal for the Philosophy of Science* 49 (1998), p. 367.

[83] The term seems to have been introduced by the great astronomer Sir Arthur S. Eddington in his Gifford Lectures of 1928, published as *The Nature of the Physical World* (Cambridge: Cambridge University Press, 1928). Steven F. Savitt, ed. *Time's Arrows Today: Recent Physical and Philosophical Work on the Direction of Time* (Cambridge: Cambridge University Press, 1995), p. 1.

[84] Unless, of course, a substantivalist view of time is adopted, according to which the flow and direction of time are brute properties of time itself.

[85] John D. Barrow, *The World within the World* (New York: Clarendon Press, 1988), pp. 126-7; 366-8; Kitty Ferguson, *The Fire in the Equations* (Grand Rapids, MI: Eerdmans, 1994), pp. 119-22.

together' to emit radiation at various times so the result is that a sphere of radiation converges on a point. There is an intuitive asymmetry here, but it cannot be explained by Maxwell's equations, since both situations are equally permissible under time-symmetry.[86]

So if the direction of time is to be determined by physical processes, it must come from a time-asymmetric process or law. Examples, which confront us daily, are the processes of ageing or weathering, or trails of wreckage. The preferred theoretical candidate for explaining this kind of asymmetry is the Second Law of Thermodynamics. In its popular form, the Second Law states that in an isolated system, one across the boundaries of which neither matter nor energy can flow, which is not in internal thermal equilibrium, entropy will always increase. The direction of increasing entropy is the future. Now, if the popular form of the Second Law were indeed a law of nature, then entropic asymmetry could ground the direction of time. But there are good reasons to regard the popular version of the Second Law as false.

The work of Boltzmann and Gibbs succeeded in giving thermodynamics a statistical interpretation at the microscopic level, resulting in a probabilistic formulation of the Second Law. According to the probabilistic formulation, there is always a non-zero, non-infinitesimal probability that, during any given interval, a particular closed system will experience a reverse in entropy. While the probability of a system in a state of low entropy experiencing an increase in entropy is very high, there is likewise a high probability that a system in a state of low entropy was, at an earlier time, in a state of high entropy.[87]

Hence, it turns out that entropic asymmetry is not time-asymmetric and cannot give any ontologically grounded answer to the direction question.[88] This

---

[86] Several attempts have been made to provide a theoretical explanation for the asymmetry of these (and similar) phenomena, or, conversely, to account for the apparent asymmetry while retaining an underlying time-symmetry. In general, physicists are deeply attached to the notion that all fundamental laws and processes are time-reversal invariant—that is, time-symmetric. See Paul Davies, *About Time: Einstein's Unfinished Revolution* (New York: Simon and Schuster, 1995), chapter 9, 'The Arrow of Time,' pp. 196-218, and references cited there. So far, neither theoretical nor experimental investigations have been able to offer an explanation of time-asymmetry, or an account of why time-asymmetry should be regarded as apparent rather than real. And if my arguments here are correct, no such explanation will ever succeed.

[87] It will almost certainly be the case in any relatively short period of observation that increased entropy coincides with the positive direction of time. This is because there are exponentially many more high entropy states than low entropy states, so statistically it is vastly more probable that a system, determined to be in a low entropy state, will be evolving toward a state of high entropy. But given enough time one will see a decrease of entropy coinciding with the positive direction of time. 'Enough time,' however, has been shown theoretically to be many orders of magnitude greater than the age of the universe, so practical applications of the Second Law are not in danger of being reversed. But since it cannot be the case that the categorical ground for the direction of time possibly reverses, entropy will not do the job.

[88] Tooley, *Causation*, pp. 221-3; Reichenbach, *The Direction of Time*, pp. 108-12.

result convinced Einstein, among others, to claim that temporal asymmetry is a purely subjective, human phenomenon, a position which accorded well with Einstein's positivism as applied to interpretations of STR.[89] Several attempts have been made to rescue the Second Law and entropic asymmetry as the grounds of temporal direction, most notably Reichenbach's hypothesis of the branch structure of the universe.[90] But finally, given the inability of statistical thermodynamics to establish time-asymmetry of entropic processes, Reichenbach concludes:

> It follows that we cannot speak of a direction for time as a whole; only certain sections of time have directions, and these directions are not the same.... To say, however, that the universe consists of separate time threads pieced together in opposite directions does have a meaning, because time *order* can be defined in classical mechanics and does not presuppose entropy. We may therefore speak of a *supertime* which orders the [entropy] curve even in sections of equilibrium, where the entropy remains practically constant, or at saddle-points, where the entropy gradient reverses its direction. Supertime has no direction, only an order, whereas it contains individual sections that have a direction, though these directions alternate from section to section.[91]

It is strange indeed that Reichenbach, who defends a causal theory of dynamic time, is led to two conclusions that seem to undermine his entire project. For if there is indeed a supertime that is ordered but has no direction, then it follows that the A-theory is incorrect with respect to cosmic or metaphysical time (whichever actually corresponds to Reichenbach's supertime). Further, if supertime is not ordered, then even the B-theory cannot be correct, since it is ordered by the non-symmetric earlier-than relation. And if time can reverse directions, how can that be consistent with a causal theory in which, as we shall see, the direction of causation determines the direction of time? (Of course, if backwards causation is possible, this would not rule out a world in which oppositely directed causal processes existed, thus giving rise to oppositely directed times. However, this is not at all what is envisioned in Reichenbach's time reversal circumstances.)

Towards the end of his investigation of these matters, Lawrence Sklar asks, 'If this asymmetry of distributions of microscopic conditions is that which in the world represents at the macroscopic level the grand asymmetry of entropic increase, isn't that enough to say that by itself it constitutes the asymmetry of "time itself?".' But he concludes, inconclusively, 'This final stage of the Boltzmann thesis is neither proven nor disproven at present.'[92]

---

[89] David Layzer, *Cosmogenesis: The Growth of Order in the Universe* (New York: Oxford University Press, 1990), p. 51.
[90] Reichenbach, *The Direction of Time*, pp. 117-43.
[91] Ibid., pp. 127, 129-30.
[92] Lawrence Sklar, *Physics and Chance: Philosophical Issues in the Foundations of Statistical Mechanics* (Cambridge: Cambridge University Press, 1993), pp. 384, 404.

So it seems that, in the end, the most promising candidate among physical processes or laws for answering the direction question is incapable of doing so.[93]

Traces and 'time's arrow' The second general class into which answers to the direction question fall comprises of those theories which claim that time's passage leaves indelible traces, and that the direction of its passage is determined by properties of those traces.[94] We never perceive such occurrences as ripples leaving the shore of a pond, merging into circular wavelets, and converging on a depression in the surface of the pond from which emerges a stone, flying towards the shore. We frequently find written traces of the past, but never of the future. And it is the past that we remember, not the future. (Even granting the possibility of precognition, there would seem to be a phenomenological difference between memory and such precognition. At least, those individuals who claim such powers seem to have no trouble telling the one from the other.)

Do such traces then give us an account of the direction of time? No, for we would still need some explanation of why we observed the kinds of trace that allegedly determine time's arrow. To see this, we must give a clearer characterization of a trace. A trace is a state of affairs that bears reliable marks of having come into being as the result of an event or a process.[95] An example might be the state of magnetization on a cassette tape, or the state of footprints in the sand of a beach. Such traces, however, can be shown to be traces of entropic processes. But this fact presents us with a dilemma. Either our perception of time's arrow is identical with increased entropy, or it is not. If the former, then all the objections to increased entropy determining the direction of time, discussed above, come to bear. This horn of the dilemma is not appealing for another reason; as Sklar notes, the requisite identification is certainly counter-intuitive:

---

[93] At present there are a few other attempts to ground the direction of time in some physical fact, process or law. One example is the breaking of time symmetry in the production and decay of the neutral K meson (K^0), the only violation of time symmetry known in particle physics. A second is the collapse of a black hole, which, unless 'white holes' exist, would seem to break time symmetry. A third is the collapse of the wave function in Quantum Mechanics, another putative time-asymmetric phenomenon. Whether any of these will turn out to be fundamental physical asymmetries is as yet unclear. See Davies, *About Time*, pp. 208-18; Roger Penrose, 'Singularities and Time-Asymmetry,' in *General Relativity: An Einstein Centenary Survey*, ed. S. W. Hawking and W. Israel (Cambridge: Cambridge University Press, 1979), p. 582; Anthony Leggett, 'Time's Flow and the Quantum Measurement Problem,' in Savitt, *Time's Arrows Today*, pp. 97-106. I would argue that even if any of these phenomena can be shown to be fundamentally time-asymmetric, it is because of the nature of causation involved in the phenomenon, and so a correct account of the direction of time will still be causal.

[94] J. J. C. Smart, 'The Temporal Asymmetry of the World,' *Analysis* 14 (1954), pp. 79-82.

[95] Bernard Mayo, 'Professor Smart on Temporal Asymmetry,' *Australasian Journal of Philosophy* 33 (1955), p. 39. Smart concurs with this characterization in 'Mr. Mayo on Temporal Asymmetry,' *Australasian Journal of Philosophy* 33 (1955), p. 125.

> We know from perception what it is for one state of a system to be temporally after some other state. And we know what it is for one state to have a more dispersed order structure than another state. We also know that these two relations are not the same . . . . So, whatever the relation of temporal afterness is, it is not an entropically defined relation.[96]

So we are left with the remaining horn of the dilemma, that time's arrow is not identical with increased entropy. But if we choose this horn, we are left with no explanation at all of the relation between traces of entropic processes and the direction of time. If we have no explanation of the relation in question, then we have no explanation of the direction of time either.

Are there perceptual traces of non-entropic processes? Karl Popper suggests that such processes, which he called de facto irreversible processes, involve a dispersal of order.[97] His example is the one alluded to earlier of ripples spreading out in a pond from the point of impact of an object striking the surface. Of course, we might object that this would be an entropic process due to the conversion of some of the kinetic energy of the ripples into heat energy as they encounter the edge of the pond. Popper could respond that such processes are equally possible in a perfectly inelastic medium where no kinetic energy would be converted to heat energy. But it is a contingent fact that there are no perfectly inelastic media in the actual world, and so as a matter of contingent fact, in the actual world, there are no non-entropic irreversible processes. Thus, in the actual world, the direction of time cannot be determined by traces of irreversible processes.[98]

The third class of answers to the direction question comprises different forms of causal explanations. Since, as we have seen, causation creeps into most answers given in the other two classes, and since we have reached the conclusion that causation gives the best answer to the flow question, there is a strong presumption that causation will provide the grounds for answering the direction question as well.

Causation and 'time's arrow'[99] At this point, however, I must make more precise just what I mean by the 'direction question.' Philosopher of science John Earman has written,

> Very little progress has been made on the fundamental issues involved in 'the problem of the direction of time.' By itself, this would not be especially surprising since the issues are deep and difficult ones. What is curious, however, is that despite all the spilled ink, the controversy, and the emotion, little progress has been made towards clarifying the issues. Indeed, it seems not a very great

---

[96] Lawrence Sklar, 'Time in Experience and in Theoretical Description of the World,' in Savitt, *Time's Arrows Today*, pp. 224-5.
[97] Karl Popper, 'The Arrow of Time,' *Nature* 177 (1956), p. 538.
[98] This is all that is needed here to respond to Popper's suggestion. Tooley offers stronger responses in *The Facts of Causation*, pp. 224-8 showing why no such analysis is possible.
[99] Much of the discussion in this section follows Tooley, *TT&C,* pp. 267-82.

exaggeration to say that the main problem with 'the problem of the direction of time' is to figure out exactly what the problem is supposed to be![100]

I have argued that time is dynamic—that there is an ontological difference between the past and the present on the one hand, and the future on the other. And I have argued that a causal theory of time offers an account of why time 'flows.' In this context, a cluster of issues about the topology of time arises. Why is it that time flows always and inexorably towards the future? Could there be eddies or reversals in the flow of time? Could there be a stoppage in its flow, and if there could be, could time start flowing again? Is it possible that there be separate 'tributaries' that converge in a single flow (fusion of temporal series)? Or could the single flow divide into separate streams (fission of temporal series)? It is the first of these that I shall take to be the principal direction question. In the following chapter some of the other topological issues will be addressed.

In this section I shall argue that certain features of causation are sufficient to answer the direction question. I begin with consideration of the temporal relations of 'simultaneous with' and 'earlier than' (or, for simplicity, simultaneity and [temporal] priority). I shall assume in this section a non-relativistic conception of time; difficulties related to the relativity of simultaneity, or the conventionality of simultaneity, stemming from STR, will be treated in the next chapter.

Simultaneity and priority possess the following formal properties: simultaneity is transitive, reflexive and symmetric, while priority is transitive only. If we combine these properties with the formal properties of causation entailed by a realist account as described above, and add one additional assumption, a perspicuous account of the direction of time emerges. Recall that a realist conception attributes to causation the formal properties of transitivity, irreflexivity and asymmetry. The one additional assumption that is needed, one that is incorporated in most realist accounts, is this: causes precede their effects (reflected in [P3] below). The formal properties, together with this assumption, give the following principles that form the basis of my argument that causation determines temporal direction:

    P1. If A is prior to B, and B is prior to C, then A is prior to C.
    P2. If A is prior to B, and B is simultaneous with C, then A is prior to C.
    P3. If A causes B, then A is prior to B.

Before proceeding with the argument, however, I must defend P3 against two serious objections: I must show that neither simultaneous causation nor backwards causation are possible.

Simultaneous causation Consider first the case of simultaneous causation. This is an important objection to a causal theory of time, for if it were possible for effects

---

[100] John Earman, 'An Attempt to Add a Little Direction to "The Problem of the Direction of Time",' *Philosophy of Science* 41 (1974), p. 15.

to be simultaneous with their causes, then causation could not account for the flow of time, let alone for its direction.

Kant, to take a notable example, believed that effects could be simultaneous with their causes:

> The great majority of efficient natural causes are simultaneous with their effects, and the sequence in time of the latter is due only to the fact that the cause cannot achieve its complete effect in one moment. The time between the causality of the cause and its immediate effect may be a *vanishing* quantity, and they may be thus simultaneous, but the relation of the one to the other will always still remain determinable in time. If I view as a cause a ball which impresses a hollow as it lies on a stuffed cushion, the cause is simultaneous with the effect. But I still distinguish the two through the time-relation of their dynamical connection. For if I lay the ball on the cushion, a hollow follows upon the previous flat shape; but if (for any reason) there previously exists a hollow in the cushion, a leaden ball does not follow upon it.[101]

It is widely thought that Kant holds that there is a categorical feature of cause and effect that determines their logical order, although they may be—and often are—simultaneous and hence unordered in time. Yet in the passage quoted he claims, 'the relation of the one to the other will always still remain determinable in time.' How are we to interpret this claim? It might be that Kant means that a cause and effect are perceptually simultaneous: they both occur in the specious present, which has some finite duration. But when Kant speaks of the time between cause and effect as being 'vanishingly small,' and when, just a few paragraphs later, he claims, 'Between any two instants there is always a time,' it seems he leaves the door open to a more sophisticated interpretation. Kant believed that time was dense, and so any causal action must span an interval of several instants, say $[t_1, t_2]$. But given the density of time, this interval includes as a proper part the half-open interval $[t_1, t_2)$, including t_1 but not t_2. Thus, though the interval is, mathematically speaking, a 'vanishing quantity,' it remains finite. So if the cause begins at t_1 and the effect at t_2, the interval during which the causal action occurs will approach zero as a limit, but will never vanish completely. Now, whether this is an accurate interpretation of Kant is somewhat irrelevant; it seems to point to the most obvious response to undermine the claim of the possibility of simultaneous causation.[102]

Further, from what we know of atomic physics, we can conclude that causal powers are not transmitted instantaneously between two ordinary objects. There is always some elasticity. Even though the nucleus of an atom is extremely close, by ordinary measurement, to its electron shell, it is still some finite distance away. And since causal signals cannot travel faster than the speed of light, an

---

[101] Immanuel Kant, *Critique of Pure Reason*, unabridged ed., tr. Norman Kemp Smith (New York: St Martin's Press, 1929), A203/B248-9.
[102] Mellor examines four variations of this approach, only to reject them all, but for reasons different than mine. Mellor, *The Facts of Causation*, pp. 220-29.

effect will be felt by the electron shell some finite time before the effect is felt by the nucleus. Hence it would seem that no physical effect could be simultaneous with its cause.

Such, at least, would seem to be the obvious response. But it fails for two reasons. First, while atoms are divisible, it might turn out that some class of subatomic particles are indivisible and inelastic (for example, point particles), and certain effects are transmitted instantaneously between them. If, say, quarks were atomic (inelastic and indivisible), then the effect of momentum transfer caused by a collision between two quarks would indeed be instantaneous; the cause and the effect would be simultaneous. So the fact that physics tells us that all the causes with which we are familiar are not simultaneous with their effects is a contingent fact. And second, even if it should turn out that physics shows that in our universe all causes necessarily precede their effects by some finite time, that would be at best a nomological necessity. In order for the causal theory of time, as I am defending it, to be consistent, the non-simultaneity of cause and effect must be at least a metaphysical necessity. So some other line of argument is needed.

Happily, another line of argument is readily available. Having argued that the world is a dynamic one characterized by postulates (C1) through (C4) about causation, and having accepted principles (P1) and (P2), we may proceed to show that simultaneous causation is not possible. Let us assume a variation of (P3), namely:

P3*. A causes B, and possibly A is simultaneous with B.

Contra Kant, I do not believe it possible to rule out a priori the Humean dictum that anything can cause anything.[103] Consequently, if A and B are simultaneous, and we have reason to believe that there is a causal relation involved, it is an open question whether A or B is the cause. Let us assume that whenever we observe A, we observe B simultaneously, but that we sometimes observe B without a simultaneous A. Now, we might assume that causal laws are deterministic, in which case we would conclude that our observations justify the hypothesis that A is a sufficient cause for B's occurrence, while B is not a sufficient cause for A's occurrence, and so A is the cause of B, not vice versa. But we live in a world where there seem to be indeterministic (or probabilistic) laws. In this case, the fact that some Bs occur in the absence of an A is insufficient to rule out B as a cause of A. Hence we are still unable to identify A as the cause and B as the effect.

But this epistemological puzzle reflects an underlying metaphysical problem. If we know that A and B co-occur, we need some metaphysical argument to identify one as the cause and the other as the effect. However, the following situation seems logically possible. Assume that A simultaneously causes B, and B simultaneously causes C. Given the Humean dictum that what causes what is logically contingent, it is possible that, rather than C, B has as its effect

---

[103] David Hume, *Treatise of Human Nature*, 2nd ed., ed. P. W. Nidditch (Oxford: Clarendon Press, 1978), 1.3.15.

not-A. But this situation is of course an impossible causal sequence. The conclusion must be that the assumption of the possibility of simultaneous causation is faulty.[104]

Backward causation As I noted above, standard arguments against backward causation typically invoke arguments against causal loops, and then move to the desired conclusion. As I also noted, Tooley has shown the illegitimacy of this move. However, at this point, we are in a position to offer a non-question-begging argument against backward causation.

Let us assume that the standard arguments against causal loops are successful. The remaining problem is the possibility of oppositely directed causal sequences which themselves are causally independent. In such a scenario, for example, as time moves dynamically from past to future, what might be observed is a baseball at rest in the lap of a spectator in the right field stands suddenly leaping into the air, traveling in a certain trajectory to make contact with a batter's bat, and then flying into the pitcher's hand while the bat swings rapidly away from the point of contact with the ball, coming to rest on the batter's shoulder. Given the time-symmetry of Newton's laws of motion, there is nothing nomologically impossible in this scenario. How can this obviously backwards scenario be ruled out?

The following line of argument, similar to the argument above against simultaneous causation, shows the impossibility of this scenario. Let us begin by considering two worlds, W and W*, both of which are assumed to include backwards causation. In W the following are true:

(i) A causes B, and B causes C.
(ii) A is later than B, and B is earlier than C.
(iii) The laws of nature are such that B cannot cause not-A.

In W, backwards causation is assumed to be possible by (ii). Now compare W to W* in which the following are true.

(iv) A is a cause of B, but B precedes A in time.
(v) The laws of nature are such that B causes not-A.

In this world, too, backward causation is assumed to be possible by (iv). Now since the laws are not logically necessary, the laws of both W and W* are logically possible. That is, both (iii) and (v) are logically possible. So it is logically possible in W* that B have not-A as an effect at a later time than the occurrence of

---

[104] This argument is similar to that of Richard Swinburne, *The Christian God*, pp. 82; 245. Hume argues that, if an effect could be simultaneous with its cause, then everything would happen simultaneously (*Treatise*, 1.3.2.), but this is a dubious claim. If simultaneous causation were possible, then possibly everything *could* happen simultaneously, but that does not entail that everything *would* happen simultaneously.

B. But this results in the manifestly impossible state of affairs where A is the cause of its own non-being. The only solution is to reject the assumption (iv) that backward causation is possible. However, since what is logically impossible is invariant across possible worlds, assumption (ii) is also impossible, and so W is not a logically possible world.

Returning now to the argument for the direction of time, since simultaneous and backward causation have been ruled out, (P3) is to be accepted. Then from (P1) through (P3), together with the formal properties of causation and temporal relations, these additional theses follow:[105]

> T1. If A causes B, and B causes C, then A is prior to C.
> T2. If A causes B, and B is simultaneous with C, then A is prior to C.
> T3. If A is not prior to B, then A is not prior to any cause of B.
> T4. If A is prior to or simultaneous with B, then A is prior to any effects of B.

I believe that (T1) through (T4), together with (P1) through (P3), suffice to determine temporal direction. It should also be apparent that any world in which these seven theses hold is also a world in which there can be causal laws satisfying (C1) through (C4). Further, although I will not do so here, it can be shown that a world such as this will give the correct answers to the counterfactuals associated with conventional accounts of causality, including the argument from preventability.

The Distance Question: Time's Metric

A theory of time should say something regarding the metric of time, or the basis for determinations of temporal distance. Whether or not (physical) time has a metric in any given temporal world is a contingent matter. In Chapter 1 we noted that the metric of time in any temporal world will depend upon the causal laws that obtain in that world; if the laws are such that no regular physical processes are possible, then there will be no metric that can be applied to time in that world. This would mean that quantitative temporal relations would not be possible, but certain qualitative temporal relations would still be possible, such as the determination that t_1 was 'more past' than t_2 (the 'earlier than' relation). Such judgments, it can easily be seen, may be grounded in (T1) through (T4), together with (P1) through (P3)—the very principles that determine temporal direction.

In a temporal world with laws permitting regular physical processes, a quantitative metric is available. Generally, if a temporal world is very much like the actual world, there will be very many possible metrics from which to choose; the purpose that the metric will serve will determine which is the appropriate one to choose. The record of attempts to determine the metric of time in human history

---

[105] Tooley goes in a somewhat different direction in developing causality, simultaneity and actuality postulates: *TT&C*, pp. 112-16.

offers numerous examples.¹⁰⁶ In an agrarian culture, the appropriate metric will serve to inform farmers when to plant their fields; in an industrial society it will determine the beginning and the end of the workday. In a technological society, greater precision may be needed, which is provided by the regularity of an atomic clock.¹⁰⁷

Each of these examples relies on a physical process, the regularity of which is causally determined—the earth's rotation on its axis or its revolution around the sun, state transitions at the atomic level. Since the causal processes involved are governed by natural law, metrics derived from these processes could all be termed intrinsic. By way of contrast, the division of the year into 12 months, or the week into seven days, represents the imposition of an artificial metric onto physical time, and thus represents extrinsic metrics.

What is important to note here, though, is that whatever process is selected as the appropriate intrinsic metric in any given situation, the regularity of the process is causally determined. Once again, the great explanatory value of a causal theory of time is apparent, as no new grounding, no novel properties, no brute facts about time must be invoked to account for time's metric. It is not at all clear that non-causal dynamic theories have an explanation so readily available for time's metric, and just how static theories account for the metric is even murkier.

The Topology of Causal Dynamic Time

To conclude the discussion of the metaphysics of time, I shall develop a topology of time consistent with the causal theory defended above. W. H. Newton-Smith, among others, has argued that the question of the topology of time is an empirical rather than a conceptual problem and should be left to the physicists.¹⁰⁸ But I believe that opinion to be incorrect, and shall offer philosophical arguments that will establish a model for the topology of time.

---

[106] Fascinating accounts of the history of humankind's attempts to measure time may be found in Daniel J. Boorstin, *The Discoverers: A History of Man's Search to Know His World and Himself* (New York: Random House, 1983), pp. 4-78. The account of thinkers' struggle to reach a conception of time as something that could be measured is told by Wolfgang Achtner, Stefan Kunz and Thomas Walter, *Dimensions of Time: The Structures of the Time of Humans, of the World, and of God*, tr. Arthur H. Williams, Jr (Grand Rapids, MI: Eerdmans, 2002), pp. 27-110.

[107] By definition, one second is the duration of 9,192,631,770 cycles of microwave radiation emitted by a cesium atom transitioning from its second lowest to its lowest energy state.

[108] W. H. Newton-Smith, *The Structure of Time* (London: Routledge and Kegan Paul, 1980).

> **A Model of Time**
>
> The argument of this section fleshes out a model of the topology of dynamic time. Features of the model:
> - Fission or fusion of times is not possible.
> - If multiple, parallel times are possible, then in principle they are either independent, or a single simultaneity suffices to establish temporal correlations.
> - The future is 'branching,' but only in the modal sense (possible futures).
> - Temporal relations are analyzed in terms of causal connections.
> - The sum total of 'what is real' grows by accretion.

Linear Time

The first general model I shall consider is the simplest. The Linear Model might be called the *mille-feuille* model of space-time, since it envisions a 'stack' of 'slices' of space, one on top of the other. The 'vertical' dimension of the stack represents time, so the temporal series is the total stack with the top 'slice' representing the entire world at the present.

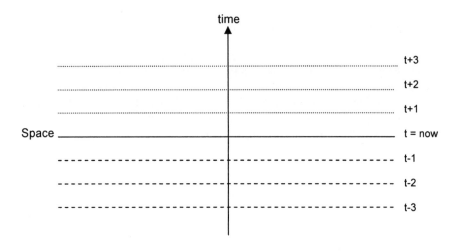

Figure 2.2 Schematic representation of a linear model

Figure 2.2 schematizes this model (with two spatial dimensions suppressed). In this figure the horizontal lines represent 'slices' of space at successive times. A solid horizontal line represents the 'slice' of space that exists at the present, as it is the only actual space. The ontological asymmetry of past and future is represented by the different broken line styles. The past is represented as a dashed line, while the future, which is not real, is represented by a dotted line. Thus, as time passes, what is 'real' grows by the addition of new 'slices.'[109]

A richer schematic representation of this model, which illustrates causal features, is possible (Figure 2.3). The reality of the past is illustrated by the lines on the left half of the diagram, showing the present effects of past causes. Here it is important to note that the times represented by the space-time slices are not durationless instants. If such were the case, slices would have zero 'thickness' and no matter how many such slices were piled on top of each other, could never yield any duration.[110] Instead, the slices should be construed as moments of arbitrarily short, but non-zero, duration.

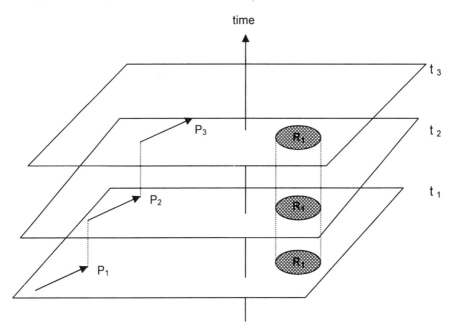

Figure 2.3 Schematic representation of causal relations in a linear model

---

[109] I will refrain from entering the debate between 'presentists,' who believe that only the present moment is real, and other thinkers such as Tooley, who say that both the past and the present are real. This latter use of 'real' is somewhat idiosyncratic: something is real iff it is or has been actual; something is actual iff it exists now (although for Tooley, 'actual as of time t' is an undefined primitive; *TT&C*, pp. 13-19; 39-42).

[110] The debate as to whether time is discrete or not has continued since Aristotle. While 'continuous' or 'dense' time raises difficult questions (the familiar problems of the continuum), so too 'atomistic' or 'quantized' time raises very counter-intuitive problems. For discussion, see, for example, Richard Sorabji, *Time, Creation and the Continuum: Theories in Antiquity and the Early Middle Ages* (Ithaca, NY: Cornell University Press, 1983); Craig, *Tensed Theory*, pp. 239-44.

Imagine a causal chain caused by a state of affairs obtaining at time t_1 at spatial point P_1 through an effect at t_2 at point P_2 and a subsequent effect at t_3 in the present at P_3. On a causal theory of dynamic time, though t_1 and t_2 are in the past at t_3, the points and causally efficacious states of affairs that were located at P1 and P2 are 'real' (in the idiosyncratic sense described in footnote 109) but not actual at t_3.

Consider also space-time region R_1, and imagine that R_1 is empty of any particulars during the entire interval (t_1, t_2, t_3). Traditionally, R_1 continues to exist through the interval because of the conservation laws. It is possible, on the causal theory of time, to interpret the conservation laws as causal laws, in which case R_1 at t_1 is the cause of R_1 at t_2 that in turn is the cause of R_1 at t_3. Empty regions of space thus have causal properties.[111] Generalizing, we can say that every complete temporal slice of the world is causally connected to every other.

This illustrates two interesting points of the theory. The first is that space is substantival, not relational, and the second is that temporal relations are to be understood in terms of causal connections, not causal connectibility. We shall look at these features in order.[112]

It seems that the most satisfactory causal theory of time will assume substantival space. Why? For one reason, it is possible for there to be empty physical worlds that are temporal. If empty space-time regions cannot enter into causal relations, then a causal theory of time cannot be correct for empty worlds. So if the arguments for a causal theory of time are sound, then space must be substantival.

Are there positive arguments to be given in favor of substantival space? Regardless of what one makes of Newton's classic arguments for substantival space,[113] other lines of argument may be adduced. One fairly straightforward argument focuses on the need to provide a categorical ground for certain claims, such as the claim that possibly a material object could be located in a presently empty region of space. Such modal claims express empirical possibilities, not merely logical possibilities, and it seems correct to require that these claims have a categorical basis. Relational theories of space cannot ascribe empirical properties to empty space, and so have great difficulty supplying the categorical basis for empirical possibilities. Substantival accounts of space avoid these difficulties altogether. In addition, substantival space will provide certain conceptual resources used later in discussions of relativity theory.

The analysis of temporal relations in terms of causal connectibility rather than in terms of actual causal connections is not uncommon. However, such

---

[111] Tooley, *TT&C*, pp. 258-64..

[112] For a thorough exploration of these issues, see ibid., pp. 258-67.

[113] Newton's views are discussed in many places. Among the most helpful are John Earman, *World Enough and Space-Time*, pp. 7-12; Jennifer Trusted, *Physics and Metaphysics: Theories of Space and Time* (New York: Routledge, 1991), pp. 92-115; William Lane Craig, *Time and the Metaphysics of Relativity*, Philosophical Studies Series 84 (Dordrecht: Kluwer Academic Publishers, 2001), *passim*.

theories are unable to provide a satisfactory analysis of temporal relations in terms of causal relations alone. This is because the truth-makers for empirical modalities, analyzed in terms of counterfactuals, will have no categorical basis within a modalized account of causal relations. Only an account that is formulated in non-modal terms of actual causal connections can do so.

The Linear Model, then, can demonstrate certain features of a physically realistic interpretation of causal time. But important questions remain, and to explore them I shall look at another model—or, rather, at a more complex extension of this model.

Branching Time?

The Linear Model simply assumed that the topology of time is linear. Although this assumption seems commonsensical, should we accept it? Might it not be the case that time is a branching structure? Using the analogy that time is like a flowing river, how can we be sure that several times do not converge in one time, like tributaries of a river? Or that time does not split to flow around an island, and then reunite? Or that time does not divide into separate streams, not to be reunited? Figure 2.4 illustrates such possible topologies.

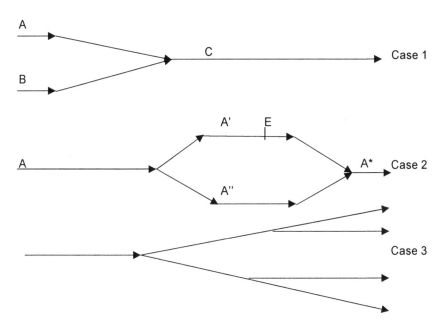

Figure 2.4 Non-linear topologies

Are there reasons for us to believe that such examples of converging or diverging temporal series are not possible?[114] Such arguments would be metaphysical in nature, not empirical; indeed, it is difficult even to imagine what empirical evidence for non-linear temporal topologies would consist in. I believe that there are such metaphysical arguments.[115] Consider first the possibility of converging temporal series as in Case 1. Assume that two entities, A and B, exist in two discrete time series, the series then merge, and an entity C continues in the single merged series. Now, is it possible that a single entity should have two distinct past histories—that is, that C should be identical to A and B? No, for this reason. A, B and C must be temporal entities. If time is contingent, then A and B came into existence in their discrete times before they merged at C. But it seems to be a necessary truth that a single entity cannot come into existence at two separate times in its history. (See Chapter 9 for an analysis of temporality and a defense of the contingency of time.) So either A or B continues to exist as C, and the other ceases to exist, or else both cease to exist and a new entity C comes into existence. Either alternative yields the result that C is not identical to A and B. It might be objected that the claim is not that temporal entities merge, but that the time series merge. However, on a causal theory of time, if time exists, then the relata of the causal relation also exist. So A, B and C could be material objects, empty regions of substantival space, or immaterial souls, and the results would be the same.

Now for Case 2, in which a single time series diverges and recombines. If any entities came into existence during the period when the times were separate, then the merging of the series would pose the same problems as in Case 1. If we imagine a single entity which splits when the times split, and then is recombined with itself, we avoid the impossibility of a single contingent entity with two beginnings, but we still have problems of identity. Suppose that A splits into A′ and A″ which later merge into A*. Suppose further that event E occurs in the time series of A′ but not in that of A″. (Some such premise is necessary, for if the history of A′ is identical with the history of A″ then is it even meaningful to assert that there had been a branching in time?) It seems to be a necessary truth that the history of A* either contains E or it does not; it cannot contain contradictory properties. Therefore either A* is the continuent of the persisting entity A-A′, and the entity A-A″ has ceased to exist, or vice versa; or else both A-A′ and A-A″ cease to exist, and A* is a new entity which has come into being. So it seems that

---

[114] Richard Swinburne once believed such topologies were possible: 'Times,' *Analysis* 26 (1965), pp.185-91. He abandoned this belief as early as the first edition of his *Space and Time* (London: Macmillan, 1968 [2nd ed., 1981]), chapter 10. Linear topologies are defended by Brian Leftow, *Time and Eternity* (Ithaca, NY: Cornell University Press, 1991), pp. 21-33; J. R. Lucas, *A Treatise on Space and Time* (London: Methuen, 1973), pp. 35-42; and Murray MacBeath, 'Time's Square,' in *The Philosophy of Time*, ed. Robin Le Poidevin and Murray MacBeath (Oxford: Oxford University Press, 1993), pp. 183-202.

[115] The arguments are analogous in interesting ways to arguments regarding identity in cases of fission and fusion of entities in a single time series. See Peter van Inwagen, *Material Objects* (Ithaca, NY: Cornell University Press, 1990), chapter 1.

we would have no grounds to claim that the same temporal series has split and recombined.

Case 3 involves a single time that splits into divergent series that never recombine. Is this alternative possible? This alternative, unlike the first two, does have its serious advocates, those who adopt Everett's Many-Worlds interpretation of QM.[116] And the argument against it is not as straightforward as the previous two cases. The first argument against Case 3 is epistemological: if time did indeed split, how would we ever know it? The arguments against merging time series in Cases 1 and 2, applied to Case 3, would rule out any causal contact between the divergent series. Anyone in any time series would, looking back, see only a single linear history, not one that was equipped with off-ramps. So in the absence of independent arguments, by Ockham's Razor we should discount this possibility. But it might be possible to give a stronger argument against this possibility. Based on the postulates of causation discussed above, the causal relation is to be analyzed in terms of the transfer of probabilities. Now if time is causal, and it splits into two discrete series, then the state of affairs that obtained at the instant of the split causally gave rise to two numerically distinct states of affairs. The states of affairs constituting the effects must be complete states of affairs, not a single one under two partial descriptions. But given the postulates, it would seem prima facie impossible for a single state of affairs to do this. So if time is dynamic and causal, then we have reason to reject Case 3.

The conclusion, then, is that time has a linear topology, with no branching or merging. But it remains an open question whether there can be only one time, or whether two (or more) parallel (i.e. non-intersecting) times are possible. Let us assume first that two parallel times exist. Either there is at least one moment at which the two coincide (that is, at which there is a causal interaction between the two series), or there isn't. Suppose that moment m in series O coincides with moment n in series P. To say that the moments coincide is to say that the same event E occurs (or the same state of affairs S obtains) in both. If E (or S) is located at m in O and at n in P, then m and n are simultaneous. Then, from that point of simultaneity, irrespective of the intrinsic metrics of O and P (or even the lack of an intrinsic metric), all points in O can be put in a one-to-one correspondence with all points in P.[117] That is to say, from a single simultaneity, common A-properties and B-relations can be assigned to the two series. If the two series are isomorphic in

---

[116] For example, Frank J. Tipler, 'The Many-Worlds Interpretation of Quantum Mechanics in Quantum Cosmology,' in *Quantum Concepts in Space and Time*, ed. R. Penrose and C. Isham (Oxford: Clarendon Press, 1986), pp. 204-14.

[117] This assertion rests on an argument that if time is linear and dense, then any time interval would contain an infinite number of points of ordinality ω and cardinality $\aleph_0$. For the details of the argument, see Craig, 'The Finitude of the Past and the Existence of God,' in *Theism, Atheism, and Big Bang Cosmology*, William Lane Craig and Quentin Smith (Oxford: Clarendon Press, 1993), pp. 4-30.

this way, then by principles of set theory and transfinite mathematics, they are the same series under different descriptions.[118]

If, however, there is no moment at which the two coincide, then there could never be any evidence of one in the other, and the claim that a completely isolated temporal series existed must be made on a priori grounds. And it is not at all clear how such an argument could even get started.

I shall argue in several places below that God is a temporal being, and that his time is to be taken as metaphysical time, the causal succession of states of affairs that underlies all possible times. Metaphysical time is not to be construed as a distinct parallel time, since from the moment of creation until the present the moments of physical time coincide with moments of metaphysical time. Perhaps most importantly, if (as I shall claim) God's providential act of sustenance is the causal act by which God maintains in existence the entire universe and all it contains, together with the properties and laws which make causal time, then God's time will be A-theoretic and his 'now' will coincide with the present of physical time.

We can now see what should be said about the possibility of three or more time series. They cannot be conceived of as parallel and independent, for the same reason that we must reject the notion of two parallel and independent times. However, if we understand metaphysical time to be infinite, then there could be infinitely many non-overlapping times. If God created many temporal worlds, each would be related to metaphysical time, and hence indirectly to each other. Of course, we might never know about these other times, but unlike the case of parallel times, they would not be inaccessible in principle. Should God so choose, he could reveal their existence to us. (Presumably this would be the case if the biblical accounts of God's creation of an angelic realm are to be taken literally.)

This completes my account of the topology of time. But there are some fascinating implications of the branching model with respect to our conception of the future. To these we turn next.

The Branching Future

As noted in the previous chapter, the pre-analytical view of the future is that many things about it remain open, that there are many ways that the world could go, and which way turns out to be the actual world is as yet undetermined. There are, however, several significantly different competing views. Several of these are represented in Figure 2.5.

While three of the four models are consistent with what has already been said about the topology of time, they differ significantly with respect to the future. Figure 2.5(a), the Minkowski Model, is the model corresponding to STR where

---

[118] Cantor's Principle of Equivalence asserts that if a one-to-one correspondence can be established between the members of two sets, then the two sets are equivalent. See also Leftow, *Time and Eternity*, pp. 24-6; Swinburne, *Space and Time*, 2nd ed., chapter 10, pp. 165-75.

space-time is conceived as a four-dimensional block universe. In this model there is only one possible future, and the ontological status of the future is no different than that of the past. In the Distinguished Branch Model, (b), and in the Branching Future Model, (d), possible ways the world could have gone in the past are indicated with dashed lines; possible ways the world could go in the future are indicated with solid lines. Both (b) and (d) incorporate modal notions into the depiction of the future, but on the Distinguished Branch Model, even though there are possible futures, it is already 'fated' which branch will be the actual future. In the Branching Future Model, however, there is genuine contingency; the future is truly open. The Many-Worlds Model, (c), is included here simply for comparison, since we already ruled out its topology. There are no dashed lines on this model; the ways the world could have gone in the past, a world did go. Those alternative worlds are as real as the actual world, but are theoretically inaccessible to us.[119]

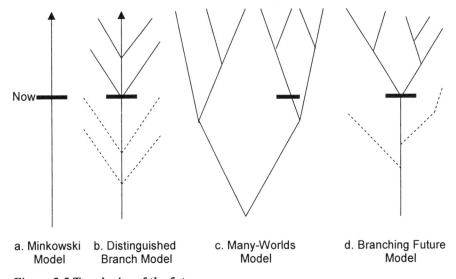

a. Minkowski Model   b. Distinguished Branch Model   c. Many-Worlds Model   d. Branching Future Model

Figure 2.5 Topologies of the future

Source: Adapted from Storrs McCall, *A Model of the Universe*, Fig. 1.4.

Of interest to us in this section is the Branching Future Model.[120] In the Branching Future Model, the temporal structure of the universe resembles a tree, with the past

---

[119] Realism about other worlds in interpretations of QM is similar to the well-known realism about possible worlds in David Lewis's ontology, although it seems that each single Lewis world contains a complete set of Everett worlds. Both ontologies seem hopelessly bloated.

[120] Storrs McCall has developed this model in detail in *A Model of the Universe: Space-Time, Probability, and Decision* (Oxford: Clarendon Press, 1994); see also his earlier papers, 'Objective Time Flow,' *Philosophy of Science* 43 (1976), pp. 337-62, and 'Counterfactuals Based on Real Possible Worlds,' *Noûs* 18 (1984), pp. 463-77.

being the trunk and the future the branches.[121] The branches represent possible ways the world could go but, as time passes, branches representing unrealized possibilities are eliminated, so that there is a singular past. The present is the point at which the 'choice' among possibilities is made, and the branches are cut off.

Now, McCall is a realist about possible futures; in his model, they are as real as the present or the past. As indicated, I think we have good reasons to avoid such a profligate ontology. But I believe that McCall's model can be stripped of its modal realism and will then offer a plausible model.[122] McCall deploys his theory to give solutions to a number of vexing philosophical problems from relativity theory and quantum indeterminacy to personal identity through time, moral freedom and free will. It is not at all clear to me that major strengths of his view would not remain if stripped of modal realism, but this is a matter I shall not pursue. Rather, I shall briefly consider troubling questions that arise from his theory. If the world truly is dynamic—and McCall claims that it is under his model—then in what sense are future possibilities real? They are in some sense more real than merely real possibilities; they really are there now. But if the world is dynamic, and time is unidirectional, then they can have no causal influence on the present, and it is not clear just what the claim comes to. Further, if the topology of time is conceived as actually splitting at each node, then the arguments in the previous section against branching time would apply, showing that such a picture of time is not possible, at least not if one assumes a causal theory of dynamic time.

A uniquely troubling aspect of the realism about possible futures—in addition to the difficulties that plague all theories which regard the future as real—is the question of what happens to the branches when they drop off. Are they still 'there,' constituting the trunks of other trees, in the same way as Everett's many worlds are there after a measurement of a quantum state has been made? Do they then cease to be real, since they are not a part of our trunk? McCall offers no answers to these questions, and I see no reason why we cannot help ourselves to much of McCall's model without at the same time accepting such cumbersome metaphysical baggage.

So the Branching Future Model shows us a picture of time in which the past is singular and fixed, the future is branched and open, and the present is where the possibilities of the future are either fixed or cut off. Such is the structure of time in any world that contains indeterminate laws, free creatures, and true contingency.

---

[121] Reminiscent of Jorge Luis Borges's felicitous comparison of the future to 'a garden of forked paths.'
[122] J. R. Lucas also proposes a branched model of temporal modalities, but does not ascribe reality to them: *A Treatise on Space and Time*, pp. 262-72.

Conclusion

If the arguments offered above are correct, we have established that time is dynamic and causal in nature, and that a realist causal theory of time accounts for the flow and the direction of time, as well as for a temporal metric. We have thus a significant metaphysical theory of time that must play a part in any theory of God's temporal mode of being. First, however, we must be assured that this metaphysical theory of time is not invalidated by the results of modern physics. The next chapter is devoted to that effort.

But before turning to that effort, I must briefly note that the causal theory of time that I have just outlined and defended should commend itself strongly to theists. Traditional theism uniformly agrees that the continued existence of the contingent universe is due to God's voluntary exercise of causal sustaining power. So the most basic relation in the universe for the theist is that of causation. With this relation already in place it should be quite easy for the theist to incorporate the causal theory of time.

Chapter 3

Physics and Time

The previous chapter developed and defended, on metaphysical grounds, a causal theory of dynamic time. But time is a subject that is inextricably intertwined with theories of physics. Physical theories place a constraint on a theory of time, requiring it to exhibit consistency with empirical observations and acceptable interpretations of those theories. In particular, the three towering theories of twentieth-century physics—the Special Theory of Relativity (STR), the General Theory of Relativity (GTR), and Quantum Mechanics (QM)—are generally regarded as calling into question the conceptual framework of a causal theory of dynamic time.

In this chapter I shall investigate objections to a causal theory of dynamic time raised on the basis of these three theories, and will show that there are adequate replies to each of the objections. This is not the place—nor do I have the space—for in-depth treatments of the objections and the responses. Fortunately, however, that is not necessary. For my purpose will be served simply by pointing out that the responses are logically and nomologically possible. The reader need not be convinced that they are true, or even more plausible than not. The possible truth of the proffered responses suffices to show that the causal theory of dynamic time is not incompatible with well-confirmed empirical results in the physical sciences. (Readers having the required level of technical detail to reach a judgment on the technical adequacy of the responses are invited to pursue the literature cited in the notes.)

The Special Theory of Relativity

Albert Einstein's publication in 1905 of what now is known as the Special Theory of Relativity changed forever both the face of physics and our understanding of the world in which we live. The Newtonian picture of absolute space and absolute time gave way to a much more complex picture of the structure of the universe. In the century since its publication, STR has been shown to yield very accurate predictions for a number of varied phenomena; its empirical adequacy has been well confirmed. The picture STR gives of space and time, and of the behavior of time in relativistic reference frames, raises serious problems for the causal theory of dynamic time drawn in the previous chapter. This section will explain the problems and suggest how dynamic time may be harmonized with STR.

Problems

Traditional interpretations of STR make the claim that simultaneity is relative to reference frames. It follows from this claim that there can be no absolute (or cosmic) time. Consequently, STR is thought to establish the ontological parity of past, present and future—that is, the static theory of time. Interpretations of this kind are generally classed as Einsteinian interpretations. An examination of these interpretations of STR will show why many physicists adopt a static view of time.[1]

STR and Dynamic Time

Under commonly held assumptions, STR shows that simultaneity is relative to reference frames. But that means that the set of events which are simultaneous with a particular B-series point is relative to the observer. Hence there is no absolute truth about what is happening 'now.'

It is not difficult to see why this result spells doom for any theory of dynamic time. For if one event is present with respect to one reference frame, but the same event is future with respect to another, then the future must have the same ontological status as the present and the past; hence, time is B-theoretic.

Further, since simultaneity relations vary with reference frames, there is no absolute time, and so there is no cosmos-wide 'truth of the matter' as to what time it is now.

---

[1] It is not necessary in what follows to rely at all on the mathematics of relativity. Indeed, it is my belief that over-reliance on mathematics at the expense of seeking physically realistic interpretations of the equations has often resulted in metaphysical silliness. I shall say a bit more about this in the section on GTR. Good introductions to STR and GTR, which discuss the physical interpretations of the theories and include enough mathematics to whet the appetite but not so much as to choke the reader, include Albert Einstein's surprisingly readable *Relativity: The Special and the General Theory*, tr. Robert W. Lawson (New York: Bonanza Books, 1961); Hermann Bondi, *Relativity and Common Sense: A New Approach to Einstein* (New York: Dover, 1964); Paul Davies, *About Time: Einstein's Unfinished Revolution* (New York: Simon and Schuster, 1995). More advanced discussions include Erwin Schrödinger, *Space-Time Structure* (Cambridge: Cambridge University Press, 1950); Murad D. Akhundov, *Conceptions of Space and Time: Sources, Evolution, Directions*, tr. Charles Rougle (Cambridge, MA: The MIT Press, 1986); Stephen Hawking and Roger Penrose, *The Nature of Space and Time* (Princeton, NJ: Princeton University Press, 1996); George F. R. Ellis and Ruth M. Williams, *Flat and Curved Space-Times* (Oxford: Clarendon Press, 1988). Discussions by respected philosophers of science, which examine metaphysical issues normally left untouched by the physicists, include Michael Friedman, *Foundations of Space-Time Theories: Relativistic Physics and Philosophy of Science* (Princeton, NJ: Princeton University Press, 1983); two volumes by John Earman, *World Enough and Space-Time: Absolute versus Relations Theories of Space and Time* (Cambridge, MA: MIT Press, 1989), and *Bangs, Crunches, Whimpers, Shrieks: Singularities and Acausalities in Relativistic Spacetimes* (New York: Oxford University Press, 1995), hereafter *Bangs*; and William Lane Craig, *Time and the Metaphysics of Relativity*, Philosophical Studies Series 84 (Dordrecht: Kluwer Academic Publishers, 2001), hereafter *TMR*.

STR entails the denial of absolute time because it denies absolute simultaneity. And if absolute simultaneity goes, absolute time will have to go as well, regardless of our intuitions to the contrary. The problem is illustrated geometrically in Figure 3.1.

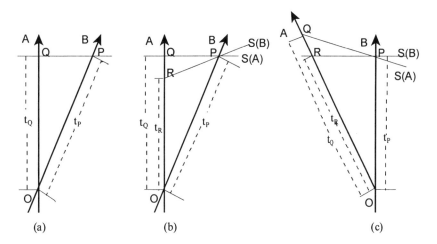

(a) Observer A determines point Q on his world-line to be simultaneous with P on B's world-line. (b) Observer B determines point R on A's world-line to be simultaneous with P on B's world-line. For B, event R precedes event Q. S(A) and S(B) are the simultaneity-lines of A and B respectively. (c) The same situation as depicted in B's rest frame.

Figure 3.1 The relativity of simultaneity

Source: Adapted from Ellis and Williams, *Flat and Curved Space-Times*, p. 87.

We cannot here delve into the mathematics of STR and derive the equations upon which these graphical representations are based. However, the point of interest is clearly illustrated. Two different observers in reference frames moving relativistically with respect to one another will arrive at different judgments about the B-relations of two events. The shocking result is that while A experiences R simultaneously with Q, B will experience R as earlier than Q. In A-theoretic terms, if R is 'now' for both observers, then R is 'now' for A but still future for B. So if simultaneity is relative, then in at least some cases the future is real. And if the future is judged to be real in any case, then one of the cardinal claims of dynamic time cannot be correct.

The principle underlying this argument is quite plausible. Call it the transitivity of reality principle:

> TR. For any two observers O_1 and O_2 and any event E, if E is real to O_1, and O_1 is real to O_2, then E is real to O_2.

Thus the relativity of simultaneity together with (TR) entails the reality of the future.

It is but a short step from this conclusion to the denial of absolute time. The key move in this step is Einstein's Principle of Equivalence, which says that the laws of physics are equivalent in any reference frame. This means, among other things, that there is no privileged reference frame—no absolute space or absolute frame of rest—with respect to which notions such as cosmic time can be defined. So if R is determined to be present in one reference frame and determined to be future in another, so be it. Nothing more can be said. In the absence of absolute time, there is no truth *simpliciter* as to whether R is present or future. The truth of temporally indexed propositions is relative to a reference frame

The implications for the causal theory of dynamic time are ruinous. There is no well-defined present marking the ontological boundary between the real past and the possible future, and no possible way to maintain causal connections as the general reduction base for temporal relations. In short, if simultaneity is indeed relative, not only must we abandon a causal theory of time, but also dynamic time altogether.[2]

Responses

I shall offer three responses to the problems raised by STR The first is to the effect that STR does not deal with metaphysical time per se, but rather with other temporal or non-temporal concepts. Thus, STR is about physical time, or about the behavior of light waves or physical bodies in light-connected relations, which may be analyzed in non-temporal terms. So the correct conclusion is not that there is no absolute time, but that the measurement of time is not absolute. The failure to keep separate the concepts of metaphysical time, which grounds all causal succession, of physical time, which obtains in the actual universe, and of the measurement of physical time, which is affected in ways predicted by the relevant laws, has led to all sorts of counter-intuitive and downright false philosophical claims about time.

The second argument takes on STR's treatment of physical time, but will suggest that it is possible to reformulate the mathematics of STR along the lines first outlined by Hendrick Lorentz, which is empirically equivalent to Einstein's formulation but does allow for absolute time and absolute simultaneity.

And, third, I will introduce two lines of astronomical evidence, specifically the anisotropy of the cosmic background radiation and the anisotropy

---

[2] This is the conclusion reached by, among others, Robert Weingard, 'Relativity and the Reality of Past and Future Events,' *British Journal for the Philosophy of Science* 23 (1972), pp. 119-21; David Malament, 'Causal Theories of Time and the Conventionality of Simultaneity,' *Noûs* 11 (1977), pp. 293-300; J. J. C. Smart, 'The Reality of the Future,' in *Essays Metaphysical and Moral* (New York: Basil Blackwell, 1987), pp. 94-9. At least two philosophers take this result to establish not only the falsity of the A-theory, but also the truth of determinism: Hilary Putnam, 'Time and Physical Geometry,' *Journal of Philosophy* 64 (1967), pp. 240-47; and C. W. Rietdijk, 'A Rigorous Proof of Determinism Derived from the Special Theory of Relativity,' *Philosophy of Science* 33 (1966), pp. 341-4.

of polarization of electromagnetic radiation across the universe, to support the possibility that there is in fact a privileged reference frame with respect to which absolute simultaneity may be defined.

Responses: STR and Dynamic Time

Three responses are available to the challenge STR presents to dynamic theories of time:
1. The definition of simultaneity under STR, which causes the difficulty for dynamic time, is conventionalist in nature. As such, it describes the behavior of clocks (or of light) in certain conditions, and hence our measurements. But this conventionalist definition does not call into question the metaphysics of time.
2. With the assumption of substantival space (defended in the previous chapter), it is possible to define notions of absolute rest and absolute simultaneity. This entails rejecting a core assumption of STR—namely, the constancy of the one-way velocity of light in all inertial reference frames in favor of a weaker assumption, the constancy of the average round-trip velocity of light.
3. Empirical support for the possibility of a privileged reference frame is found in certain observed anisotropies of electromagnetic radiation in the universe.

Conventionalist definitions The problems raised by STR for dynamic time begin with Einstein's conventionalist definitions of time and of simultaneity. For Einstein, time is 'what is measured by clocks,' an oft-repeated comment which reflects his early positivism.[3] Time is defined operationally, by readings on the clock used to determine quantitative temporal measurements. But the metric of time is a function of the laws of nature that obtain in a particular temporal world. So if the relevant laws affect the mechanism of the clock, then we would expect different results from quantitative measurements taken in different conditions that affect the clock's mechanism differently. (The famous time-dilation factor, which becomes significant at different points in a gravitational field or under conditions of acceleration, is a clear example.)

Since time is what is measured by clocks, Einstein stipulates that two events are simultaneous iff they occur at the same time as indicated by a clock at the locations of the two events. This presents no problem if the two events are close together. But if the events are separated widely in space, then a more elaborate method determining simultaneity is needed. First, two accurate clocks must be synchronized, and then moved apart (infinitely slowly to avoid the time-dilation effects of acceleration!). Once the clocks are in place, a synchronizing signal is sent from one to the other, and the second reflects the signal back to the first. The total round-trip time of the signal is measured and divided by two to determine the time required for the signal to go one way. But this rests on a crucial

---

[3] 'Certainly the original arguments in favor of the relativistic viewpoint are rife with verificationist presuppositions about meaning, etc. And despite Einstein's later disavowal of the verificationist point of view, no one to my knowledge has provided an adequate account of the foundations of relativity which isn't verificationist in essence.' Lawrence Sklar, 'Time, Reality, and Relativity,' in *Reduction, Time and Reality*, ed. R. Healey (Cambridge: Cambridge University Press, 1981), pp. 140-41.

assumption, one for which Einstein offered no argument: that the velocity of light was constant in both directions. To challenge this claim might seem audacious, since we all learned at our mother's knee that the speed of light, c, is constant in every direction. But the claim is open to question, as we shall see.

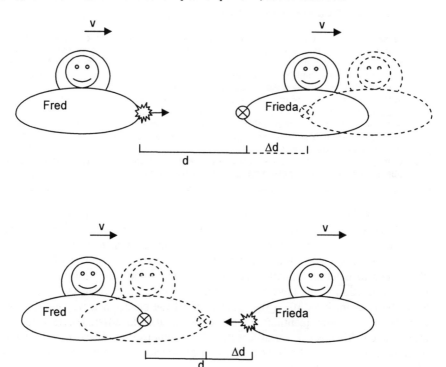

Two space ships moving at constant velocity v are at rest with respect to each other. (a) Fred's signal to Frieda travels a total distance of d + Δd, while (b) Frieda's return signal to Fred travels d − Δd. Since Fred and Frieda form an inertial reference frame, c should (according to the one-way assumptions of Einsteinian interpretations of STR) be the same in (a) and (b).

Figure 3.2 Average Round-Trip Velocity of Light

Consider the thought experiment illustrated in Figure 3.2. Two space ships are traveling at a high velocity v in tandem, d kilometers apart, but are at rest with respect to each other. Fred, in the trailing ship, sends a signal to Frieda in the lead ship (a). But by the time the signal gets to Frieda, her ship has moved beyond where it was when the signal was sent. The distance covered by the signal is d + Δd. If the signal is immediately reflected back to Fred, by the time it gets to Fred, he too will have advanced, so that the distance covered by the signal is d − Δd (b). Surely the time required for the signal to get from Fred to Frieda takes longer than from Frieda to Fred. But STR requires us to assume that since the two ships form a rest frame relative to each other, the one-way velocity of light will be c, not c + v or c − v (as in classical physics). True enough, the round-trip time will be 2d/c, but it is not clear that the signal arrives at Frieda and is reflected at ½ (2d/c).

Let t_1 be the time the signal leaves Fred; t_2 the time it arrives at Frieda and is reflected back; and t_3 be the time the reflected signal arrives back at Fred. Einstein's one-way assumption yields the equality of the intervals: $(t_2-t_1) = (t_3-t_2)$. Consequently, $t_2 = t_1 + ½(t_3-t_1)$. But the constant ½ reflects the one-way assumption. If this assumption is incorrect, then some other ratio will be needed.

Let ε be that constant. It turns out that there is no independent way to establish that $ε = ½$.[4] The choice of $ε = ½$ is pure convention.[5] If this conclusion to the thought experiment is correct, what it shows is that the method for synchronizing clocks is purely conventional and is based on the behavior of light, and has precious little to do with time as such. Einstein has abolished absolute time by fiat, and leaves us with nothing more than time as measured by clocks at rest in a particular reference frame.

An additional aspect of the conventionality of the formulation of STR is Einstein's assumption that space-time is Minkowskian. Craig notes:

> How could Einstein, in section 1 of his paper, already assume that spacetime had the structure determined by [STR], since he was still at that point attempting to discover the metric of space and time? To justify the definition of distant simultaneity based on clock synchronization via light signals by appealing to the metric of relativistic spacetime would plainly be question-begging.[6]

Conventionalism is a form of anti-realism, and since Einsteinian interpretations of STR rely on conventionalist assumptions, it is quite dubious to claim that the results of the theory should be interpreted realistically.

Neo-Lorentzian Interpretations If we replace the assumption of the constant one-way velocity of light with the weaker assumption that the average round-trip velocity of light is constant, then an interesting consequence follows. Given our earlier acceptance of substantival space, this allows for the retention of absolute time, irrespective of the relativistic motions of different clocks. And it also allows

---

[4] See Craig, *TMR*, pp. 30-42.
[5] Of course, Fred and Frieda are at rest relative to one another. The thought experiment assumes that there is an inertial reference frame with respect to which their velocities are indeed *v* and not 0. This is the assumption of substantival space. In rejecting the notion of the aether, Einstein did away with the Newtonian concept of absolute space. But I argued in the previous chapter for substantival space, and so have reason to disagree with Einstein here.
[6] Craig, *TMR*, p. 35.

for a definition of absolute simultaneity in terms of actual spatial connections without fear of distortions of relativistic space-time.

The basic idea is this: space-time points P_1 and P_2 are at the same spatial location at different times iff P_1 causes P_2 (or vice versa), and there is no third point P_3 which is causally related to either one of P_1 or P_2 but not the other. This allows a definition of rest with respect to absolute space and, in turn, a definition of absolute simultaneity as simultaneity with respect to an inertial reference frame at rest with respect to absolute space.[7]

The principle of the constant average round-trip velocity of light[8] then opens the door to a reformulation of the mathematical equations of STR. The treatment of STR offered by Tooley, which I shall describe, achieves just this sort of reformulation. The mathematics of STR even today are expressed using the Lorentz transformations, which allow calculation of the space-time coordinates in an inertial reference frame of space-time points given in a different inertial frame. The transformations must be modified if the one-way assumption is replaced by the average round-trip assumption for the velocity of light; the resulting transformations are known as the 'ε-Lorentz transformations.' These equations allow one to take account of variations in the one-way velocity of light.[9] While this reformulation brings along with it a certain theoretical baggage (for example, a necessary principle of the conservation of space), the resultant reformulation is consistent, entails the same empirical consequences as the Einsteinian formulation, and defines relations of absolute simultaneity.[10]

Tooley also argues that the modified version of STR has greater explanatory power. To note just one application where the explanatory power is greater, we can look at Quantum Mechanics. As early as June of 1980, Karl Popper noted that recent experimental results seem to confirm that 'action at a distance' was indeed part of QM. Popper suggested that this result had implications for STR, viz., that Einstein's assumption of the relativity of simultaneity had to be replaced by a neo-Lorentzian reformulation of STR that

---

[7] Michael Tooley, *Tine, Tense and Causation* (Oxford: Oxford University Press, 1997), pp. 340-44, hereafter *TT&C*. The simplification above follows Robin Le Poidevin, review of *Time, Tense and Causation*, by Michael Tooley, *The British Journal for the Philosophy of Science* 49 (1998), pp. 367-8.

[8] To be rigorous, the principle must include several additional stipulations, so that in the end it becomes something like this; 'The average round-trip velocity of light, propagated in a vacuum in a closed path, as measured within (note: not 'relative to') a given inertial reference frame, will be constant.'

[9] The interested reader can find descriptive details in Tooley, *TT&C*, pp. 348-54, and in Craig, *TMR*, pp. 171-94. For mathematical details see John Winnie, 'Special Relativity without One-Way Velocity Assumptions,' *Philosophy of Science* 37 (1970), pp. 81-99, 223-38.

[10] See Tooley, *TT&C*, pp. 354-60 for details. Tooley also argues that the modified version of STR has greater explanatory power.

included absolute simultaneity.[11] For without absolute space and absolute simultaneity, the observed results of experiments in QM are inexplicable. Quentin Smith remarks, 'There has been much hand-waving in this direction since then, but Tooley is the first to achieve a rigorous physical theory of EPR-based absolute simultaneity.'[12]

It seems, then, that since a neo-Lorentzian interpretation of STR is empirically equivalent to an Einsteinian interpretation and has greater explanatory power than an Einsteinian interpretation, we should actually prefer the neo-Lorentzian formulation. So since we have a version of STR that allows for absolute simultaneity, the alleged objection to dynamic time based on STR disappears.

Supporting empirical evidence Thirdly, there is positive empirical evidence that may suggest the possibility of determining a privileged reference frame for the universe. The foundations of this concept date to the 1965 discovery by Arno Penzias and Robert Wilson of the so-called cosmic background radiation, predicted by George Gamow in 1946. This microwave radiation has the spectral characteristics of black body thermal radiation at 2.76°K, and is amazingly isotropic in all directions. Data from the Cosmic Background Explorer satellite (COBE), launched in 1989, established the homogeneity of the background radiation to better than one part in 20,000 on all angular scales greater than 7°.[13] This background radiation, of course, was major confirmatory evidence for the Big Bang, and its isotropy seemed to indicate either that the Earth was at the center of the universe—an unthinkable coincidence—or that there was no privileged point or axis with respect to which one could define cosmic time.

Recently, more precise empirical observations have called this conclusion into question, for it appears that there is a very small anisotropy in the background radiation. COBE measurements show small fluctuations in the background radiation of the order of $5 \pm 1.5 \times 10^{-6}$ over angular scales of 10° to 90°.[14] This

---

[11] Karl R. Popper, *Quantum Theory and the Schism in Physics*, from *Postscript to the Logic of Scientific Discovery*, ed. W. W. Bartley, III (New York: Routledge, 1982), pp. 27-30. The issue has its roots in the Einstein-Podolsky-Rosen (EPR) thought experiment, proposed in 1937. The point of EPR was to show that the Copenhagen interpretation of QM entailed what Einstein called 'spooky action at a distance.' In 1964 J. S. Bell developed an equation that showed that EPR was in principle empirically testable. Alain Aspect's series of experiments in the late 1970s provided the empirical confirmation of Bell's inequality. See Peter Kosso, *Appearance and Reality: An Introduction to the Philosophy of Physics* (New York: Oxford University Press, 1998), pp. 133-51; Craig, *TMR*, pp. 223-33.
[12] Quentin Smith, review of *Time, Tense and Causation*, by Michael Tooley, *The Philosophical Review* 108 (1999), p. 127.
[13] G. F. Smoot *et al.*, 'Preliminary Results from the COBE Differential Microwave Radiometers: Large Angular Scale Isotropy of the Cosmic Microwave Background,' *Astrophysical Journal Letters* 371 (1991), L1-5.
[14] William Lane Craig and Quentin Smith *Theism, Atheism, and Big Bang Cosmology* (Oxford: Clarendon Press, 1993), p. 71. In February 2002, the first analyses of results of

small anisotropy is, however, sufficient to allow cosmologists to determine with some accuracy the motion of earth relative to that location with respect to which the universe is expanding isotropically. The reference frame of fundamental particles at rest with respect to the expansion of the universe would be the privileged reference frame needed to establish a standard for cosmic time.

> This special state of affairs, this carefully selected view of the cosmos, singles out the reference frame . . . [which] we can use to define a *cosmic* time, a time by which to measure historical change in the universe. Fortunately, the Earth is moving at only 350 kilometers per second relative to this hypothetical special clock. This is about 0.1 percent of the speed of light, and the time-dilation factor is only about one part in a million. Thus, to an excellent approximation, Earth's historical time coincides with cosmic time.[15]

More recently, a second set of empirical data has been interpreted in a way that might point to a preferred axis in the universe. An analysis of the polarization of electromagnetic radiation from over 160 distant galaxies led to the startling observation that:

> . . . the additional polarization rotation is *anisotropic* in nature as it depends systematically on the *direction* that the plane-polarized electromagnetic wave moves through space. This global, anisotropic dependency of the polarization rotation revealed itself as we systematically searched through the totality of all directions on the sky as seen from Earth.[16]

The researchers caution that the anisotropic axis, which runs through the constellation Aquila, Earth, and the constellation Sextans, represents only a direction and not a location in the universe. Still, the reason for this universal asymmetry of direction is puzzling, since it cannot be due, for reasons related to the nature of polarized electromagnetic radiation, to any effect that depends upon physical matter. The researchers surmise that 'it is Vacuum itself that flaunts a form of electromagnetic anisotropy.'[17] Stripping off the personification (all too common among cosmologists), what this comes to is a supposition that there is a privileged direction in space. While this is insufficient to establish a privileged reference frame for the entire universe, since such a frame would need location as well as directional orientation, it does suggest the possibility that such a frame exists.

---

data from the Wilkinson Microwave Anisotropy Probe were released, which confirmed to a greater degree of precision the results of the COBE observastions. For technical details, see http://lambda.gsfc.nasa.gov/product/map/map_bibliography.html
[15] Paul Davies, *About Time*, pp. 128-9.
[16] Borge Nodland, 'A Glimpse of Cosmic Anisotropy,' <http:// www.cc.rochester.edu/college/rtc/Borge/overview.html>, p. 2. The research was first published by B. Nodland and J. P. Ralston, 'Indication of Anisotropy in Electromagnetic Propagation over Cosmological Distances,' *Physical Review Letters* 78 (April 1997), p. 3043.
[17] Ibid., p. 4.

In summary, while the standard Einsteinian interpretations of STR pose problems for a causal theory of dynamic time, adequate responses are available, and it is possible to formulate the theory in terms of definitions and mathematical equations which are completely consistent with both empirical observations and theoretical constraints of dynamic time.

The General Theory of Relativity

If Einstein's 1905 publication of STR was challenging enough, both in its mathematics and in its overthrow of traditional views of space and time, his papers of 1916 deriving the General Theory of Relativity are far and away more challenging. In one sense, all GTR did was generalize STR to contexts involving acceleration and gravity. But Einstein struggled at this from 1905 through November of 1915, finally deriving a system of equations describing the gravitational field in geometrical terms. The equations, known as the Einstein Field Equations (EFE), employ very difficult mathematical techniques, and result in a geometry of space (actually space-time) that is curved, unlike the flat geometry of STR. GTR has proved a very fruitful theory and is regarded today as the best-confirmed scientific theory of the twentieth century.[18]

GTR and Dynamic Time

GTR is a much more complex theory than is STR, but it has been confirmed experimentally to a very high degree of accuracy. Three features of GTR pose difficulties for a theory of dynamic time.
1. Acausalities. Certain solutions of EFE allow for closed time-like curves, in which causation has direction but no order. Thus acausalities undermine causal theories of dynamic time.
2. Singularities. Elements of EFE 'blow up' (become undefined) under certain conditions, which have become known as singularities. Black holes and the Big Bang singularity are examples. The problem is that at singularities it is not at all clear what happens to time.
3. Topologies with handles. Some solutions to EFE allow for wormholes providing shortcut connections between distant space-time locations. Such topologies undercut causal dynamic theories of time.

Now, it is correct that if GTR is true, then STR—strictly—is not. For the existence of matter/energy within the universe described by EFE causes the curvature of space-time in the Riemannian geometry of GTR and destroys the equivalence of inertial observers postulated in the flat space-time of the Minkowskian geometry of

---

[18] For detailed discussions of GTR, from both the standpoint of physics and of philosophy, see Craig, *TMR*, chapter 10, pp. 195-241; Earman, *Bangs*; Ellis and Williams, *Flat and Curved Space-Times*, chapters 5-7, pp. 197-289; Friedman, *Foundations of Space-Time Theories*.

STR.[19] So it would have been tempting to eliminate the preceding discussion of STR entirely. However, theoretical approaches today regard STR as the 'first term of the expansion' of EFE in GTR. Just as Newtonian mechanics are a first approximation and accurate for non-relativistic velocities and non-astronomical distances, so STR is an accurate approximation for isolated or medium-sized objects moving at uniform velocities or at low accelerations. So if STR threatens dynamic time, the threat is transmitted to GTR. Having offered responses to STR, though, we must still consider the implications of GTR for dynamic time and offer responses to a new set of problems. Three problems arise in GTR: acausalities, the presence of closed time-like curves; the existence of space-time singularities; and the possibility of space-time topologies with handles (wormholes).

Acausalities: Closed Time-like Curves

There are many possible solutions to EFE, each of which represents a model of the universe. In general, a solution is defined as a triple <M, g_{ab}, T^{ab}> where M is a continuously differentiable four-dimensional manifold, a collection of space-time points together with a topology; g_{ab} is a metric-field tensor defined everywhere on the manifold, which specifies metric (for example, distance) and geometric relations among the points of the manifold; and T^{ab} is the stress-energy tensor which represents the distribution of matter and energy in the universe and imparts curvature to the space-time. The question of what elements of a particular model represent substantival space-time is an interesting one, and it is doubtful that the usual answer—that the manifold M gives space-time—is correct, since the manifold has no spatial or temporal structure without the metric tensor. Without the metric, one could not define spatio-temporal relations, so there could be no A-properties or B-relations, no light cones, and no distance relations. We accepted substantival space in the first place because it was a necessary precondition to an explanation of the causal relations that determine time. Without a metric, a manifold will not have the requisite structure to allow for causation, so the manifold alone cannot represent substantival space-time.[20] The point of this is that, even after more than 80 years of working with GTR, physicists are not agreed upon the physical interpretation of the elements of the theory. Consequently, announcements of scientists to the effect that certain metaphysical or physical points have been proved by relativity theory should not be taken as conclusive arguments.

---

[19] Earman, *Bangs*, p. 195. I should note, however, that there is one possible solution to EFE which incorporates a Minkowski manifold plus a metric tensor, but at the expense of being devoid of matter and energy.

[20] Carl Hoefer, 'The Metaphysics of Space-Time Substantivalism,' *Journal of Philosophy* 93 (1996), pp. 5-27, claims that it is the metric field tensor that represents substantival space-time. John Earman interprets the metric as giving the properties of the space-time points contained in the manifold, and so thinks that if substantivalism is correct, space-time is represented by the manifold plus the metric: *World Enough and Space-Time*, chapter 11. The point at issue above does not depend upon an answer to this question.

Through experience, skill and intuition, physicists have discovered models and families of models that represent solutions to EFE. Some are physically realistic, some clearly are not, and many are quite puzzling. Some of the boldest claims regarding time and GTR come from a class of solutions in which closed time-like loops are possible. One such family was discovered by the mathematician Kurt Gödel (Figure 3.3). In such worlds time travel is possible, indeed, ultimately inevitable.

The direction of time at the axis is along the vertical axis. But as one moves away from the axis, the angular acceleration affects the direction of time-like trajectories until the time-like direction is horizontal. At this region of Gödel space-time, a time-like trajectory would be closed, and time would loop back on itself.

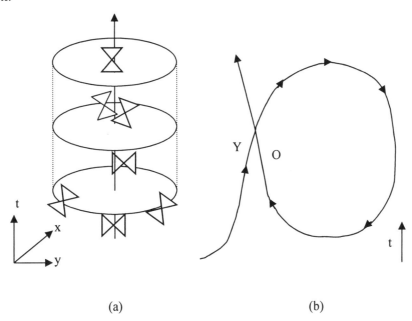

(a) (b)

(a) A rotating universe. The small local light-cones are vertical in the axis, but 'tip over' until they become horizontal as one moves away from the axis, resulting in a closed time-like trajectory. (b) An individual world-line in such a universe. The time-like line comes back to itself, so that it is possible for a young person (Y) to meet her old self (O).

Figure 3.3 Gödel's rotating universe

Source: Adapted from Ellis and Williams, *Flat and Curved Space-Times*, p. 228.

George Ellis and Ruth Williams interpret this scenario to imply that a young woman could possibly meet her much older self, as shown in Figure 3.3 (b).[21] But this interpretation is quite problematic. A point on a closed time-like trajectory in a Gödel world would have the B-properties of earlier/later than, at least with respect to its immediate neighbors, but the temporal ordering of the points would be impossible, since each point both precedes and succeeds itself. So it is not at all clear that any meaning such as 'younger' or 'older' could be given to segments of the closed trajectories.[22]

Be that as it may, scenarios such as this have spawned a good bit of science fiction. But are such universes possible? Well, yes and no. Gödel's models are mathematically possible solutions to EFE, certainly. But mathematical possibility is a long way from physical possibility. The mere fact that a Gödel solution to EFE incorporates the possibility of closed time-like loops does not entail that the modality is physical possibility. In fact, whether of the self-sustaining type or of the more pathological self-undercutting type, closed causal loops such as those envisioned in a Gödel world are impossible, as I argued in the previous chapter.

In point of fact, a Gödel solution is physically unrealistic on other grounds, viz. the stress-energy tensor requires all matter to be distributed uniformly throughout the universe. But neither that fact, nor the fact that such a model would violate any realistic analysis of causation, keeps Ellis and Williams from commenting that:

> It is in principle possible for an observer on any galaxy in this space-time to travel from any event in the galaxy's history to any previous event in its history, by accelerating far enough away from its world-line and then back. There is no evidence that this can occur in the real universe, but on the other hand this possibility (which raises various causal paradoxes) *has not been disproved observationally or experimentally.* We do not claim it is likely that the real universe is like this, but merely point out that curved space-time models exist where this is a theoretical possibility.[23]

It is difficult to know how to take this comment, even though it comes from two respected theoretical physicists. What exactly would it be to prove observationally or experimentally the impossibility of closed time-like curves? I suspect that the reason for their unwillingness to rule out the physical possibility of such space-time models stems more from their deep respect for the power of the mathematical tools used to formulate the models than to reflective thought about the actual physical or metaphysical implications of the models. My claim should be clear. We have already arrived at certain conclusions regarding the dynamic, causal nature and the topology of time. There is nothing within GTR that forces us to

---

[21] Ellis and Williams, *Flat and Curved Space-Times*, p. 228.
[22] This is Earman's complaint in *Bangs*, pp. 163-4.
[23] Ellis and Williams, *Flat and Curved Space-Times*, p. 229, my emphasis.

accept mathematically possible solutions to EFE that entail closed time-like loops as physically possible.

Space-Time Singularities

There is another feature within GTR that presents problems to a dynamic theory of time, the notorious singularities that were such an embarrassment to Einstein. Prior to the 1960s physicists largely ignored singularities, since Einstein himself believed they were mere artifacts of physically non-realistic models, or confined to the very earliest epochs of the history of the universe. With the publication of the Hawking and Penrose theorems in 1970, the problem returned with a vengeance, since the theorems show that singularities plague large classes of models, not merely a few. A singularity may be characterized as a point in space-time at which the Einstein Field Equations defining the behavior of space-time either 'blow up,' becoming undefined, or else oscillate. All physically realistic models of GTR predict singularities will occur at the Big Bang, at the gravitational collapse of stars (black holes), and possibly at the 'Big Crunch.'

To get an idea of what happens when the equations 'blow up,' consider the example of the time-like trajectory of an astronaut falling into a black hole, as depicted in Figure 3.4. The astronaut begins at A, some distance from the black hole. As he falls, he will approach the Schwarzchild radius r_s, the 'event horizon.' Nothing that has fallen within the event horizon can ever escape from the black hole, and no observer outside the horizon can observe anything within the horizon. The time axis t measures Schwarzchild time, which is time as measured by an observer far from the black hole. The astronaut's own watch records his proper time τ, which is measured along the time-like curve. The curve of τ is asymptotic at r_s, meaning that while the Schwarzchild time goes to infinity at B, proper time τ as measured along the curve does not. As the astronaut falls through the event horizon, the Schwarzchild time begins to decrease, even though proper time τ does not. Strangely, although t goes to infinity at B, the net Schwarzchild time is finite, as is the astronaut's proper time τ. In this example, several seeming absurdities appear. Inside the event horizon the astronaut's clock will appear to him as if it is running as always, but if *per impossibile* it could be observed from outside the horizon it would appear to be running backwards. Also, the net travel time as observed from a distance (Schwarzchild time) for the astronaut to fall into the singularity is finite, even though the time went to infinity part-way through the motion. Such absurdities are examples of the bizarre behavior of space-time at a singularity, and even if the precise nature of the problem is not apparent, the sort of problems indicated by the rough term 'blow up' is.[24]

So it is clear that there is a problem posed by singularities for dynamic time, but it is unclear just what the problem is. What interpretation should be given to the behavior of 'time' in the equations when they 'blow up'? In

---

[24] Stephen Hawking relates the account of the many bizarre experiences of the astronaut in *A Brief History of Time* (New York: Bantam, 1988), p. 87.

Minkowski space-time, a singularity is said to exist at point p if the electromagnetic energy density goes to infinity as one approaches p along any path. But when this concept is extended to the curved space-time of GTR, it is not clear what the 'blow-up' consists in. For the singularities of GTR are no longer unbounded increases in a physical quantity, but in space-time itself.[25]

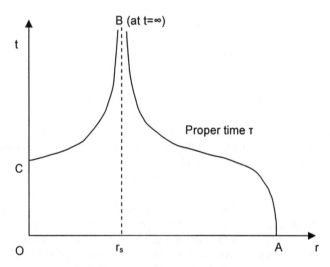

The distance from the black hole is measured along the r axis; r_s designates the Schwarzchild radius. The time axis, t, measures Schwarzchild time. The astronaut's proper time τ, the time kept on his watch, is measured along the time-like curve.

Figure 3.4 Astronaut falling into a black hole

What can be said about singularities? Three things, I think, which are not all that enlightening but which suffice to deflect the problem. The first, in what is becoming a familiar argument, is that the formulations of the problem are in terms of physical time as measured by clocks that are susceptible to the relativistic effects of gravity and acceleration. There is no reason to believe that the 'blow-up' of EFE at singularities has anything to do metaphysical time, nor is it clear how such formulations affect a causal theory of time. At the worst we might conclude that in certain regions of space-time the laws are such that the metric of physical time is locally altered, or that the causal relations observed elsewhere behave differently. (On a singularist account of causation, this may not be not a grave problem.) Second, as Earman has argued, singularities seem to be part of the boundary of space-time and are not properly a part of space-time itself. At present,

---

[25] Earman, *Bangs*, p. 28.

the theoretical nature of singularities is poorly understood, and it is not clear that singular points can be attached to the manifold.[26] Earman is not terribly optimistic that mathematical solutions to the problems will be forthcoming any time soon. Hawking notes that the presence of singularities shows that GTR is an incomplete theory.[27] The hope is that GTR can be combined with QM in a theory of quantum gravity, and while small successes have been achieved in certain carefully defined cases, the hoped-for unified theory does not seem imminent.

My third point relative to singularities is a bit more technical, and must begin with some definitions. If a space-time (comprising the two elements M, $g_{a,b}$) is constructed under GTR so as to rule out any 'pathologies' which would prevent a global Laplacian determinism, the space-time is called globally hyperbolic. Now any space-time contains space-like hypersurfaces, which are the relativistic analogue of constant time slices in Newtonian space and time. A globally hyperbolic space-time will contain a Cauchy surface, a hypersurface that partitions the space-time into past and future domains. By definition within GTR, and fully coherent with a causal theory of time, a Cauchy surface cannot be intersected more than once by any future-directed time-like or causal curve.[28]

It follows that any globally hyperbolic space-time will contain a Cauchy surface, and is thus determinately partitionable into past and future. Further, it follows that a globally hyperbolic space-time cannot contain any closed time-like curves (due to the presence of the Cauchy surface), and also, for the same reason, cannot contain naked singularities. A naked singularity is a space-time point observable in principle at which EFE become undefined. This would mean that, in Earman's words, 'naked singularities are sources of lawlessness' due to the breakdown of determinism and/or predictability.[29] In 1969 Roger Penrose offered the Cosmic Censorship Conjecture: Nature abhors a naked singularity. If there are singularities in our universe, he suggested, they are all conveniently hidden in black holes where they are unobservable. Efforts to prove cosmic censorship as a theorem of GTR have failed, although it has not been disproved either. However, it has been proved that cosmic censorship holds for globally hyperbolic space-times.[30]

---

[26] Ibid., chapters 2-3.

[27] Stephen Hawking, 'Classical Theory,' in Hawking and Penrose, *The Nature of Space and Time*, p. 20.

[28] See Craig, *TMR*, pp. 202-41 for an extended discussion of the foliation of space-time consistent with the reality of cosmic time. Russian physicist Aleksandr Friedmann in 1922 discovered a viable solution to EFE, which has been extended by others. Currently, those models that are physically realistic are the FRW (Friedmann-Robinson-Walker) family of solutions to EFE. There are natural symmetries in FRW models that guide a preferred foliation as described above.

[29] Earman, *Bangs*, p. 65. Much of the discussion in this section is based on chapters 2 and 3 of Earman's work.

[30] It should be noted that the debate about the Cosmic Censorship Conjecture rages on among cosmologists and theoretical physicists. In the late 1990s, Stephen Hawking conceded that the conjecture had been disproved. But his concession seems to have been

The significance of this is that all standard models of the Big Bang are globally hyperbolic, meaning that in any physically realistic model devised so far, cosmic censorship holds.[31] According to the mathematical formulations of GTR, all instantaneous information relevant to the evolution of a globally hyperbolic space-time is specified on the Cauchy surface, so the pathological lawlessness of singularities will not affect the causal structure of time in such a universe, and apparently—happily—ours is such a universe. And to anticipate the next section slightly, Hawking states:

> One can predict what will happen in U from data on the Cauchy surface, and one can formulate a well-behaved quantum field theory on a globally hyperbolic background. Whether one can formulate a sensible quantum field theory in a nonglobally hyperbolic background is less clear. So global hyperbolicity may be a physical necessity.[32]

I conclude that our degree of warrant for accepting a causal theory of dynamic time is greater than that for rejecting a causal theory on the grounds of the possibility of space-time singularities, since even if there are real singularities in the actual universe, they would be censored and unobservable.

Topologies with Handles

Talk of wormholes in space-time sounds more like science fantasy than genuine physics. Yet cosmologists are serious about the idea.[33] A wormhole is a 'handle' in the topology of space-time, connecting two widely separated times and places. Theoretically, a wormhole would allow causal connections between otherwise causally unconnected regions of space-time, and—theoretically—allow for time travel.

As we have seen, all known physically realistic models of space-time are globally hyperbolic, allowing for the definition of an absolute 'now' by means of a hypersurface. The topography of this Cauchy surface could be quite complicated, containing bumps and valleys, ripples and dips. But given the definition of a Cauchy surface—that no time-like causal trajectory intersects it more than once—it will not allow for folds or twists. It would have the topological properties of a sheet or surface. The ripples could account for the relativistic effects on local time and space, but there would still be one universal 'now' defined by the surface. By contrast, a space-time that allowed for handle topologies would not be globally

---

premature. A recent web search for 'cosmic censorship' recorded over 10,000 hits, many links to academic papers and on-line refereed journal articles, as well as moderated discussion lists, which deal with the topic.

[31] Earman, *Bangs*, p. 69, points out that this means that the Big Bang singularity is not naked either.

[32] Hawking, 'Classical Theory,' pp. 9-10.

[33] For example, Hochberg and Kephart postulated that at Planck time (about 10^{-43} second after the Big Bang), the universe was riddled with wormholes, thus (perhaps) offering a solution for the smoothness problem. See Earman, *Bangs*, pp. 147-8.

hyperbolic. (Of course, other amazing conditions are required for the existence of wormhole, such as a hyperspace within which the topological handle can form, and just the right amount and kind of exotic matter to hold the wormhole open.) [34]

At the same time, however, I cannot ignore the great predictive successes of STR and GTR. Is it possible to construct a model of time that is both dynamic and causal, and still allows for a space-time that conforms to the more common physical interpretations of GTR?

Storrs McCall has suggested that in such a universe the past, represented by the trunk of a tree, would be a single four-dimensional manifold.[35] (A manifold, I would add, which is modeled by a physically realistic solution to EFE.) The present would be a Cauchy surface, a space-like hypersurface of simultaneity that is the 'leading edge' of reality, the place where possible but non-actual future branches are cut off, and the real space-time structure grows by addition. If a causal theory of dynamic time is correct, then the 'advance' of this cutting edge of reality is brought about by causation.

Relativity: Summary

In summary, I want to emphasize a more general point that I have made before. The mere fact that a solution to a mathematical equation exists is no reason to believe that the solution is physically realistic. When I want to know how long to cut a piece of plywood for a peaked roof on a doghouse, I can use the Pythagorean Theorem and get two answers: 35 inches, and −35 inches. It takes very little deliberation to decide that I should not cut the plywood negative 35 inches from the edge. But that is a mathematical possibility. Of course, it is much more difficult to give a physical interpretation to the mathematical expression of a solution to EFE, but the principle is the same: not all mathematical possibilities are physical possibilities.

And as noted in the previous chapter, mathematical abstractions of the laws of nature or of descriptions of physical processes are formalized in time-symmetric terms; prediction and retrodiction are equivalent (in most cases). So when physicists derive their equations and find solutions, there is the to think of the solutions in non-modal terms—as mapping actuality as opposed to possibility. For example, when cosmologists depict light-cones, with the present at the point where two oppositely oriented cones touch at their vertices, the temptation is strong to regard the future light-cone as being just as real as the past light-cone (see Figure 3.5).

The temptation is strengthened when one asks what ontological status is to be assigned to events outside the light-cones of an observer O, such as event E in Figure 3.5. Is it real, or not? Not being in the past or future light-cones of O, is

---

[34] See Kip Thorne, *Black Holes and Time Warps* (New York: W. W. Norton, 1994), pp. 483-521.
[35] Storrs McCall, 'Objective Time Flow,' *Philosophy of Science* 43 (1976), p. 342.

it simultaneous with O at the present? Is it undefined? There is no clear answer at all unless one assumes static time and the reality of a four-dimensional space-time.

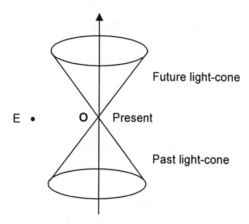

Figure 3.5 Past and future light-cones of an observer O

If the dynamic theory of time is correct, then of course such realism about the future must be rejected. I think that the correct claim is that the light-cones show regions that are causally or epistemically accessible, given the laws of nature that obtain in the actual universe. We should not think that the future light-cone represents actual reality, but rather represents the region of causal accessibility, or possibility.

I conclude that, in spite of the titillation of some of the bizarre interpretations of STR and GTR, there seems to be at this time no reason whatsoever to accept the claim that relativity has proved that time is static, or that determinism is true, or that time travel will work, or that backward causation is possible. In fact, as I have shown, a causal theory of dynamic time is compatible with both STR and GTR if certain relatively simple and plausible assumptions are granted.

Quantum Mechanics

Relativity theory had to do with very large scales, QM with the very small. From its beginnings in the 1920s, QM has made claims that strike common sense as just as outlandish, if not more so, as claims of STR or GTR. And if the latter are more generally accepted, perhaps it is only because they had a head-start of a decade or two. Curiously, some aspects of QM pose a threat for STR as well, notably the notion that non-locality, rather than local hidden variables, is really at work in certain quantum effects. But, if taken at face value, non-locality would seem to necessitate that absolute simultaneity holds in the universe.[36]

While many other aspects of QM are counter-intuitive and raise difficulties for a critical realist approach to science, only the difficulties posed for a causal theory of dynamic time will be surveyed here.

Problems

I shall consider two difficulties for dynamic time that are raised by QM. First is the suggestion by Hawking that at some epoch earlier than Planck time (10^{-43} seconds after the Big Bang) quantum effects merged space and time into his well-known notion of 'imaginary time.' Second, some interpretations of QM regard time at the quantum level as either 'fuzzy' or quantized (i.e. discrete), thus posing conceptual difficulties for a dynamic, causal theory of time.

QM and Dynamic Time

Two difficulties for dynamic time arise from QM:
1. Quantum cosmology might imply that there is such a thing as imaginary time. If correct, this would call into question the whole notion of dynamic time.
2. On a more general level, QM raises conceptual difficulties for dynamic time via such notions as 'fuzzy' time, quantized time, and Heisenberg uncertainty.

Quantum cosmology and imaginary time Hawking's claims about imaginary time emerged in a famous paper coauthored with James Hartle in 1983.[37] Hartle and Hawking attempted to combine QM with a physically realistic solution to EFE and derive the wave function of the universe. This quantum cosmology approach to the Big Bang is enticing, for it is evident that the conditions of the universe in early epochs of its expansion are subject to the laws of QM rather than GTR. Instead of starting with a Big Bang singularity, Hartle and Hawking postulated certain initial conditions in which the beginning of the universe was a hemispherical Euclidian four-space, with the time dimension being imaginary. From that paper Hawking went on to make several claims. Among them were that since the proposed

---

[36] It is beyond the scope of this book to discuss the proper interpretation of Bell's theorem; yet I will merely note that there is a renewed interest in some form of global hidden variable theory as opposed to non-locality. I thank Michael Friedman for calling my attention to this in conversation.

[37] J. Hartle and S. W. Hawking, 'Wave Function of the Universe,' *Physical Review D* 28 (1983), pp. 2960-75. My comments here are based upon the following papers: William Lane Craig, '"What Place, Then, for a Creator?" Hawking on God and Creation,' and Quentin Smith, 'The Wave Function of a Godless Universe,' both in *Theism, Atheism, and Big Bang Cosmology*, pp. 279-300 and 301-37; William Lane Craig, 'Hartle-Hawking Cosmology and Atheism,' Quentin Smith, 'Quantum Cosmology's Implication of Atheism,' and Robert J. Deltete and Reed A. Guy, 'Hartle-Hawking Cosmology and Unconditional Probabilities,' all in *Analysis* 57 (1997), pp. 291-5, 295-304, and 304-15 respectively.

spacetime, being hemispherical, has no edge or point (singularity), the boundary conditions of the universe are that it has no boundary conditions. He also claimed that time is imaginary at the beginning of the universe, so there is no temporal beginning.[38]

The threat to dynamic time here is hard to assess. It is not at all clear what imaginary time comes to, or even if the term has meaning. Nevertheless, it does seem that some response must be given since if time can be imaginary, that would constitute a prima facie case against a causal theory of time.

As a response, I'd begin by noting that Hawking's speculations depend on an as-yet nonexistent theory of quantum gravity and cannot be construed in realist terms. Most quantum physicists and philosophers of science would agree with Roger Penrose:

> The theory [QM] has, indeed, two powerful bodies of fact in its favor, and only one thing against it. First, in its favor are all the marvelous agreements that the theory has had with every experimental result to date. Second, and to me almost as important, it is a theory of astonishing and profound mathematical beauty. The one thing that can be said against it is that it makes absolutely no sense![39]

Of all the physical theories in general use today, QM seems far and away the most likely candidate to be labeled operationalist rather than realist. There is no question that QM yields accurate predictions, and equally no question that much of QM defies physical interpretation.

Hawking's use of the notion of imaginary time must be regarded as non-realist. He asserts, 'Imaginary time may sound like science fiction but it is in fact a well-defined mathematical concept.'[40] Of course, the use of imaginary numbers—those involving the square root of a negative number—is an important and well-defined tool in trigonometry and has important applications in electrical engineering. But it surely is a mistake to confuse the usefulness of a formal technique with reality. The astronomer Arthur Eddington, as far back as 1920, suggested using imaginary numbers for the time coordinate of spacetime, but noted that 'it can scarcely be regarded as more than an analytical device' which 'certainly does not correspond to any physical reality.'[41] Hawking himself sounds a bit like an operationalist when he says, 'Only if we could picture the universe in terms of imaginary time would there be no singularities. . . . When one goes back to the real time in which we live, however, there will still appear to be singularities.'[42] So it would seem that even he regards imaginary time as nothing more than a mathematical fiction. But he only adds to the confusion in the following passage:

---

[38] Hawking, *A Brief History of Time*, pp. 133-6.
[39] Roger Penrose, 'Gravity and State Vector Reductions,' in *Quantum Concepts in Space and Time*, ed. R. Penrose and C. Isham (Oxford: Clarendon Press, 1986), p. 129.
[40] Hawking, *A Brief History of Time*, p. 134.
[41] Cited by Craig, '"What Place, Then, for a Creator?": Hawking on God and Creation,' *British Journal for the Philosophy of Science* 41 (1990) p. 291.
[42] Hawking, *A Brief History of Time*, p. 144.

This might suggest that the so-called imaginary time is really the real time, and that what we call real time is just a figment of our imaginations. In real time, the universe has a beginning and an end at singularities that form boundaries to space-time and at which the laws of science break down. But in imaginary time, there are no singularities or boundaries. So maybe what we call imaginary time is really more basic, and what we call real is just an idea that we invent to help us describe what we think the universe is like.[43]

In Hawking's articulation of his theories, he offers no physical interpretation of imaginary time, leaving the reader wondering if the concept is not physically unintelligible. And he frequently asserts a four-dimensional ontology that spatializes time: 'The use of imaginary numbers has an interesting effect on space-time: the distinction between time and space disappears completely. . . . there is no difference between the time dimension and directions in space. . . . time is imaginary and is indistinguishable from directions in space.'[44]

To assert that time and space are indistinguishable simply because certain mathematical representations treat the two entities the same is to fail to recognize a type distinction. That measures of time and space are both determinable quantities has no relevance in deciding the nature of what is being measured. Even most B-theorists, who are not uncomfortable with the notion of 'spatializing time,' will not go so far as to assert the indistinguishability of time and space.

In brief, there is nothing in Hawking's development of the Hartle-Hawking cosmology which should be regarded as anything more than instrumentalist. The basic failure is the mistaken notion that one can move directly from mathematics to physical interpretation with no intervening philosophizing.

'Fuzzy' time, quantized time   The second sort of difficulty related to time that stems from QM has to do with interpretations of certain of phenomena of time at the quantum level. The phenomena are interpreted variously as showing that time is 'fuzzy,' or in some sense indeterminate, or that space-time at the quantum level is foam-like. Other phenomena are interpreted as showing that quantum time comes in discrete units (chronons). The Heisenberg Uncertainty Principle and the seemingly intractable measurement problem call into question the relation between causation and time. And the possibility of non-locality—instantaneous action at a distance—does the same. There is no need here to describe the various experiments in question; such descriptions would either be too superficial to be of any value, or so technical that they would not belong in a chapter like this.

Nevertheless, there clearly are conceptual difficulties that QM poses for causal dynamic time. For example, if time is 'fuzzy' at the quantum level, then we seem to be faced with a dilemma. Either the present is a fuzzy region in which, however briefly, time and causation do not behave as they do on the macro-level, or else there are mechanisms at play in this phenomenon that seem to violate all known laws of physics.

---

[43] Ibid., p. 139.
[44] Ibid., pp. 134-5.

I must confess to some bewilderment when it comes to quantum phenomena. Watching demonstrations of quantum tunneling, or versions of the double-slit experiment, is an exercise that stirs wonder and amazement. It seems clear that Aspect's experimental confirmation of Bell's theorem proves conclusively that one cannot have local hidden variables and retain determinate properties of quantum particles, such as spin. The only alternative seems to be non-locality, instantaneous action at a distance. This would seem to strike at the heart of a causal theory of time, requiring not only simultaneous causation, but also simultaneous causation at non-spatially contiguous locations. And the list could go on.

Once again I shall resort to an oblique response. Perhaps the central internal difficulty in QM is the measurement problem. The measurement problem can be stated in several ways, but central to it is the notion that when a measurement is made, an interaction occurs between the measuring device and the system being measured. All the weirdness at the quantum level (superpositions, collapsing wave functions, Heisenberg uncertainties) somehow results in a determinate measurement, and the interactive process of measurement is not at all understood.[45] Earman notes that the measurement problem in QM indicates that the theory is incomplete: 'The measurement problem in QM shows, prima facie, that the theory is empirically inadequate in the worst way: it cannot account for the fact that measurement procedures yield definite outcomes.'[46] The fact that QM cannot explain how its empirical results are obtained in any one particular case is an argument for the incompleteness of the theory. And, as I have already claimed, there is no reason to regard QM in realist terms. So since QM has no real explanation for the phenomena that allegedly call into question the nature and direction of causation and time at the micro-level, I do not feel compelled to offer an explanation of my own. Is this the coward's way out? Perhaps, but better to note that only fools rush in to give realist interpretations of such wondrous mathematical constructs.

Conclusion

On balance, investigation of the metaphysics of time favor a causal theory of time in which time is dynamic and the past and present are real and a multiplicity of futures are possible. Study of the salient physical theories reveals various interpretations that might challenge this theory of time, but there are quite reasonable responses to each of them. I conclude, then, that the theories and empirical results of modern physics, while often used in arguments for static time, cannot alone decide the issue against dynamic time. And since there are strong

---

[45] Kosso, *Appearance and Reality*, pp. 163-76; Ian Stewart, *Does God Play Dice? The New Mathematics of Chaos*, 2nd ed. (Malden, MA: Blackwell, 1997), pp. 329-56.
[46] Earman, *Bangs*, p. 222. He opines earlier (p. 4), 'my own conviction is that neither clever semantical rules nor extravagant metaphysics will suffice and that eventually new physical principles will have to be recognized.'

philosophical reasons for preferring dynamic time, we are justified in retaining that belief. In Part II, then, I shall show how the theory of time that I have just developed and defended yields an understanding of God's temporal mode of being.

PART II
SCRIPTURE AND TRADITION

Chapter 4

The Evidence from Scripture

All theoretical works presuppose certain foundational truths. This is a work in philosophical theology, and more explicitly philosophical theology in the Judeo-Christian tradition. Hence certain foundations will be assumed and not argued for. (Although I believe there are good arguments for each of the assumptions, they are far beyond the scope of this book.) Specifically, the existence of and rationality of belief in God, the evidential force of Scripture, and the weight of the Judeo-Christian theological tradition, function as 'control beliefs.'[1] Thus in Part II biblical evidence and traditional formulations relating to God's temporal mode of being will be examined, along with contemporary formulations and arguments.

The present chapter deals with the evidence from Scripture. This evidence is of two kinds: philological and exegetical. The philological evidence takes into account the semantics of the different Hebrew and Greek words used for the concepts of time and eternity, while the exegetical evidence looks at texts that have implications for the topic at issue.

Philology: Hebrew and Greek Words for Time and Eternity

The Hebrew and Greek texts of the Old and New Testaments form the basis of the Judeo-Christian tradition. While neither systematic nor dogmatic theology, a work in philosophical theology in this tradition must consider the contributions made by these basic documents. Specifically, the question is whether the biblical texts require either an atemporal or a temporal view of God's mode of existence, or whether the evidence is neutral between the views.

Only philosophical analysis can reveal the nature of time. But study of the semantics of Hebrew and Greek terms used in the Tanakh[2] and in the New Testament for time and eternity can reveal how time was perceived by the

---

[1] I use the term in the sense developed by Nicholas Wolterstorff, *Reason within the Bounds of Religion*, 2nd ed. (Grand Rapids, MI: Eerdmans, 1984), especially pp. 72-80. His discussion clearly illustrates the function such control beliefs play in *any* intellectual inquiry.
[2] 'Tanakh' is an acronym formed from the first letters of the Hebrew designations of the three divisions of the Hebrew Scriptures, and is widely used by Hebrew-speaking Jews to refer to their Scriptures. Although the order of the books in the Tanakh is different from the order of the canonical books in the Old Testament in English Bibles, the contents are the same. I shall refer to this collection as the Tanakh so as to avoid possible pejorative connotations some might associate with the term 'Old Testament.'

respective cultures. Also, by noting how these terms were used in reference to God, we can see God's temporal mode of being and his relation to time were conceived by the formative communities of faith.

In what follows I shall demonstrate that an examination of significant Hebrew and Greek words used to speak of time and eternity shows that none of them must be taken as referring to time as static or substantial, or to eternity as timeless. To the contrary, the basic concepts in both languages seem to be that of duration or sequence (with eternity being endless, limitless time). Hence the biblical data do not require an idea of an atemporal God or of timeless eternity, and in fact might seem to support a concept of eternity as everlasting, entailing duration and succession, rather than the more traditionally accepted notion of eternity as timelessness.

Significant Hebrew and Greek Words used for Time and Eternity	
Hebrew	Greek
'ôlām (עוֹלָם)	aion (αἰών)
nētsakh (נֶצַח)	kairos (καιρός)
'ēt (עֵת)	chronos (χρόνος)
'attâ (עַתָּה)	
yôm (יוֹם)	
mo'ēd (מוֹעֵד)	

Before beginning the detailed study I shall make a few comments concerning methodology. From roughly the beginning of the twentieth century, biblical theology was dominated by philology grounded almost exclusively in an etymological approach.[3] This is perhaps understandable, coinciding as it did with the birth of modern comparative linguistics and the discovery of a wealth of material from Indo-European and cognate Semitic languages—Akkadian, Babylonian, Arabic, Aramaic and Ugaritic. But, in the process, one of the fundamental principles of linguistics was ignored: usage, not etymology, determines meaning.[4]

The assumption of the older methodology was that the theoretical framework of a people could be 'read off' their lexical stock for different concepts. The fallacy, of course, was that by analyzing the etymology of words, a philologist

---

[3] The approach is pervasive in at least the first volume (German publication in 1933) of the monumental ten-volume *Theological Dictionary of the New Testament*, ed. Gerhard Kittel, tr. Geoffrey W. Bromiley (Grand Rapids, MI: Eerdmans, 1964 *seq*.). The influence of the method can be seen throughout the remaining volumes, and also in the companion *Theological Dictionary of the Old Testament*, ed. G. Johannes Botterweck and Helmer Ringgren, tr. John T. Willis and David E. Green (Grand Rapids, MI: Eerdmans, 1974 *seq*.), begun in 1970 and still in progress; hereafter *TDNT* and *TDOT* respectively.

[4] See the thorough critique of this methodology by James Barr, *The Semantics of Biblical Language*, (Oxford: Oxford University Press, 1961); M. Silva, *Biblical Words and their Meaning: An Introduction to Lexical Semantics* (Grand Rapids, MI: Zondervan, 1983).

could assign the word to a particular theoretical role quite independent of the psychological, theological or philosophical presuppositions of the philologist herself. But invariably such studies arrange the lexical material into categories derived from—where? Simon de Vries comments:

> We are altogether justified in rejecting any method that proceeds deductively on the basis of a psychological theory rather than inductively on the basis of contextual evidence. . . . Unfortunately, theologians have at times been all too eager to misuse philological data for the purpose of a facile systematic arrangement.[5]

Thus, although I shall begin this chapter with word studies, the purpose is to show that conclusions urged by others are untenable even on the grounds of their own methodology. Ultimately it is the statements of the Bible that must be considered if one desires to demonstrate a unique biblical doctrine of time.[6]

Hebrew Words

Much of what has been written concerning the biblical view of time has relied on the work of Conrad von Orelli, published in 1871.[7] But Orelli was strongly influenced by Kantian and Hegelian philosophy as well as by the etymological method. Much of what has been written about 'the Hebrew view of time'—for example, that the Hebrews were the first ancient culture to have a realistic view of history, or that the Hebrews conceived of time as linear while other cultures, such as the Egyptians and especially the Greeks, conceived of time as cyclical, or that the Hebrews viewed time as substantial rather than chronological—is due to Orelli's influence. But even a cursory study will show that attributing to the Hebrews a doctrine of time unique among ancient peoples is mistaken. While my purpose in what follows is thus to a large degree corrective, I do not intend to offer a detailed critique of the conclusions reached by Orelli and his followers. Such critiques are readily available in the works already cited by Barr and de Vries, among others. Where I do discuss etymologies, they are included to give insight into how usage developed. The primary focus of the study below is on usage in context.

---

[5] Simon J. de Vries, *Yesterday, Today and Tomorrow: Time and History in the Old Testament*. (Grand Rapids, MI: Eerdmans, 1975), pp 32-4, hereafter *YTT*.

[6] A notable example of proper methodology in compiling a 'theological wordbook' is the *Theological Wordbook of the Old Testament*, 3 vols., ed. R. Laird Harris, Gleason L. Archer, Jr., and Bruce K. Waltke (Chicago: Moody Press, 1980); hereafter *TWOT*. I have benefited from their treatment of specific words in what follows.

[7] Conrad von Orelli, *Die hebräischen Synonyma der Zeit und Ewigkeit genetisch und sprachvergleichend dargestellt* (Leipzig: A. Lorentz, 1871). Orelli's methodology, conclusions and influence are thoroughly discussed by de Vries, *YTT, passim*, and by James Barr, *Biblical Words for Time*, rev. ed. (Naperville, IL: Alec R. Allenson, 1969), chapter 4, hereafter *BWT*.

'ôlām (עוֹלָם)

Etymology This word is probably derived from root *'lm*, 'to hide'; hence, 'what is hidden in the distant past or future.'

Usage This word occurs 440 times in Hebrew (and another 20 times in the short Aramaic portions of the Tanakh) to indicate continuance into the indefinite future; and some 20 times to indicate 'a time long ago' (which leaves open the possibility of, but does not require, a limitless past). *'ôlām* never occurs independently but only as the object of a preposition indicating direction ('since,' 'until,' etc.), as an adverbial accusative of direction, or in a genitive construct.

Perhaps the basic meaning is 'most distant times,' either past or future, depending upon the syntactical indicators. However, certain usages favor a meaning of 'indefinite duration, perpetuity,' such as Exodus 21:6, 'He will be his servant in perpetuity [*lᵉ'ôlām*],'[8] or Leviticus 25:34, 'The pastureland belonging to their towns is not to be sold; it is their land in perpetuity [*lᵉ'ôlām*].' The word does not in itself contain the notion of limitless duration, as shown by its use to refer to a definite period in the past, such as Joshua 24:2: 'Long ago [*mē'ôlām*], your forefathers, including Terah the father of Abraham and Nahor, lived beyond the River.' The basic meaning of the word, then, would be somewhere in the range between 'perpetuity, indefinite duration,' and 'remote, distant times.'

In the Qumran material *'ôlām* is quite common, with much the same meanings as seen in the canonical books of the Tanakh. A common phrase is *lᵉkôl qasēy 'ôlāmîm*, 'for all the ages of eternity.'[9]

Given the semantics of *'ôlām* as seen in non-theological uses, when *'ôlām* is used attributively of God it would seem that a notion of everlastingness or perpetuity (sempiternity—everlasting duration in time—rather than atemporality) is intended. Thus in Genesis 21:33, '[Abraham] called upon the name of the Lord, the Eternal God [*ēl 'ôlām*]' (New International Version), 'Everlasting God' would be a better translation.

The plural *'ôlāmîm* is also found. It might be tempting to conclude from this word that the Hebrews conceived of sequence (successive ages) in eternity, but it may be more plausible to interpret it as a plural of extension: 'The plural of extension is used to denote a lengthened period of time in עוֹלָמִים [*'ôlāmîm*] eternity.'[10]

---

[8] Barr, *BWT*, p. 73 n.1. The New International Version sensibly translates the expression 'for life.' *The Holy Bible, New International Version* (New York: New York International Bible Society, 1978).
[9] Barr, *BWT*, p. 124-5.
[10] W. Gesenius and E. Kautzsch, *Hebrew Grammar*, 2nd English ed., tr. A. E. Cowley, (Oxford: Clarendon Press, 1910), p. 397.

nētsakh (נֵצַח)

Etymology   The Hebrew verbal root denotes a wide range of meanings from 'be perpetual' through 'excel' (and even 'oversee' as a Piel participle[11]). The Arabic cognate connotes both 'pure' and 'reliable.' The basic idea seems to be that which is enduring and strong as well as that which is pure or brilliant.

Usage   In three of the 42 occurrences of the noun *nētsakh*, the word means 'victory' or 'majesty,' as in 1 Chronicles 29:11: 'Yours, O LORD, is the greatness and the power and the glory and the majesty/victory [*nētsakh*].' This meaning is apparently derived from the verbal root's sense of 'to excel.' The remaining 39 occurrences have the sense of 'continual' or 'perpetual.' This is seen in such examples as 2 Samuel 2:26, 'Abner called out to Joab, "Must the sword devour forever [*halānētskh*]?"' or Jeremiah 15:18, 'Why is my pain unending [*nētsakh*]?' In Isaiah 34:10 the plural is used in parallel to *'ôlām*: 'It will not be quenched night and day; its smoke will rise forever [*'ôlām*]. From generation to generation it will lie desolate; no one will ever [*nētsakh n^etsakhim*] pass through it again.' Here, as for *'ôlām* above, we should understand the plural as a 'plural of extension,' thus, 'continual continuation' or some such connotation. There is no connotation of timelessness, but rather of long duration.

Although *nētsakh* is never used attributively of God, note Psalm 16:11: 'You have made known to me the path of life; you fill me with joy in your presence, with eternal [*nētsakh*] pleasures at your right hand.' But, given the semantic range that we have seen for *nētsakh* in other usages, it would beg the question to read this as timeless eternity rather than as unending duration.

'ēt (עֵת)

Etymology   The same verbal root *'nh* exhibits four clearly distinct semantic ranges: (i) 'to answer, respond, speak, shout'; (ii) 'to be occupied, busy with'; (iii) 'to afflict, oppress, humble'; and (iv) 'to sing.' Such diverse meanings may well indicate that distinct homonymous roots crept into Hebrew from different languages. While the standard Hebrew lexicon suggests that *'ēt* derives from (i),[12] it seems much more likely that the derivation is from (ii), 'to be occupied'—that is, a filled temporal extension. Nevertheless, uncertainty plagues the etymology.

Usage   The word *'ēt* occurs 296 times in the Tanakh, making it one of the more common Hebrew words for time in general. It is used of recurring or regular events, such as the rains in Ezra 10:13: 'it is the rainy season [*'ēt*].' The word is

---

[11] The Piel stem generally conveys an intensive form of the root; hence 'to excel greatly' leads to the concept of 'to oversee.'

[12] Francis Brown, S. R. Driver, and Charles A. Briggs, ed., *A Hebrew and English Lexicon of the Old Testament* (Oxford: Clarendon Press, 1907), s.v. עָנָה.

also used of nonrecurring events which are considered to have come at the appropriate time, such as death in Ecclesiastes 7:17: 'why die before your time [*'ēt*]?' A third group of usages refers to the suitable times God has designed for all of life's activities, best exemplified in the well-known poem from Ecclesiastes 3: 'To everything there is a season, and a time [*'ēt*] to every purpose under the heaven.' The poem continues repeating *'ēt* some 16 times.

While these usages have been adduced jointly as evidence for a Hebraic view that conceives of time as 'opportunities' or 'seasons,'[13] this would seem to force a particular meaning on the general concept of time. In contexts such as Amos 5:13, 'Therefore the prudent man keeps quiet in such times [*'ēt*, singular], for the times [*'ēt*, singular] are evil,' the word refers to an extended but general period as indicated by the English translation, 'times.' A reasonable interpretation takes *'ēt* to refer generally to nonspecific units of time. However, the word frequently occurs in the expression 'at that time,' either to distinguish the event in question from other events, or to establish a temporal relation between events. Finally, the use of *'ēt* to designate a particular time of the day (Exodus 9:18: 'At this time [*'ēt*] tomorrow. . .') confirms that the word is perhaps as flexible as the word 'time' in English.

'attâ (עַתָּה)

Etymology This adverb is derived from the same root *'nh* as *'ēt*.

Usage Found frequently in the Tanakh (425 times), *'attâ* generally seems to refer to the present, as in Genesis 19:9: 'This fellow came here as an alien, and now [*'attâ*] he wants to play the judge.' But often the sense of *'attâ* expresses a logical or situational relation rather than a strictly temporal one. An example of *'attâ* introducing the next point in consequence is in Isaiah 5:3, after a parable about God working hard planting and cultivating a vineyard and receiving only bad fruit: 'Now [*'attâ*] you dwellers of Jerusalem and men of Judah, judge between me and my vineyard.' In certain prophetic contexts it is used proleptically as, for example in Isaiah 43:19: 'Behold, I will do something new; now [*'attâ*] it will spring forth.' So, rather than serving as a temporal indexical as well as a logical or situational connective (as the English 'now,' and to some extent the Greek νῦν), *'attâ* seems to have the basic meaning of 'in this situation' or perhaps 'as a result of this situation.' (In fact, Hebrew does not express the temporal indexical 'now' by a single word. See below under *yôm* for the use of *hayyôm hazzeh*, 'this day' or 'this time,' for what comes closest to a temporal indexical in Hebrew.)

---

[13] *TWOT*, vol. 2, p. 680. De Vries argues that *'ēt* cannot have been the primary Hebrew term for expressing pure time relationships. He reserves that function for *yôm* (*YTT*, p. 41).

yôm (יוֹם)

Etymology The root is common in Akkadian and Ugaritic as well as other Semitic languages; its origin is obscure but doubtless very ancient.

Usage Yôm is the fifth most common substantive in the Tanakh, occurring 2,304 times in Hebrew (and 16 times in Aramaic). It is 'the most important concept of time in the OT by which a point of time as well as a sphere of time can be expressed.'[14] While the word often refers to the diurnal period of daylight, in contrast to the night (e.g. Genesis 8:22), it also refers to the 24-hour day (e.g. Exodus 20:8-10). It is used of a year (e.g. 1 Samuel 27:7: 'David lived in the Philistine territory a year [*yamim*, plural] and four months.') as well as of a general period of time (e.g. Genesis 2:4, where *yôm* (singular) refers to the six days of creation, however long they were). In general, however, the plural is used of an indefinitely extended duration, while the singular is used of a unique event, even if that event (as creation) extended over more than one literal day.

De Vries has analyzed the rich nuances of *yôm* when combined with adjectival or adverbial expressions,[15] but much of the interest of this study is in understanding the eschatological perspective of the Hebrews and adds nothing to what has already been said. However, we should note that the expression *hayyôm hazzeh*, 'this day,' could serve the function of the indexical 'now.' An example is 1 Kings 2:6: 'Go back to your fields in Anathoth. You deserve to die, but I will not put you to death now [*hayyôm hazzeh*].'

Perhaps the most reasonable conclusion we can reach regarding the usage of *yôm* in the Tanakh is this: as the day was certainly the primary unit of temporal measure for primitive peoples (including Semitic and Hebrew peoples), it is understandable that *yôm* would be the most common temporal designator, one which acquired great flexibility as the need grew to express a wider variety of temporal nuances. That no particular doctrine of time can be constructed from the many uses of *yôm* is, then, hardly surprising.

mô'ēd (מוֹעֵד)

Etymology Mô'ēd derives from the common Semitic root *w'd*, the original meaning of which is 'recurrence.'[16] The nominal derivative *mô'ēd* thus refers, in its basic notion, to something recurrent.

Usage The nominative form *mô'ēd* occurs 223 times in the Tanakh, designating either a determined time or place. While the purpose of the designation may not be stated, the use of *mô'ēd* clearly implies that there was such a purpose. The time or place so designated is generally a recurring one. In Genesis 1:14 the heavenly

---

[14] *TWOT*, vol.1, p. 370.
[15] De Vries, *YTT, passim*; see his summary, pp. 39-53.
[16] *TDOT*, vol. 4, p. 135.

bodies are said to have been placed by God to 'serve as signs to mark seasons [*mô'ēd*] and days and years.' Consequently, in Leviticus 23:2, God speaks of the seven annual religious festivals as 'my appointed feasts [*mô'ēd*], the appointed feasts [*mô'ēd*] of the Lord, which you are to proclaim as solemn assemblies [*mô'ēd*].' However, *mô'ēd* also refers to events that are clearly 'designated' or 'appointed' though not recurring, such as the birth of a child (Genesis 17:21), the coming of a plague (Exodus 9:5), or a fixed temporal interval, as in 1 Samuel 13:8: 'He waited seven days, the time set [*mô'ēd*] by Samuel, but Samuel did not come.'

The fact that *mô'ēd* is not used interchangeably with *'ēt* would indicate that the Hebrews did not view time as cyclical, although they certainly understood the cyclic recurrence of seasons. Nor did they view all times as fixed or appointed, such as would have been the case if their worldview were more fatalistic. But since humans, as well as God, can determine the purpose for which a particular *mô'ēd* is set, it would be illegitimate to conclude that God must be atemporal or outside of time in order to appoint times.

Other Evidence from Hebrew

Other words are found in the Hebrew vocabulary stock for time, such as *'ad* (עֵד), 'time,' or the late Aramaic loan word *zᵉman* (זְמָן), 'seasons' (occurring in parallel in Daniel 2:21), but they add little or nothing to our understanding of the Hebrew conception of time gained from the more common words discussed above. Quite clearly, the Hebrew vocabulary for time reflects a phenomenological perspective and does not, qua lexical stock, furnish a basis for drawing conclusions about the Hebrews' metaphysics. Hence George Knight is not warranted in concluding that the Hebrews viewed time 'spatially,' or that 'To the Hebrews time was identical with its substance.'[17]

Other sorts of evidence that have been adduced to ground a presumed 'Hebraic view of time' include the morphology and syntax of the Hebrew verb system,[18] or the lack of distinct Hebrew words for 'past and 'future.'[19] However,

---

[17] George A. F. Knight, *A Christian Theology of the Old Testament* (Richmond, VA: John Knox Press, 1959), p. 102; cf. p. 314.

[18] The verbal system of classical Hebrew is based on aspect (*Aktionsart*) rather than tense—that is, on modes of action rather than time of action. Thus the Perfect tense is used for completed action, the Imperfect for uncompleted action, and the participle for continuing action. Now it is true that, generally speaking, the Perfect (completed action) corresponds to the English past tense, and so forth. That the equivalence is not exact, however, is shown by two features of Hebrew: (1) Often, especially in prophetic material and notably in relation to God's actions, an act of God which is still in the future at the time of the prophet's speaking (writing) is described in the Perfect. This so-called 'prophetic Perfect' does not mean that the prophets regarded the event as already in some sense completed (that is, existing timelessly in the space-time continuum) but rather that they regarded it as being 'as good as done,' since God had given his word. (2) In narratives, we find the use of the '*waw*-consecutive,' where the narrative begins using the Perfect, but then the next verb is preceded by the Hebrew word *waw*, 'and,' and succeeding verbs are in the Imperfect. This

all such efforts depend upon the same proposition—namely, that a language reflects a people's psychology that in turn determines the people's metaphysics.[20] However, it is far from clear that this proposition is true; an argument could be made from the perspective of critical realism that it is reality which forms psychology which in turn shapes linguistic expression. Be that as it may, it has long been noted that the Hebrews (at least prior to Philo in the first century AD) never produced anything approaching speculative philosophy; nor, apparently, did they think in explicitly philosophical categories about time and God's relation to time—certainly they employed no special vocabulary for these categories.

Greek Words

Unlike the Hebrew of the Tanakh, the Greek of the New Testament can be viewed in the context of a well-attested linguistic heritage. However, there are differences of usage between the classical Greek of Homer or Plato and the Koinē Greek of the New Testament era, and it would beg the question to assume that the New Testament writers were conversant with Greek philosophical analyses of time and chose their words in light of what Empedocles, Plato or Aristotle had written. On the other hand, it would be equally question-begging to assume that the New Testament writers had a distinctly different concept of time than that current in the Greek-speaking world in which they lived, as Oscar Cullmann did in his influential study, claiming that the Hebrew concept of time was linear, while the Greek concept was cyclical.[21] The purpose of the present section is not to treat the

---

reflects, most likely, the ability of the storyteller to place himself mentally in the story, to adopt the perspective of the time of the narrative rather than the time of the telling. That these two features are quite different from any English equivalent says more about the imaginative ability—or customs—of the Hebrews than it does about their psychology or their metaphysics of time. See G. L. Klein, 'The "Prophetic Perfect",' *Journal of Northwest Semitic Languages* 16 (1990), pp. 45-60; G. S. Ogden, 'Time, and the Verb היה in O.T. Prose,' *Vetus Testamentum* 21 (1971), pp. 451-69; Bo-Krister Ljungberg, 'Tense, Aspect, and Modality in Some Theories of the Biblical Hebrew Verbal System,' *Journal of Translation and Textlinguistics* 7:3 (1995), pp. 82-96.

[19] De Vries suggests, 'Linguistically speaking, the Hebrews seem to have been more two-dimensional than three-dimensional with respect to time. That is, many of their words for the past and for the future were essentially the same' (*YTT*, p. 40). But this is overstated. Certainly the Hebrews had and used the linguistic equipment to distinguish the 'already' from the 'not yet,' a distinction which seems to be a fundamental, ineliminable feature of human experience of the world.

[20] See, for example, Mircea Eliade, *The Myth of the Eternal Return, or, Cosmos and History*, tr. Willard R. Trask (Princeton, NJ: Princeton University Press, 1971), who contrasts the 'primitive' view of time with the 'biblical' view that understands time in a realistic manner quite different from other ancient cultures.

[21] Oscar Cullmann, *Christ and Time: The Primitive Christian Conception of Time and History*, tr. Floyd V. Filson (Philadelphia: Westminster Press, 1950), pp. 37-68. See also Barr's criticism of Cullmann's methodology in *BWT*, pp. 50-85.

linguistic history of Greek words for time and eternity, nor to give an analysis of different Greek concepts of time,[22] but to demonstrate that the language of the New Testament does not require any particular view of time, eternity, or God's temporal mode of being.

aion (αἰών)

Usage Translated variously as 'age,' 'long time,' 'eternity,' or 'world,' this word and its adjectival form αἰώνιος, generally rendered 'eternal,' 'everlasting,' or 'forever,' occur 121 times and 73 times respectively in the New Testament. In formulaic usages we find 'from the beginning of the world [ἀπ' αἰῶνος]' (e.g. Acts 3:21; 15:18), or 'for ever and ever [εἰς τὸν αἰῶνα τοῦ αἰῶνος]' (e.g. Hebrews 1:8), indicating an indefinite past or future. Αἰών is frequently used to indicate the time or duration of the world, as in Matthew 13:39: 'The harvest is the end of the world [αἰῶνός].' In Luke 1:70, 'as he said through his holy prophets long ago [ἀπ' αἰῶνος],' the reference is clearly to historical times. In such usages the focus is on a lengthy but limited duration. (By metonymy the employment of αἰών, meaning 'epoch, age,' was extended to mean the 'world' itself, so that αἰών is frequently synonymous with κόσμος, as in John 9:32, 'Since the world [αἰῶνος] began, nobody heard of anyone opening the eyes of a man born blind.')

Interestingly, in addition to its employment to indicate the duration of the world, αἰών is also used to speak of God's being. Several texts speak of God planning or ordaining something 'before this world/age [αἰών],' thus seeming to contrast God's being with the duration of this world (which is a clear meaning of the word). In Ephesians 3:11, 'according to his eternal [αἰώνων] purpose,' the plural is used, perhaps in the sense of 'the entirety of time'—that is, ternity. God is even called 'the eternal [αἰωνίου] God' in Romans 16:26. But to conclude on the basis of these theological uses that αἰών must mean timeless eternity, in the absence of any other clear uses of αἰών for timelessness, would be to beg the question.

Beyond theological contexts, αἰών seems to be used in contexts that would clearly require 'forever' as the meaning, as, for example, in 1 John 2:17: 'The world [κόσμος] and its desires pass away, but the one who does the will of God lives forever [εἰς τὸν αἰῶνα].' But these uses of αἰών in themselves do not require that the word mean 'timeless eternity,' since αἰών can also clearly mean 'long duration.'

This brief survey establishes that no conclusions can be reached regarding either the temporality or the timelessness of eternity on the basis of the use of αἰών. However, I should note here that Plato, in an apparent departure from

---

[22] For such studies, see John F. Callahan, *Four Views of Time in Ancient Philosophy*, rev. ed. (Cambridge, MA: Harvard University Press, 1979); Richard Sorabji, *Time, Creation and the Continuum: Theories in Antiquity and the Early Middle Ages* (London: Duckworth, 1983).

conventional usage, distinguished between αἰών as the timeless, ideal eternity in which there are no days or months or years, and χρόνος, the time of the world created as a moving image of eternity (*Timaeus* 37d). Under Plato's influence, Philo of Alexandria said that χρόνος is the βίος of the κόσμος αἰσθεντός, while αἰών is the βίος of God and the κόσμος νοντός,[23] an idea traces of which can perhaps be discerned in the thought of the later Neoplatonists and the Neoplatonic theology of the early Church. However, there is no clear instance of the Platonic meaning of the word in the New Testament, and attempts to fix the meaning of the word in the New Testament by referring to Plato's unconventional usage are surely exercises in misguided methodology or special pleading.

kairos (καιρός)

Usage Almost all of the 86 occurrences of καιρός in the New Testament are translated by 'time' or 'season' (that is, a temporal interval). The reference may be to a precise moment, as in Matthew 11:25, 'At that time [καιρῷ] Jesus answered,' or to an imprecise interval, parallel in meaning to αἰών, as in Luke 18:30, '[He] will not fail to receive many times as much in this age [καιρῷ] and, in the age [αἰῶνι] to come, eternal [αἰώνιος] life.' Cullmann's study[24] claimed that καιρός generally or always had the meaning 'decisive moment' (in contrast to χρόνος, which meant time in general), but this cannot be sustained on the basis of a study of the use of the word.[25] The point is that καιρός does not indicate 'time' as defined by its contents (a staple of those who, like Cullmann, desire to demonstrate a distinctive biblical doctrine of time), which could possibly lead to a substantial view of time. If a substantial view of time is to be defended in philosophical theology, it cannot be on the basis of the use of καιρός in the New Testament.

chronos (χρόνος)

Usage This word occurs 53 times in the New Testament, and is virtually indistinguishable from καιρός in usage. Thus the sense of 'precise time' is seen in Matthew 2:7: 'Then Herod called the Magi secretly and asked them carefully the time [χρόνον] that the star appeared.' And the sense of 'imprecise interval' can be seen in Matthew 25:19: 'After a long time [χρόνον] the master of those servants returned.' In Acts 1:7, '[Jesus] said to them: "It is not for you to know the times (χρόνους) or dates (καιροὺς) the Father has set by his own authority",' the

---

[23] Philo of Alexandria, *De Fuga et Interventione* 57, cited by Hermann Sasse in *TDNT*, vol.1, p. 198.
[24] Cullmann, *Christ and Time*, pp. 39-44. Gerhard Delling claims the sense of 'decisive moment' is clear in classical Greek from the time of Sophocles, and sees in the New Testament use the meaning 'fateful and decisive point' (*TDNT*, vol.3, pp. 455-9). But that the word is used of times that turn out, in fact, to be fateful and decisive, does not mean the word itself means 'fateful and decisive times.'
[25] Barr, *BWT*, pp. 50-85.

pairing of καιρός and χρόνος is probably due to the use of parallelism in spoken Aramaic; nevertheless, it is clear that the two terms are interchangeable. Any attempt to derive a philosophical conclusion from contrasting uses of καιρός and χρόνος is untenable.

Other Evidence from Greek

As in Hebrew, other words are used in the Greek New Testament to express temporal concepts. Among them are the common words ἡμέρα, 'day,' or ὥρα, 'hour' (both of which occur together in Matthew 24:36, probably in the general sense of 'time'), or the *hapax legomenon* στιγμῇ, 'moment,' in Luke 4:5, but they do not add to the discussion. I would suggest that, even given a rich tradition of philosophical reflection on the nature of time in Greek philosophy, the usage of the different words for time (certainly in the Koinē Greek of the New Testament) is imprecise and phenomenological. None of the differing words used for time in Classical or in Koinē Greek served as technical terms with distinct metaphysical connotations. Consequently no philosophical conclusions may be drawn simply from noting word usages. It is methodologically unwarranted to attempt to read off a philosophical view from the vocabulary stock of a language in the absence of affirmative evidence for the view.[26]

Exegesis: Central Texts that Speak of Time and Eternity

As James Barr emphasizes in his study of the semantics of biblical words for time:

> It must be clear that no kind of theology can be built upon such material as the occurrence of this word in each of these places, or the occurrence of different words in two places. A valid biblical theology can be built only upon the statements of the Bible, and not upon the words of the Bible.[27]

It should be apparent from the discussion above that common expressions such as 'eternal God' do not establish that any particular concept of God's relation to time is *the* biblical concept. Nevertheless, it might be claimed, even though the vocabulary of the Tanakh and of the New Testament cannot do so, there are certain biblical texts which do clearly present a particular view of God's relation to time. In response to such claims, I will argue that careful contextual study of crucial texts relating to God's eternity shows that timeless duration is not the required interpretation, and that some sort of temporal succession is either presumed or allowed for God's temporal mode of being. My aim here is to lay the foundation for the claim to be advanced in the next chapter that the medievalists were more heavily influenced by Neoplatonism than by biblical exegesis, and thereby to

---

[26] Ibid., pp. 135-58.
[27] Ibid., p. 154.

Hebrew Texts

There are two classes of Hebrew texts to consider. First will be those that speak of God as eternal, followed by those that speak of God as 'repenting' or 'changing his mind.'

An example of the first class occurs in Psalm 145:13, where the Psalmist writes, 'Your kingdom is an everlasting [*kol-'ōlāmîm*, literally, 'of all *'ōlāms'*] kingdom, and your dominion endures through all generations.' The use of the plural *'ōlāmîm* was discussed above; here I wish to call attention to the parallel expression 'all generations.' Clearly, generations are successive, and while placing a concept of succession in parallel with the common word for eternity does not prove that the Hebrew concept of eternity was of succession, it at least makes that interpretation possible.

Psalm 90:2 shows a formulaic use of *'ōlām*, in which the idea of eternity is unmistakable: 'Before the mountains were born, or you brought forth the earth and the world, from everlasting to everlasting [*mē'ōlām ad-'ōlām*] you are God.' If the Tanakh contained evidence that the Hebrews believed time began with creation, then this formula in this context would indicate timeless eternity. But such evidence is totally absent, and in light of the semantics of *'ōlām* discussed above, it would be unwarranted to read such a meaning into this formula.

Only in the Book of Ecclesiastes do we find something approaching a philosophical conception of time and eternity. Shortly after the well-known poem that begins 'To everything there is a time . . .' (Ecclesiastes 3:1-8), there is a very enigmatic saying: 'I have seen the burden God has laid on men. He has made everything beautiful in its time. He has also set eternity [*'ōlām*] in the hearts of men; yet they cannot fathom what God has done from beginning to end' (Ecclesiastes 3:10, 11). While it is not uncommon to find interpreters who take 'eternity' in this text as a reference to the realm of the atemporal, in contrast to the realm of the temporal (the present world), such an understanding does not accord good sense to the saying. Barr suggests taking *'ōlām* in the sense of 'perpetuity,' thus translating the text as 'He has done everything well in its time; he has also set perpetuity in their heart [i.e. in the heart of man]—yet so that man cannot find out the action which God does from the beginning to the end.' Barr glosses the verse:

> The reference to perpetuity would mean the consciousness of memory, the awareness of past events. The predicament of man is that he has this awareness, and yet cannot work out the total purpose of God. This avoids the artificiality of the old rendering 'eternity,' understood as the consciousness of another world beyond this one.[28]

---

[28] Ibid., p. 124, footnote 1.

Thus even here, in the only context in the Tanakh where 'eternity' stands as an independent object, a sempiternal rather than an atemporal understanding fits better.

One further passage, Ecclesiastes 9:11, is worthy of note: 'The race is not to the swift, or the battle to the strong, nor does food come to the wise or wealth to the brilliant or favor to the learned; but time [*'ēt*] and chance happen to them all.' If there remains an inclination to regard *'ēt* as 'filled time' or 'substantial time,' in line with some kind of fatalistic understanding, this text should put an end to it. The most natural understanding of the verse is that time in general is characterized by contingency, uncertainty and vicissitudes; that time does not merely unveil a preexisting space-time continuum, but is that which brings the contingent future to reality.

The second category of texts to be considered is of those that speak of God 'changing his mind.'[29] The Hebrew word used in these texts is *nhm* (נָחַם), generally translated as 'repent, reconsider, change one's mind.' One class of these texts often serve as proof-texts of immutability—for example, Numbers 23:19, 'God is not a man, that he should lie, nor a son of man, that he should change his mind [*nhm*],' or I Samuel 15:29, 'He who is the Glory of Israel does not lie or change his mind [*nhm*]; for he is not a man, that he should change his mind [*nhm*].' At the most, it seems, such texts only support a concept of God who does not change, not a God who cannot change. This is a somewhat weak understanding of the nature of immutability that is generally taken as a necessary property of deity.

But there is a second class of texts that say exactly the opposite. One example is Exodus 32:14: 'Then the Lord relented [*nhm*] and did not bring on his people the disaster he had threatened.' Another is Jeremiah 18:7-10, where God himself says:

> If at any time I announce that a nation or kingdom is to be uprooted, torn down and destroyed, and if that nation I warned repents of its evil, then I will relent [*nhm*] and not inflict on it the disaster I had planned. And if at another time I announce that a nation or kingdom is to be built up and planted, and if it does evil in my sight and does not obey me, then I will reconsider [*nhm*] the good I had intended to do for it.

This class of texts seems to indicate that a real change takes place in God's intentions based on temporal changes in creation. However, many theologians who assume divine immutability disagree:

> On the surface, such language seems inconsistent, if not contradictory, with certain passages which affirm God's immutability. . . . When *nāham* is used of God, however, the expression is anthropopathic and there is no ultimate tension. From

---

[29] Significant discussions of these texts may be found in H. Van Dyke Parunak, 'A Semantic Survey of *NHM*,' *Biblica* 56 (1975), pp. 512-32, and Robert B. Chisholm, 'Does God "Change His Mind"?,' *Bibliotheca Sacra* 152 (1995), pp. 387-99.

man's limited, earthly, finite perspective it only appears that God's purposes have changed.'[30]

But the natural reading of the texts supports a temporal God. The best interpretation will emerge if it is possible to harmonize these two classes of *nhm* texts.

In the texts in question, *nhm* occurs in the Niphal or the Hithpael stem.[31] In these stems the root falls into one of four semantic categories: (1) the sense 'to experience emotional pain or weakness'—for example, Genesis 6:6-7; (2) the sense 'to be comforted' or 'to comfort oneself'—for example, Genesis 24:67; (3) references to God's 'relenting from' a course of action already underway—for example, Deuteronomy 32:36; and (4) references to God's 'retracting' or 'deviating from' an announced course of action—for example, Exodus 32:12. In a detailed study of occurrences in categories (3) and (4), Robert Chisholm discerns two distinct uses, and distinguishes different statements or contexts of intention. 'Some are decrees or oaths that are unconditional and bind the speaker to a stated course of action. Others, which may be labeled announcements, retain a conditional element, and do not necessarily bind the speaker to a stated course of action.'[32] This distinction is clear in non-theological contexts—that is, contexts in which the subject is a human being—as well as in theological contexts. Chisholm concludes:

> The texts analyzed. . . clearly show that God can and often does retract announcements. Two of the passages [Joel 2:14 and Jonah 4:2] even regard this willingness to change His mind as one of His most fundamental attributes. . . . If He has decreed a certain course of action or outcome, then He will not retract a statement or relent from a declared course of action. Verses stating or illustrating this truth must not be overextended, however. . . . They should not be used as proof texts of God's immutability. . . . If God has not decreed a course of action, then He may very well retract an announcement of blessing or judgment. In these cases the human response to His announcement determines what He will do. Passages declaring that God typically changes His mind as an expression of His love and mercy demonstrate that statements describing God as relenting should not be dismissed as anthropomorphic.[33]

Admittedly, as much of our language about God is used analogically or anthropomorphically, interpreting these texts in this way is not doing violence to the text. But not all language about God is anthropomorphic, and a natural understanding of these texts is quite plausible, once a commitment to strong

---

[30] *TWOT*, vol. 2, p. 571.
[31] The Niphal stem generally conveys a reflexive, reciprocal or passive sense, often resembling the semantics of the Greek middle voice. The Hithpael stem is an intensive reflexive, reciprocal, or passive, often indicating 'an action performed *with regard to* or *for* oneself.' Gesenius and Kautsch, *Hebrew Grammar*, §54f.
[32] Chisholm, 'Does God "Change His Mind"?,' p. 389.
[33] Ibid., p. 399.

immutability is abandoned. <u>So if these texts count as evidence, they support a temporal view of God's being.</u>

Greek Texts

There are several texts in the New Testament which speak of God acting before creation, such as Ephesians 1:4, 'He [God] chose us in him [Christ] before the foundation of the world,' or 1 Peter 1:20, 'He [Christ] was chosen before the creation of the world.' Such texts are often cited in support of a timeless realm in which God exists. But such an understanding would only be defensible if the New Testament clearly assumed that time was coextensive with creation—an idea which, if entertained at all, was never addressed in the New Testament. A more natural understanding of these texts would be that God's time is inclusive of the time of creation, but is not limited by creation.

1 Timothy 1:17 is a classic text: 'Now unto the King eternal [τῶν αἰώνον], immortal [ἀφθάρτῳ], invisible, the only God. . . .' Doesn't this verse clearly and explicitly state God's eternity? Eternity, yes; timelessness, no. The precise wording is instructive. 'Eternal' is literally 'of the ages' (τῶν αἰώνον); hence God is the 'King of the ages' or, taking the plural as discussed above, 'King of the entirety of time.' Further, God is said to be 'immortal,' which more often has the sense of 'incorruptible' (ἀφθάρτῳ)—that is, something not subject to change or decay with the passage of time (see 1 Peter 3:3, 4: 'Your beauty should not come from outward adornment, such as braided hair and the wearing of gold jewelry and fine clothes. Instead, it should be that of your inner self, the unfading [ἀφθάρτῳ] beauty of a gentle and quiet spirit. . . .'). To take this verse as implying timeless existence is to read more into the text than is justified. There is no hint of timelessness here; it must be imported.

Finally, note must be taken of Revelation 10:6. Many see in this verse one of the strongest statements of God's eternity: 'And he [an angel] swore by him who lives for ever and ever [τῷ ζῶντι εἰς τοὺς αἰῶνας, literally, 'the one living unto the ages of the ages'] who created the heavens and all that is in them, the earth and all that is in it, and the sea and all that is in it.' Granted, this is indeed a strong statement of God's eternity, but it certainly does not necessitate an understanding of timelessness; 'the longest possible duration' or some such paraphrase would be at least equally likely. But the verse continues, translating literally what the angel swore: 'that time no longer shall be [ὅτι χρόνος οὐκέτι ἔσται].' Now this, taken in isolation, could well seem to be a clear statement that time (as measured or as sequential duration) will cease, but the context invalidates that interpretation. The speech of the angel continues in verse 7, 'But in the day when the seventh angel is about to sound his trumpet, the mystery of God will be accomplished, just as he announced to his servants the prophets.' While the interpretation of the apocalyptic literature of Revelation is difficult (and well beyond the scope of this work), it seems about as clear as anything in Revelation that the seventh angel referred to in verses 6 and 7 is the same as in 11:15: 'The seventh angel sounded his trumpet, and there were loud voices in heaven, which said, "The kingdom of his world has become the kingdom of our Lord and of his Christ, and he will reign for ever and ever [εἰς τοὺς αἰῶνας τῶν αἰώνων].'" But several things happen in what seems to be a chronological sequence between the angel's statement that 'time shall be no more' and the blowing of the seventh trumpet, and many more things happen in sequence following the trumpet, such as seven angels pouring out seven successive bowls of judgment (described in chapter 16)—surely actions which can only be interpreted as chronologically successive events. Thus the interpretation of the angel's statement in 10:6 that takes it as announcing the end of time and the beginning of a timeless eternity cannot stand.

A better interpretation is readily available. Delling notes that 'behind 10:6 there stands the Christian expectation that the eschatological events of judgment and salvation will be fulfilled as proclaimed.'[34] The New International Version translates the statement as 'There will be no more delay!' Barr acknowledges, 'It is true that χρόνος sometimes amounts to "delay",' but continues, 'It is rather the context within the Apocalypse as a whole that shows that the sense here is not an abolition of time and its replacement by timelessness, but "no more time" from the words of the angel until the completion of the divine purpose.'[35] Thus what some would take to be the verse in the Bible that most clearly indicates a timeless eternity as distinct from the present temporal age is seen, by a careful consideration of the context, not to mean that at all.

Conclusion

In summary, neither the lexical stock nor often-cited texts of either the Tanakh or the New Testament convey any particular metaphysical view of time to the reader. Perhaps the Hebrews and the first-century Christians had determinate concepts of time and eternity and of God's relation to both, but, if so, they did not communicate them in the texts. Attempts to derive 'the' biblical view of time from the vocabulary or the text of the Bible are doomed from the outset. James Barr concludes his study of the biblical words for time (and various attempts to derive philosophical and theological conclusions from them) in this way:

> The position here developed means in effect that if such a thing as a Christian doctrine of time has to be developed, the work of discussing and developing it must belong not to biblical but to philosophical theology.
>
> In fact we may observe that the origin of the question lies in many ways within philosophical rather than biblical material. . . . If, then, many or most of the modern discussions start out in fact from some problem or difficulty found in modern theological-philosophical discussion, we may regard it as probable that satisfactory results can be reached only by seeking clarification within that area, and not by expecting the Bible to answer the problem for us. If, on the other hand, Eichrodt is right in suggesting that there is in fact no biblical notion of time which

---

[34] *TDNT*, vol. 9, p. 592.
[35] Barr, *BWT*, p. 80 n. 2.

differs widely from our own, this may be a reason for theology to avoid being forced into developing such a theological doctrine of time, or at any rate to avoid claiming that any such doctrine developed rests on a certain biblical basis.[36]

I would concur with Barr. This chapter demonstrates that there is no clear, univocal biblical view of time or eternity, or of God's relation to time. If pressed, perhaps we could conclude that the *nhm* texts tend to favor a temporal interpretation. But the evidence is far from conclusive. Any satisfactory theory must be based on metaphysics and philosophical theology. Consequently, as we move in the next chapter to consider the traditional view, that of an atemporal God, we will see that the medieval philosophical theologians either depended directly on philosophical thought or read the Scriptures in light of philosophical—specifically Neoplatonic—presuppositions.

---

[36]Ibid., pp. 156-7.

Chapter 5

The Medieval Consensus: God is Atemporally Eternal

The past three decades have seen a renewed interest in philosophical theology as a specialty distinct from both philosophy of religion and systematic theology, and a corresponding interest in the exegetical study of major philosophical theologians of the Middle Ages. In the Middle Ages, of course, few but the clergy had the leisure to pursue philosophy, so it is not surprising that virtually all medieval philosophy is concerned either with explicitly theological issues or with philosophical questions spawned by theological issues.

The major difficulties associated with the relation of God and time are not new; philosophical theologians have grappled with them for centuries. Most of the medieval philosophers shared the view that God's being, and eternity itself, were atemporal. There does seem to be less of a consensus regarding the nature of time. Most presuppose an intuitive dynamic theory of time, but with some twists such as Augustine's psychological theory. But, through the centuries, much stronger hints of a static theory emerge.

My purpose in this chapter, then, is to survey four major medieval philosophers in order to determine their views of time and of God's relation to it—views which established themselves so firmly in the tradition as to become virtually canonical. But I shall also show that the influence of Neoplatonism is the most probable explanation for the medieval consensus that emerges.

St Augustine

Augustine of Hippo (354-430) is a crucial figure in the tradition of philosophical theology, since subsequent medieval philosophical theologians either followed or reacted to his lead. He is also an influential figure in the history of the philosophy of time. Consequently I shall devote more space to Augustine's views than those of later thinkers. But Augustine's views on time and God's relation to it are motivated in the context of his meditations on the doctrine of creation, and must be carefully disentangled from his larger concerns. And if Augustine is not always as clear as we could wish in propounding his philosophy of time, we should remember that this is not his primary interest.

Augustine is clearly committed to God's atemporality. I shall analyze Augustine's arguments for this conclusion, and show that it is rooted not in biblical exegesis, but in the doctrines of divine simplicity and immutability.

> **St Augustine's Views of God and Time**
>
> Augustine profoundly influenced both theology and philosophy in the West for centuries. He is the first genuine philosopher of time, and his theological views of God's nature and the nature of eternity were unquestioned for nearly a millennium.
> - God is atemporal; eternity is timeless. Augustine's argument for God's atemporality is based on the following:
>   1. Time began with the creation of the world.
>   2. Since God created the world, God created time.
>   3. Therefore, God must not be in time.
> - God's foreknowledge is of timeless ideas, not of the non-existent future.
> - Time is subjective yet dynamic. Augustine is ambiguous between two interpretations:
>   A. ontological, in which case objective time flow is false; or
>   B. epistemological, in which case time flow can only be known by minds.
> - Neoplatonism clearly influenced Augustine's thinking about God's simplicity and immutability. Both entail the timelessness of God's being and of eternity.

Next I will show that Augustine's views on time, while discussed at some length, are not entirely clear. In the *Confessions* he seems to hold to a psychological view of time, with temporal becoming being mind-dependent, but elsewhere to hold to a realist, dynamic view of time. There is more than a hint of contradiction here, indicating perhaps that he changed his mind about the nature of time. Be that as it may, I shall compare Augustine's psychological theory of time to the psychological theory of contemporary philosopher Adolf Grünbaum, and then to the transcendental idealist theory of Immanuel Kant. I shall suggest that a quasi-Kantian theory offers us a possible way of interpreting Augustine.

I shall then trace the roots of Augustine's views on divine timelessness to the influence of the Neoplatonists, notably Plotinus, and suggest that within Neoplatonist metaphysics it is possible to see that Augustine's doctrines of God's timelessness and of dynamic time can be understood as consistent and coherent.

The investigation of Augustine's views will generally follow his discussion of time in the *Confessions*, where his most systematic treatment of the nature of time may be found. Recently, James Wetzel has argued that it is misguided to attempt to find a philosophical treatment of the nature of time in the *Confessions*,[1] locating the work in Augustine's somewhat mystical Christianity, and seeing it for what it claims to be: a confession of Augustine's inescapable sinfulness. Wetzel may well be correct in his religio-mystical construction of Augustine's views on time; nevertheless, it is the traditional philosophical interpretation that I shall discuss here. For better or worse, it is this interpretation that has influenced subsequent philosophers.

---

[1] James Wetzel, 'Time after Augustine,' *Religious Studies* 31 (1995), pp. 341-7. See also John M. Quinn, 'Time,' in *Augustine Through the Ages: An Encyclopedia*, ed. Allan D. Fitzgerald (Grand Rapids, MI: Eerdmans, 1999), pp. 832-8.

God Exists Atemporally

Augustine begins Book XI of *Confessions* with this question: 'O Lord, since you are outside of time in eternity, are you unaware of the things that I tell you? Or do you see in time the things that occur in it?'[2] How did Augustine come to the conclusion that God (and eternity) was outside of time? Augustine takes up the question of time in the context of his discussion of Genesis 1:1-3 in Books XI to XIII. Thus it is the doctrine of creation that prompts Augustine's reflections on time. He considers an objection to the doctrine of creation: what was God doing before he made heaven and earth? Now if the answer is that God was doing nothing, then at some point he must have decided to create, thus marking a change in the divine will, and thus destroying divine immutability.[3] Augustine's response is that it is a misconception to speak of time before creation:

> How could these countless ages have elapsed when you, the Creator, in whom all ages have their origin, had not yet created them? What time could there have been that was not created by you? How could time elapse if it never was? . . . Furthermore, although you are before time, it is not in time that you precede it. If this were so, you would not be before all time. It is in eternity, which is supreme over time because it is a never-ending present, that you are at once before all past time and after all future time. . . . Your years are completely present to you all at once, because they are at a permanent standstill. They do not move on, forced to give way before the advance of others, because they never pass at all.[4]

He continues, 'It is therefore true to say that when you had not made anything, there was no time, because time itself was of your making. And no time is co-eternal with you, because you never change; whereas, if time never changed, it would not be time.'[5]

Near the conclusion of his discussion of time in Book XI, Augustine again takes up this theme, responding to those who ask, 'How did it occur to God to create something, when he had never created anything before?':

> If a man is said never to have made anything, it can only mean that he has made nothing at any time. Let them see, then, that there cannot possibly be time without creation. . . . Let them understand that before all time began you are the eternal Creator of all time, and that no created thing is co-eternal with you, even if any created thing is outside time.[6]

Augustine's argument may be summarized as follows:

1. Time began with the creation of the universe.

---

[2] Augustine, *Confessions*, tr. R. S. Pine-Coffin (London: Penguin Books, 1961).
[3] Ibid., XI.10.
[4] Ibid., XI.13.
[5] Ibid., XI.14.
[6] Ibid., XI.30.

2. God created the universe.
3. Therefore, God created time.
4. If God created time, he could not have done so in time.
5. Therefore, God exists outside of time.

This line of argument is clear enough. While a non-theist would challenge (2), there is no question that (2) is a central belief of the Judeo-Christian tradition. Even within that tradition, however, we might wonder whether (1) is true. Its plausibility derives, I think, from Augustine's assumption that since God is immutable, there can be no succession in his being, and hence the eternity in which he has his being must be timeless.

But that leads to a second line of argument, intertwined with the first, which may be stated in modal terms:

6. Necessarily, God cannot change.
7. Possibly, anything in time can change.
8. Therefore, necessarily, God is not in time.

Here the argument from immutability is clearly spelled out. Augustine clearly understands God's immutability in a strong sense. It is not enough that God does not change; for Augustine, immutability means God cannot change. Elsewhere he writes:

> That which is changed does not retain its own being, and that which can be changed, even if it is not actually changed, is able not to be that which it had been. For this reason, only that which not only is not changed, but also is even unable to be changed in any way, is most truly said to be.[7]

The modality in (6) captures 'unable to be changed in any way.' Let us call this the strong sense of immutability:

> SI: An entity is strongly immutable iff, necessarily, none of its properties can change.

This would be opposed to a weak sense of immutability:

> WI: An entity is weakly immutable iff, necessarily, none of its non-relational properties can change.

(I shall not attempt at this point to resolve the many difficulties lurking at the edges of this distinction.) Strong immutability, I believe, depends upon Augustine's doctrine of divine simplicity and its Neoplatonic overtones. But I shall reserve comment on his Neoplatonism until later.

---

[7] Augustine, *De Trinitate* V.ii.3, cited in Brian Leftow, *Time and Eternity* (Ithaca, NY: Cornell University Press, 1991), p. 73, hereafter *T&E*.

Although Augustine does not make it explicit at this point, given his answer to the question he raised in *Confessions* XI.1 (quoted above), it is clear that he believed that God had timeless knowledge of temporal events. In this, his position is different from the views of several Greek philosophers who, like Augustine, held to something akin to a doctrine of divine timelessness. But for them, temporal events were 'so lacking in inherent dignity as to be beneath God's notice.'[8] Augustine's unique contribution is thus the notion of God's timeless knowledge of temporal events. (While this position would come to play a central role in the thinking of Boethius and his followers in their treatment of the problem of divine foreknowledge and human freedom, it is interesting that Augustine never used his position in this way. Perhaps the connection which seems so obvious since Boethius was not really all that obvious, or perhaps Augustine felt that he had said all that needed to be said on the foreknowledge problem in *De libero arbitrio*, written a year or two before the *Confessions*.)

The Nature of Time

Augustine now launches into a discussion of the nature of time itself, beginning with his celebrated disclaimer:

> What then is time? There can be no quick and easy answer, for it is no simple matter even to understand what it is, let alone find words to explain it.... What, then, is time? I know well enough what it is, provided that nobody asks me; but if I am asked what it is and try to explain, I am baffled.[9]

Yet, despite his bafflement, Augustine actually has quite a lot to say about the nature of time. He first asserts that neither the past nor the future exist: 'Of these three divisions of time, then, how can two, the past and the future, be, when the past no longer is and the future is not yet?'[10] Hence:

9. The past and the future do not exist objectively.

So Augustine proceeds to analyze the present. If we speak of a present interval of time, say, the present year, it is clear that some months are past, one is present, and the rest are future. But since the past and future do not exist, it is the present month of which we must be thinking. But in the present month some days are past, one is present, and the rest are future. So it is really one day that is present. But in the day there are hours, and in the hours there are minutes, and finally there must be an indivisible instant, for if it were divisible, part would be past, part present, and part future. Therefore, he concludes, the present is only a durationless instant, of no

---

[8] William Hasker, *God, Time, and Knowledge* (Ithaca, NY: Cornell University Press, 1989), p. 3, n. 1.
[9] Augustine, *Confessions*, XI.14.
[10] Ibid.

extension.[11] Here Augustine's view agrees with the celebrated view of Aristotle.[12] Thus:

> 10. The present is a durationless instant.

This premise, coupled with (9), gives

> 11. Therefore, temporal intervals do not exist objectively.

But Augustine then considers a counter argument to this version of presentism, which begins with a model of static time:

> Another view might be that past and future do exist, but that time emerges from some secret refuge when it passes from the future to the present, and goes back into hiding when it moves from the present to the past. Otherwise, how do prophets see the future, if there is not yet a future to be seen? It is impossible to see what does not exist. In the same way people who describe the past could not describe it correctly unless they saw it in their minds, and if the past did not exist it would be impossible for them to see it at all. Therefore both past and future do exist.[13]

Augustine's response to this counter-argument is to claim that whatever it is that is 'seen,'[14] when it is claimed that the past or the future are seen, is really present at the time of the seeing. When a person describes the past, it is not past facts he 'sees' but his memory of those facts, which is present to him. And when we 'foresee' the future, 'we do not see things which are not yet in being, that is, things which are future, but it may be that we see their causes or signs, which are already in being.'[15] Consequently the past and the future are not 'real'; they do not exist at some hidden location. He then confesses that whatever method God uses to reveal the non-existent future is beyond his understanding.

Here we see what seems to be a dynamic presentist view of time. Augustine's doctrine represents a type of A-theory of time, for the A-theory is not inconsistent with the ontological unreality of the past and the future. Augustine concludes, 'it is not strictly correct to say that there are three times, past, present, and future. It might be correct to say that there are three times, a present of past

---

[11] Ibid., XI.15.
[12] See Aristotle, *The Physics*, Loeb Classical Library, tr. Philip H. Wicksteed and Francis M. Comfort (New York: G. P. Putnam's Sons, 1929), 6.3, 239a9-19, hereafter *Physics*.
[13] Augustine, *Confessions*, XI.18.
[14] We should note Augustine's representational theory of cognition, conceived of by analogy with sight. Thus cognition is inner sight. It is possible that had Augustine entertained a different theory of cognition, or employed a different metaphor or analogy, his treatment of foreknowledge would have been different. But sight is a compelling metaphor for cognition, and was used by all four of the medieval philosophers discussed in this chapter.
[15] Augustine, *Confessions*, XI.18.

things, a present of present things, and a present of future things.'[16] Then follows a hint of what will prove to be his final solution: 'Some such different times do exist in the mind, but nowhere else that I can see.' Whatever there is to tense, then, is mental.

Augustine next confesses to being puzzled by our ability to measure and compare the lengths of different intervals of time, given his conclusion expressed in (11):

> As I said, we measure time while it is passing. If I am asked how I know this, my answer is that I know it because we do measure time. We could not measure a thing which did not exist, and time does not exist when it is past or future. How, then, do we measure present time when it has no duration? It must be measured while it is in process of passing.[17]

Thus,

12. We are able to measure temporal intervals.

But (11) and (12) generate a paradox. Clearly we can measure the length of intervals of time, but that would seem to be impossible. How is it that we can measure something that has no objective existence? We cannot measure the length of past or future intervals, since they do not exist (or exist only subjectively), and we cannot measure the length of a present (objective) interval, since the present is a durationless instant. How can we do the seemingly impossible?

Augustine leaves for a while the paradox he has created, and considers a theory he once heard to the effect that the motions of the heavenly bodies are what constitute time. But this is easily refuted, Augustine shows. First, even if the heavenly bodies ceased moving, the potter's wheel would continue to move. But, more telling, it is by time that we measure and compare the intervals of motion of the heavenly bodies; therefore, it is not their motion that constitutes time. Granted, if their motion were faster or slower, we might use our words for time (such as 'day') differently, but we could still measure the motion as being faster or slower.[18] He concludes, 'It is clear, then, that the movement of a body is not the same as the means by which we measure the duration of its movement. This being so, it must be obvious which of the two ought more properly to be called time.'[19] But buried

---

[16] Ibid., XI.20.
[17] Ibid., XI.21.
[18] Ibid., XI.23. It is interesting to compare Augustine's views here to those of Aristotle. For the latter, time measures motion: τοῦτο ἐστιν ὁ χρόνος, ἀριθμὸς κινήσεως κατὰ τὸ πρότερον καὶ ὕστερον; literally, 'This is time, a number of motion according to the before and after' (*Physics*, IV.11.219b1). For Augustine, though, as we have seen, it is the mind that does the measuring. Thus there is an ineliminable mental aspect to time for Augustine, independent of motion, whereas for Aristotle, it is motion, rather than the mental aspect, which is ineliminable.
[19] Augustine, *Confessions*, XI.24.

in this discussion there is a second indication of what will ultimately be his solution: 'I see time, therefore, as an extension of some sort.'[20]

Now follows another of Augustine's bemused reflections on his ignorance of the nature of time:

> I confess to you, Lord, that I still do not know what time is. Yet I confess too that I do know that I am saying this in time, that I have been talking about time for a long time, and that this long time would not be a long time if it were not for the fact that time has been passing all the while. How can I know this if I do not know what time is?[21]

Returning to his paradox and moving towards its resolution, Augustine once again remarks on the fact that we are able to judge comparative duration of temporal intervals. He then becomes even more explicit as to his view: 'It seems to me, then, that time is an extension, though of what it is an extension I do not know. I begin to wonder whether it is an extension of the mind itself.'[22,23]

Having suggested that time is mental extension, Augustine offers the example of measuring the duration of a sound, or of syllables, and concludes,

> I must be measuring something which remains fixed in my memory. It is in my own mind, then, that I measure time. I must not allow my mind to insist that time is something objective. . . . It is the impression that I measure, since it is still present, not the thing itself, which makes the impression as it passes and then moves into the past. When I measure time it is this impression that I measure. Either, then, this is what time is, or else I do not measure time at all.[24]

Here Augustine argues that:

> 13. When we measure temporal intervals, we are measuring mental impressions.

Hence, the resolution of the paradox contained in (11) and (12) is simple:

---

[20] Ibid., XI.23.
[21] Ibid., XI.25.
[22] Perhaps the word 'extension' in the Pine-Coffin translation is misleading. The Latin is *distentio*. Callahan comments on the distinction: 'Since the present in which the soul exists and measures is indivisible, we must not think of the distention which time is called as having a quantitative extension. Rather it is a vital activity that is without quantity, and time, strictly speaking, is without quantitative extension because any duration is measured in an indivisible moment by the coincidence in the soul of past, present, and future. Thus time may be called a distension and said to possess extension with reference to the motions that are measured by it.' John F. Callahan, *Four Views of Time in Ancient Philosophy*, rev. ed. (Cambridge, MA: Harvard University Press, 1979), p. 177.
[23] Augustine, *Confessions*, XI.26.
[24] Ibid., XI.27.

14. Temporal intervals are mental impressions.
15. Mental impressions are subjective.
16. Therefore, temporal intervals are subjective.

Consequently, even though temporal intervals have no objective existence (11), they do exist subjectively—in the mind. So the paradox posed by (11) and (12) is resolved.

Augustine makes one further move in the text just cited, claiming that what is measured when a temporal interval is measured is time itself.

17. Therefore, time is a mental impression.

From which it follows that:

18. Time is subjective.

What is subjective, of course, is in the mind. Augustine uses this conclusion to explain past and future:

> All the while the man's attentive mind, which is present, is relegating the future to the past. The past increases in proportion as the future diminishes, until the future is entirely absorbed and the whole becomes past.[25]

> But how can the future be diminished or absorbed when it does not exist? And how can the past increase when it no longer exists? It can only be that the mind, which regulates this process, performs three functions, those of expectation, attention, and memory. The future, which it expects, passes through the present, to which it attends, into the past, which it remembers.[26]

To summarize: for Augustine in the *Confessions*, time is something mental. Temporal intervals exist only in the mind (because past and future are non-existent and the present is durationless), and can only be measured by mind. The paradox disappears, since it is not truly past or future time that is being measured, but an extension of the mind. The mind has the power, as it were, to extend itself into the future by expectation, and into the past, by memory.

In this doctrine of time as mental extension we can see also the probable explanation of Augustine's insistence that 'without creation there cannot possibly be time.' Augustine never makes it clear just why there cannot be time without creation, but perhaps the explanation is to be inferred from this doctrine. In the concluding paragraph of Book XI he states, 'It is far otherwise with you, for you are eternally without change, the truly eternal Creator of minds.' Clearly Augustine thinks that the minds of human beings and other beminded creatures (angels) are different in kind from the mind of God. Since God is atemporal, time

---

[25] Ibid.
[26] Ibid., XI.28.

is not an extension of God's mind. Therefore created minds must exist if time is to exist; hence, without creation there is no time.

Subjective Time: Grünbaum and Kant

Significant problems remain in the attempt to grasp Augustine's full theory of time. We are left somewhat puzzled as to the ontological status of time, for it is one thing to say that time is mind-dependent, and quite another to explain what this means. Now, even granting Augustine's notion of degrees of being (discussed in the next section), he should still be considered a realist about the external world. This is presupposed in his comments about the movement of an object that cannot itself be time. Consequently Augustine cannot be interpreted as being a thoroughgoing idealist about time; time does bear some relation to objective reality. I have already said that Augustine's is a dynamical theory of time; he would, if he were aware of the terminology, have agreed that time is described by A-determinations. But he would also hold that events possess A-determinations only because the events bear B-relations to mental events. Could Augustine really have confused consciousness of time, as a mode of knowing, with the reality of time, as the object of knowledge?

We can understand the subjectivity of time in two ways: Augustine could intend an ontological thesis:

    A. The passage of time is wholly psychological.

Or he could intend an epistemological thesis:

    B. We only know the passage of time subjectively.

Many interpreters take Augustine in the sense expressed in (A), and cite him as the originator of the psychological theory of time.[27]

Adolf Grünbaum At this point it would be helpful to compare Augustine's view to contemporary psychological theories of time such as that of Adolf Grünbaum.[28] Grünbaum's theory 'makes nowness (and thereby pastness and futurity) depend on

---

[27] So Richard Sorabji, *Time, Creation and the Continuum: Theories in Antiquity and the Early Middle Ages* (Ithaca, NY: Cornell University Press, 1983), pp. 30-31, 167-8, hereafter *TC&C*; David Wood, *The Deconstruction of Time* (Atlantic Heights, NJ: Humanities Press International, 1989), p. 61; J. R. Lucas, *A Treatise on Space and Time* (London: Methuen, 1973), p. 14.

[28] Adolf Grünbaum, 'The Status of Temporal Becoming,' in *The Philosophy of Time*, ed. Richard M. Gale (New Jersey: Humanities Press, 1978), pp. 322-54; see also Grünbaum's *Philosophical Problems of Space and Time*, 2nd ed. (Dordrecht: Reidel, 1973). Another twentieth century proponent of a psychological theory of time, though not as well known for the view as Grünbaum, was Kurt Gödel. See Hao Wang, 'Time in Philosophy and in Physics from Kant and Einstein to Gödel,' *Synthese* 102 (1995), pp. 215-34.

the existence of conceptualized awareness that an experience is being had.'[29] Thus temporal becoming is mind-dependent. But Grünbaum's theory assumes a 'block universe'—four-dimensional Minkowskian space-time, in which events are ordered by B-relations.[30] In such space-time there is no past or future, but such A-properties come to be applied to events in the B-series when a conscious mind experiences one of the events. It is certain that Grünbaum would not accept (16), because he would reject (14). For him, temporal intervals exist objectively and are subject to B-theoretic descriptions. What is subjective for him are the A-properties a conscious mind assigns to those intervals. As McCall summarizes, 'No conscious beings, no present. No present, no flow of time, and no division into past and future.'[31]

But could Augustine accept the existence of space-time in which all physical events timelessly exist? Probably not. We have seen that Augustine rejected the static model in which past and future events existed in some secret place on an ontological par with the present.[32] He held that the present was truly ontologically—not merely psychologically—privileged. So it seems that he would be forced to reject a Minkowskian space-time, and with it also contemporary psychological theories of time. Thus (A) cannot be the proper understanding of Augustine's subjective time.

This means that Augustine must have intended something like (B), which does not deny the objective ontological status of time. But (B) sounds like a trivial claim: *all* experience is subjective, after all. Could Augustine really have devoted so much argument to support the platitude that time is something we experience and therefore it is only knowable subjectively? Well, such a claim might not be quite so trivial after all. It is possible that Augustine meant that we, as creatures occupying space and having a lower degree of being than the immutable God, couldn't but experience transience; our minds are subject to the limitations of the experience of time. While it appears that no philosopher took up Augustine's

---

[29] Grünbaum, 'The Status of Temporal Becoming,' p. 335.
[30] McCall comments:
> The mind-dependence theory, according to which the passage of time and its division into past, present, and future occupy roughly the same status as the secondary qualities of colour, sound, and taste did for John Locke, has been laid out persuasively by Grünbaum. I shall show that within the context of the Minkowskian picture of the universe that he accepts, Grünbaum's arguments are incontrovertible. Any attempt to defeat them can be turned aside. But if the Minkowskian picture is replaced by something different, a mind-independent theory of temporal becoming can be consistently introduced.

Storrs McCall, *A Model of the Universe: Space-Time, Probability, and Decision* (Oxford: Clarendon Press, 1994), p. 27. It is just this Minkowskian picture that I criticized in Chapter 3.
[31] Ibid., p. 28.
[32] Augustine, *Confessions*, XI.17.

suggestion of a psychological interpretation of time until Leibniz,[33] it was Kant who made the subjectivity of time a fundamental principle of his philosophy. Indeed, I believe that an excursus into Kant's analysis of time might prove helpful in understanding Augustine.[34]

Immanuel Kant In the *Critique of Pure Reason*, Kant set out to prove that the presentations of time and space are the transcendentally ideal a priori forms of all empirical intuition. In the sections of the 'Transcendental Aesthetic' called 'Metaphysical Exposition,' Kant attempts to give an exposition of space and time that clearly exhibits both as given a priori.[35] Although he treats space and time separately, his thought about both is clearly symmetric. However, I am here interested only in his thoughts about time. I believe that Kant not only held that time was empirically real but also that time formed a necessary precondition to any mental construction of the world. What he attempts to do is to prove (i) that time is an a priori presentation—that is, not empirically derived; and (ii) that it is in fact an intuition and not a concept, where an intuition is a singular term standing in immediate relation to an empirical object, and not a general or abstract term.

The first argument Kant offers for the a priori nature of time is this: We cannot individuate successive events without possessing an a priori idea of time. Time is already there, antecedent to the temporal individuation, and hence must be a priori:

> Time is not an empirical concept that has been derived from any experience. . . . Only on the presupposition of time can we represent to ourselves a number of things as existing at one and the same time (simultaneously) or at different times (successively).[36]

The second argument for the a priori nature of time involves the impossibility of ever conceiving of its absence—that is, non-time—although we can certainly conceive of empty or eventless time. Since, then, time is an ineluctable 'given' of all appearances, it must itself be a priori.

> Time is a necessary representation that underlies all intuitions. We cannot, in respect of appearances in general, remove time itself, though we can quite well

---

[33] Although he does not expressly acknowledge Augustine; see his fifth letter to Clarke, in *The Leibniz-Clarke Correspondence*, ed. H. G. Alexander (New York: Manchester University Press, 1956), §§27, 47-9; cf. Sorabji, *TC&C*, p. 30.

[34] The comparison of Kant's and Augustine's theories of time was suggested by Bertrand Russell, who called Augustine's 'a better and clearer statement than Kant's of the subjective theory of time.' *A History of Western Philosophy*, 2nd ed. (New York: Simon and Schuster, 1972), p. 354.

[35] Immanuel Kant, *Critique of Pure Reason*, unabridged ed., tr. Norman Kemp Smith (New York: St Matrin's Press, 1929), B38/A23.

[36] Ibid., B46/A30.

think of time as void of appearances.... In it alone is actuality of appearances possible at all....[37]

Next Kant argues that time is an intuition as opposed to a concept. In Kant's vocabulary, a concept may be thought of as a general term; an intuition as a singular term. Roughly speaking, an intuition is what is produced immediately in the mind, while a concept is what is produced mediately (by employing the concepts of understanding) in the understanding by a plurality of intuitions.[38] Remembering that Kant is concerned in the *Critique* with the conditions under which synthetic a priori judgments are possible, it becomes clear why time (and space) must be not only a priori but also an intuition. For, if time were a concept, there would be no guarantee that it was actually related to sensible objects of empirical reality, and hence could never play a role in synthetic judgments. Thus Kant must show that time is an intuition. This he does by arguing that it is a singular and immediate term:

> Time is not a discursive, or what is called a general concept, but a pure form of sensible intuition. Different times are but parts of one and the same time; and the representation which can be given only through a single object is intuition....[39]

> The infinitude of time signifies nothing more than that every determinate magnitude of time is possible only through limitations of one singular time that underlies it.... But when an object is so given that its parts, and every quantity of it, can be determinately represented only through limitations, the whole representation cannot be given through concepts, since they contain only partial representations; on the contrary, such concepts must themselves rest on immediate intuition.[40]

In these paragraphs Kant assumes that time is continuous. Time, although a succession of individual durationless moments, cannot be partitioned into independent times that are not also in continuous time. There are not many times, but a singular time. Therefore, since parts of time cannot be composed into a general term, time cannot be a concept, but must be a singular intuition.

We must take note of Kant's dictum that space is the 'form of the outer sense' while time is the 'form of the inner sense.'[41] He distinguishes form and matter in this way: 'That in the appearance which corresponds to sensation I term its matter; but that which so determines the manifold of appearance that it allows of being ordered in certain relations, I term the form of appearance.'[42] The form, then, is that which is 'the subjective condition of sensibility.'[43] Form belongs to

---

[37] Ibid., B46/A31.
[38] Cf. ibid., B74/A50.
[39] Ibid., B47A32.
[40] Ibid., B48/A32.
[41] Ibid., B37/A22.
[42] Ibid., B34/A20.
[43] Ibid., B42/A26.

the subject as pertaining to the structuring of sensibilities, and is thus a necessary precondition of all intuition. Consequently Kant makes a claim which has definite Augustinian overtones: 'This form is not to be looked for in the object in itself, but in the subject to which the object appears; nevertheless, it belongs really and necessarily to the appearance of this object.'[44] That the a priori intuitions of space and time are forms of every intuition, then, means that we can never have completely unstructured perceptions; all sensibilities are inevitably embedded in space and time.

It is clear from the start that Kant has two opposing views in mind against which he is directing the Transcendental Aesthetic.

> What, then, are space and time? Are they real existences? Are they only determinations or relations of things, yet such as would belong to things even if they were not intuited? Or are space and time such that they belong only to the form of intuition, and therefore to the subjective constitution of our mind, apart from which they could not be ascribed to anything whatsoever?[45]

The first view is the absolutist view of space and time developed by Newton; the second is the relational view of Leibniz and Wolff. Kant wants to take the debate in a different direction. He denies that intuitions of space and time are somehow in contact with noumenal reality, for that can never be known. Thus Newton is refuted in his hypothesis of absolute space and absolute time. On the other hand, space and time are more than mere relations, for in order for there to be relations, there must be relata, and Kant has shown that the intuitions of space and time remain even if emptied of all objects. Thus Leibniz is refuted in his relational view.

But where does that leave us? Kant repeatedly states that space and time are transcendentally ideal and also empirically real.[46] Now, if Kant were speaking of space and time in themselves as transcendentally ideal, several awkward conclusions would follow. First, the concepts of space and time in modern mathematics and physics would render Kant's conceptions little more than historical curiosities. The critical problem arises in Kant's examples of geometric truths; for example, that the shortest distance between two points is a straight line, or that two lines can never bound a figure—at least three lines are necessary. Clearly these 'truths' only hold in plane Euclidean geometry; even simple spherical geometry contains counter-examples. More problematic, less than 50 years after the publication of the *Critique*, Lobachevskian geometry and, shortly thereafter, complex Riemannian space, were developed mathematically (the latter being assumed by Einstein's General Theory of Relativity of 1916). Further, just over a century later, David Hilbert (ironically, another native of Königsberg) would devise mathematical systems that decisively undercut Kant's theory of geometry. The *Grundlagen der Geometrie*, published by Hilbert in 1898, opened with a motto

---

[44] Ibid., B55/A38.
[45] Ibid., B37/A23.
[46] At, for example, ibid., B52/A36.

taken from Kant: 'All human knowledge begins with intuitions, proceeds to concepts, and terminates in ideas.' But Hilbert's development of geometry established a decidedly anti-Kantian view of the subject, and it is Hilbert's work in infinite dimensional space that forms an important part of contemporary formulations of Quantum Mechanics.

The obvious question is whether these advances in mathematical theory, especially when demonstrated to have empirical application, invalidate Kant's argument that (Euclidean) geometry proves the transcendental nature of the intuition of space. I would argue that they do not. For Kant posits space and time as the forms of intuition. But, as forms of intuition, space and time are the framework in which all sensibilities arise. Now, sensibilities of the empirically accessible world will never give rise to intuitions of Riemannian or Hilbert space; even relativity theory allows that the perceptions of the immediate surroundings of an observer in a relativistic reference frame will appear to her as normal (the Principle of Equivalence). So modern mathematical constructs cannot invalidate Kant's claims; at best, they can show his understanding of geometry was seriously incomplete. If Kant were writing today, doubtless he would use other illustrations and be more careful in his pronouncements. But I conclude that Kant's use of geometry to demonstrate the transcendental nature of the intuition of space is compatible, *mutatis mutandis*, with modern mathematical theory. That is, if Kant is talking about the representations of space and time as Euclidean and linear, then his argument still stands, but this would not be the case if his use of space and time in themselves entailed their transcendentally ideality.

Second, it seems that a strong transcendental ideality interpretation does in fact leave Kant open to charges that his is merely a version of Berkeley's idealism, a charge he roundly rejects.[47] For strong transcendental ideality would reduce space to nothing but the form of outer sense, and time to nothing but the form of inner sense. It would then follow that perceptions themselves do not exist in space and time, but cannot be intuited as other than spatial and temporal. What intuitions of space and time yield would not be empirical reality but only psychological phenomena. And clearly Kant intends more than that.

Finally, when he says that 'Time and space, taken together, are the pure forms of all sensible intuition,'[48] he must be presupposing that space-time (or the spatio-temporal order), as the necessary ground of empirical sensibility, is entirely unaffected by the transcendental ideality of space and time. But this view cannot easily be harmonized with strict transcendental ideality. For while, on Kant's view, sensible intuitions can never reach the noumenal world of things-in-themselves, nevertheless they are grounded in the empirically real phenomenal world. Certainly the presentations of empirical objects are intuited according to the transcendentally ideal presentations of space and time. However, if space and time themselves are transcendentally ideal, it would seem necessary to conclude that the empirical world of sensible intuitions would collapse into transcendental ideality,

---

[47] Ibid., B69-71.
[48] Ibid., B56/A39.

and space-time, rather than being the 'matrix' within which sensation occurs, would become nothing but the mind imposing structure on mental constructs. This would leave us two steps removed from empirical reality, which Kant seems specifically to deny.[49]

Let us return now to Augustine. Despite the argument of the *Confessions* that time is subjective, it seems clear that he, like Kant, holds to the empirical reality of time, as illustrated in a well-known passage from *The City of God*:

> For if eternity and time are rightly distinguished by this, that time does not exist without some movement and transition, while in eternity there is no change, who does not see that there could have been no time had not some creature been made, which by some motion could give birth to change—the various parts of which motion and change, as they cannot be simultaneous, succeed one another—and thus, in these shorter or longer intervals of duration, time would begin? Since then, God, in whose eternity is no change at all, is Creator and ordainer of time, I do not see how He can be said to have created the world after spaces of time had elapsed, unless it be said that prior to the world there was some creature by whose movement time could pass. . . .
>
> For that which is made in time is made both after and before some time—after that which is past, before that which is future. But none could then [at the moment of creation] be past, for there was no creature by whose movements its duration could be measured. But simultaneously with time the world was made, if in the world's creation change and motion were created.[50]

Now perhaps Augustine changed his mind between the *Confessions* (AD 397-98) and *The City of God* (413-26), but he gives no indication of repudiating his earlier views. A more charitable reading seeks to harmonize the two works, and I believe that our excursus into Kant gives us some idea as to how this is possible. Augustine believed that we perceive the real, objective world; unlike Kant, he did not posit an unknowable noumenal realm beyond empirical reality. Nevertheless, in his Neoplatonic doctrine of degrees of being, Augustine opened the door to an ontology analogous to Kant's. Kant's noumenal realm that cannot be perceived is analogous to Augustine's realm of archetypal ideas that cannot be perceived. According to Kant, we can only perceive empirical reality through the pure forms of the intuition, space and time. Analogously, for Augustine, we limited creatures cannot perceive other creatures in their individual, archetypal essences, but only as temporal particulars. As for Kant, time itself would be a necessary presupposition to any perception at all.

If this quasi-Kantian understanding of Augustine is correct, we can read him as holding to a 'psychological theory' of time without denying the objective, ontological status of time as a necessary part of the created universe. With this interpretation of (B) we are not required to presuppose a static space-time, as in modern psychological theories of time.

---

[49] Ibid., B69-71.
[50] Augustine, *The City of God*, tr. Marcus Dods (New York: Random House, 1950), XI.6.

Yet, in his insistence that God knows events that to us are yet future (he does not want to speak of foreknowledge, but of knowledge *simpliciter*[51]), Augustine seems to have in mind some way in which all events exist timelessly, so that they may be known to God. He clearly wants to maintain that the future is literally non-existent, not merely non-actual. How, then, can future events be present to God in his timeless eternity? Given his insistence that God has his being in timeless eternity, we must look to Augustine's understanding of eternity to untangle this final knot in his doctrine of time. But, first, I shall, as promised, survey some of the salient themes of Neoplatonic thought that influenced Augustine.

Neoplatonic Influences on Augustine

The influence on Augustine from Plato and the Neoplatonists, especially Plotinus (AD 205-270), is well known.[52] Augustine's *De vera religione*, written in 390, only four years after his conversion, shows the clear Platonic overtones of Augustine's apologetic, notably his references to the immutable forms, which he says exist in the mind of God, and by which humans are able to gain knowledge of sensible, temporal things.[53]

Neoplatonism

Neoplatonism refers to the development of Plato's thought beginning with Plotinus (AD 205-270), and extending at least until the Emperor Justinian closed the Platonic School in Athens in AD 529 (although some historians argue that Neoplatonism continued to characterize the philosophy of the Byzantine Empire).
   Principal Neoplatonist thinkers include Plotinus, Amelius, Porphyry, Iamblichus, Proclus, and Simplicius.
   Key ideas:
- 'The One' is above the '*Noûs*' or world-intellect.
- True, pure being belongs only to the One.
- The One is metaphysically simple and immutable.
- The 'Great Chain of Being' emanates from the One.
- Eternity is conceived of as timeless.

---

[51] 'What then is foreknowledge, if not knowledge of the future? But what becoming is there in God, who transcends all time? If then God's knowledge possesses the things themselves, they are not for Him future, but present. It follows that one may not in this case speak of foreknowledge, but simply knowledge.' *De diversis quaestionibus ad Simplicianum* 2.2.2, tr. William Lane Craig in *The Problem of Divine Foreknowledge and Future Contingents from Aristotle to Suarez* (Leiden: E. J. Brill, 1988), p. 73, hereafter *PDF*.
[52] Callahan, *Four Views of Time*, pp. 176-8; Craig, *PDF*, pp. 76-8; Leftow, *T&E*, pp. 73-111; Sorabji, *TC&C*, pp. 163-73. The clear influence of Neoplatonism on Augustine should not be overstated, however; in many respects he departs from the teachings of Plotinus. Cf. Sorabji, *TC&C*, pp. 168-72.
[53] The Platonic tone is most clear in xxix, 52-xliv, 82, where Augustine attempts to show how reason can lead men to God.

Of crucial importance to understanding Augustine's doctrine of God's relation to time are the elements of Neoplatonism in his doctrines of immutability and simplicity. So we begin by picking up the thread of immutability, which was left hanging earlier.

A basic conviction of Neoplatonism, which shaped Augustine's thinking about God, is that 'true' existence is immutable existence. That God possessed 'true' existence was axiomatic to his Christian theism:

> Being is a name for immutability. For all things that are changed cease to be what they were, and begin to be what they were not. Nobody has true being, pure being, real being except one who does not change. . . . What does 'I am who am' mean but 'I am eternal . . . I cannot be changed'?[54]

> But you say, Why do they [mutable creatures] become defective? Because they are mutable. Why are they mutable? Because they have no supreme existence. And why so? Because they are inferior to him who made them. Who made them? He who supremely is. Who is he? God, the immutable Trinity, made them through his supreme wisdom and preserves them by his supreme loving-kindness. Why did he make them? In order that they might exist. Existence as such is good, and supreme existence is the chief good. Whatever is good must have some form, and though it be but a minimal good it will be good and will be of God. The highest form is the highest good, and the lowest form is the lowest good.[55]

Now a full understanding of these claims demands an understanding of Augustine's concept of degrees of being, a concept that Leftow characterizes as 'just bizarre.'[56] Nevertheless, Leftow proceeds to give a good framework for a coherent understanding of degrees of being,[57] which I shall not rehearse here, although I think he has offered a very reasonable interpretation of the concept. Suffice it to say that we can see here strong echoes of Plato's teaching concerning the Forms that are in some sense more real than any particular participating in the Form.

The same notion is found in Plotinus. According to this great Neoplatonist, the realm of intelligible essences, which constitute the archetypes for temporal particulars, is the immediate object of the *Noûs*. Plotinus views this realm as containing timeless Platonic forms as well as individual essences. The *Noûs* has a sort of timeless intuitive knowledge of essences and forms:

> Admiring this world of sense . . . let us mount to its archetype, to the yet more authentic sphere: here we are to contemplate all things as members of the Intellectual—eternal in their own right, . . .—and, presiding over all these, the unspoiled Intelligence and the unapproachable wisdom.
>
> . . . For here is contained all that is immortal; nothing here but is Divine

---

[54] Augustine, Sermon 7,7, cited in Leftow, *T&E*, p. 73.
[55] Augustine, *De vera religione*, in *Augustine: Earlier Writings*, tr. John H. S. Burleigh (Philadelphia: Westminster Press, 1953), xviii, 35.
[56] Leftow, *T&E*, p. 80.
[57] Ibid., pp. 81-9.

Mind; all is God. . . . All its content, thus, is perfect, that itself may be perfect throughout, as holding nothing that is less than the divine, nothing that is less than the intellective. Its knowing is not by search but by possession . . . ; for all belongs to it eternally and it holds the authentic Eternity imitated by Time. . . . [T]his is pure being in eternal actuality; nowhere is there any future, for every then is a now; nor is there any past, for nothing there has ever ceased to be.[58]

Thus greatest (truest, purest) being cannot change, and hence (as shown above in (6)-(8), section 5.1.1) is timeless.

19. Greatest (truest, purest) being belongs to God.
20. Greatest (truest, purest) being is immutable.
21. What is immutable is timeless.
22. Therefore, God is timeless.

There is another aspect of Neoplatonic thought concerning greatest being that must be explored—that of the unity or simplicity of greatest being. Just as greatest being cannot have temporal parts, for that implies at least the possibility of change, so too it cannot have any other kind of proper parts. Now anything with no proper parts is simple, is one. Thus Plotinus writes, 'because what the soul seeks is the One . . . it must withdraw from . . . objects of the lowest existence and turn to those of the highest . . . it must again become one. Only thus can it contemplate . . . the One. . . . As we turn toward the One we exist to a higher degree, while to withdraw from it is to fall.'[59]

Similarly impressed with the simplicity of greatest being, Augustine says such things as 'if it [the soul] serves God with the mind and a good will, it will undoubtedly be restored, and will return from the mutable many to the immutable One';[60] and 'We must then, seek the realm where number exists in complete tranquility; for there existence is, above all, unity.'[61]

But no spatio-temporal thing can possess the unity of greatest being:

The reason why corporeal beauty is the lowest beauty is that its parts cannot all exist simultaneously. Some things give place and others succeed them, and all together complete the number of temporal forms and make of them a simple beauty.[62]

But who can find absolute equality or similarity in bodily objects? Who would venture to say, after due consideration, that any body is truly and simply one? All are changed by passing from form to form or from place to place, and consist of parts each occupying its own place and extended in space. True equality and

---

[58] Plotinus, *Enneads*, 5.1.4, cited in Craig, *PDF*, p. 76.
[59] Plotinus, *Enneads* 6.9.3 and 9, cited by Leftow, *T&E*, p. 93.
[60] Augustine, *De vera religione*, xii, 24.
[61] Ibid., xlii, 79.
[62] Ibid., xxi, 41.

similitude, true and primal unity, are not perceived by the eye of flesh or by any bodily sense, but are known by the mind.[63]

In the same vein Augustine 'denies that God is properly called a substance that has properties; he is more properly called an essence because he is properties.'[64] Consequently,

23. Greatest (truest, purest) being belongs to God.
24. Greatest (truest, purest) being is simple.
25. What is simple is timeless.
26. Therefore, God is timeless.

We can now pull all these threads together. The doctrines of degrees of being, of immutability and of simplicity as found in Neoplatonism and, as incorporated by Augustine in his understanding of God, require God to be timeless.

But as it is not a settled question that the doctrines of strong immutability and simplicity are correctly applied to the God of the Judeo-Christian tradition, nor that the doctrine of degrees of being is required by an ontology faithful to biblical data, it is not at all clear that the God of the Tanakh and the New Testament is properly analyzed in Neoplatonic concepts.[65] To the degree that these concepts distort an accurate picture of God and of reality, they are misleading. They must be subjected to critical analysis and evaluation by the best philosophy and metaphysics available to us, and even though they figure large in the tradition, they cannot be allowed to determine without further argument the conclusions of a contemporary investigation in philosophical theology.

Eternity

Given the arguments summarized above, which Augustine uses to defend God's timelessness, it follows that he also took eternity to be timeless. Here, too, Augustine was not a solitary figure, but reflects the influence of Plato, the Neoplatonists, and perhaps others as well.

---

[63] Ibid., xxx, 55.

[64] Richard Swinburne, *The Christian God* (Oxford: Clarendon Press, 1994), p. 168. The reference is to Augustine, *De Trinitate*, vii, 5.

[65] Edward R. Wieringa, *The Nature of God: An Inquiry into Divine Attributes* (Ithaca, NY: Cornell University Press, 1989), p. 173. Simplicity is defended by Eleonore Stump and Norman Kretzmann, 'Absolute Simplicity,' *Faith and Philosophy* 2 (1985), pp. 353-82. James Ross objects in 'Comments on "Absolute Simplicity",' *Faith and Philosophy* 2 (1985), pp. 383-91, to which Stump and Kretzmann offer a rejoinder in 'Simplicity Made Plainer: A Reply to Ross,' *Faith and Philosophy* 4 (1987), pp. 198-201.

While Plato is somewhat ambiguous in his references to eternity, at least in the *Timaeus* he seems to regard the Forms as timeless, durationless, existing always.[66] For instance, concerning the Forms, Plato writes,

> But that which is always changeless and motionless cannot become either older or younger in the course of time .... And all in all, none of the characteristics that becoming has bestowed upon the things that are borne about in the realm of perception are appropriate to it. These, rather, are forms of time that have come into being—time that imitates eternity [αἰών].[67]

Plotinus apparently had a clearer concept of timelessness than did Plato, and made a number of attempts to clarify the notion of unextended, durationless eternity; for example:

> So it does not contain any this, that and the other. Nor therefore will you separate it out, or unroll it, or extend it, or stretch it. Nor then can you find any earlier or later in it. If then there is neither any earlier nor later about it, but 'is' is the truest thing about it, and indeed is it, and this in the sense that it is by it essence and life, then again we have got the very thing we are talking about, namely, eternity [αἰών].[68]

The denial of duration in eternity can be seen before Plotinus, however. Probably the first to grope towards this notion was Parmenides. Echoes are heard in Plutarch as well as the Christian fathers Clement of Alexandria, Origen, and Gregory of Nyssa.[69] The Jewish scholar Philo of Alexandria also reflects this Platonic tradition:

> So with God there is no future, for he has put beneath himself the very boundaries of all times. Indeed, his life is not time, but eternity (αἰών), the archetype and model for time. And in eternity nothing is past or future, but simply has being.[70]

While it is problematic to say just how much Augustine knew of the Greek fathers, it is still clear that in understanding eternity as timeless, unextended and durationless, he was operating within a firm tradition of Jewish and Christian theology as well as Greek philosophy. Thus his claim that God exists timelessly has broader grounding than simply the Neoplatonic concepts of immutability and

---

[66] Richard Sorabji argues that in the *Timaeus* Plato conceived of eternity as durationless: *TC&C*, pp. 108-12.
[67] Plato, *Timaeus*, in *Readings in Ancient Greek Philosophy: From Thales to Aristotle*, ed. S. Marc Cohen, Patricia Curd, and C. D. C. Reeve, tr. D. J. Zeyl (Indianapolis, IN: Hackett, 1995), 38A3, hereafter *Timaeus*.
[68] Plotinus, *Enneads* 3.7.6, cited by Sorabji, *TC&C*, p. 112. Sorabji cites many other references in Plotinus that emphatically deny that eternity is extended.
[69] Plutarch, *On the E at Delphi*, 393A-B; Clement of Alexandria, *Stromateis* 1.13; Origen, *On First Principles* 2.2.1; *Commentary on John* 1:29; Gregory of Nyssa, *Contra Eunomium* 1.359-64, all cited by Sorabji, *TC&C*, pp. 121f.
[70] Philo, *Quod Deus Immutabilis Sit* 6.23, cited by Sorabji, *TC&C*, p. 121.

simplicity. Whatever the sources of the idea, Augustine clearly conceived of eternity as timeless:

> Try as they may to savour the taste of eternity, [men's] thoughts still twist and turn upon the ebb and flow of things in past and future time. But if only their minds could be seized and held steady, they would be still for a while and, for that short moment, they would glimpse the splendour of eternity which is for ever still . . . . They would see that time derives its length only from a great number of movements constantly following one another into the past, because they cannot all continue at once. But in eternity nothing moves into the past; all is present.[71]

> Furthermore, although you are before time, it is not in time that you precede it. If this were so, you would not be before all time. It is in eternity, which is supreme over time because it is a never-ending present, that you are at once before all past time and after all future time. . . . Your years are one day, yet your day does not come daily but is always today, because your today does not give place to any tomorrow nor does it take the place of any yesterday. Your day is eternity.[72]

But just how do these considerations answer the question that we left hanging above? Given that Augustine maintains that the future is literally non-existent, not merely non-actual, how can future events be present to God in his timeless eternity?

If we recall Plotinus' words to the effect that the *Noûs* comprehends the essences in a sort of timeless intuition, then we can understand Augustine's view of God's knowledge in the context of Neoplatonism.[73] Augustine himself directly addresses the question in writing to Simplician:

> Hence in Latin we can call the ideas either 'forms' or 'species,' which are literal translations of the word. But if we call them 'reasons,' we obviously depart from a literal translation of the term, for 'reasons' in Greek are called *logoi*, not 'ideas.' Yet, nonetheless, if anyone wants to use 'reason,' he will not stray from the thing in question, for in fact the ideas are certain original and principal forms of things, i.e., reasons, fixed and unchangeable, which are not themselves formed and, being thus eternal and existing always in the same state, are contained in the Divine Intelligence. . . . As for these reasons, they must be thought to exist nowhere but in the very mind of the Creator. For it would be sacrilegious to suppose that he was looking at something placed outside himself when he created in accord with it what he did create. But if these reasons of all things to be created or [already] created are contained in the Divine Mind, and *if there can be in the Divine Mind nothing except what is eternal and unchangeable*, . . . then not only are they the ideas, but they are themselves true because they are eternal and because they

---

[71] Augustine, *Confessions*, XI.11.
[72] Ibid., XI.13.
[73] See Linda Zagzebski, 'Individual Essence and the Creation,' in *Divine and Human Action: Essays in the Metaphysics of Theism*, Thomas V. Morris, ed. (Ithaca, NY: Cornell University Press, 1988), pp. 119-44.

remain ever the same and unchangeable. *It is by participation in these that whatever is exists in whatever manner it does exist.*[74]

What is striking here is that the divine ideas are the medium by which God knows his creatures even after creation. In discussing this passage Craig makes a point that has often been overlooked by other commentators on Augustine:

> Hence, it is not the creatures themselves which are timelessly present to God's *scientia*, but their archetypal ideas. The creatures in many cases do not yet exist and therefore are not 'seen' by God; in this sense His knowledge of them may be said to be foreknowledge. But the divine ideas are always present to God's mind and thus are 'seen' by Him; this is the sense in which Augustine could say to Simplician that God's knowledge possesses the things themselves (*scientia dei res ipsas habet*). . . . Hence, for Augustine God's knowledge in eternity of that which is future for us in time is not really foreknowledge, but simply knowledge of the equally eternal divine ideas.[75]

Whether this sort of knowledge entails at least a 'soft' form of theological fatalism is another question, one beyond the scope of this study.[76]

Summary

Augustine's views on God, time and eternity would have a profound influence on philosophers and theologians for centuries to come. He held a dynamical, presentist theory of time in which the past and the future do not exist, but in which, although time is real, temporal becoming is subjective. He believed that God's immutability and simplicity entailed God's atemporality. And he believed that eternity was durationless and unextended.

We should not ignore the strong influence that Neoplatonism had on Augustine. Since Augustine's arguments for God's atemporality depend heavily on the Neoplatonic understanding of immutability and simplicity, his view of an atemporal God is as strong—or as dubious—as the Neoplatonic foundations.

I have already argued in Chapter 2 for a dynamic theory of time, and I should like to claim Augustine as an ally. On a causal theory of time, temporal becoming is objectively real, not merely psychological, and I argued in the second chapter that a psychological theory of time entails at least one A-series in reality— namely, the A-series in the mind of the person experiencing time. I have tried here to show how Augustine's psychological theory could be harmonized with a realist dynamic theory of time, using Kant as a point of comparison. However, the

---

[74] Augustine, *De diversis quaestionibus ad Simplicianum*, 46.2.21-32, 53-64, cited by Craig, *PDF*, p. 77, my emphasis.
[75] Craig, *PDF*, p. 77.
[76] Hasker believes that, by the time of writing *De diversis quaestionibus*, Augustine had moved from his earlier libertarianism to a soft determinism. See Hasker, *God, Time, and Knowledge*, pp. 4-6.

validity of that reading of Augustine may be questioned; hence Augustine remains a troublesome ally at best.

Boethius

Anicius Manlius Severinus Boethius (480-524) is notable for attempting to reconcile the problem of God's foreknowledge and human freedom by invoking God's atemporality, and for offering a famous definition of eternity. Further, Boethius is the first to assign fixed terms to certain concepts, distinguishing clearly between everlastingness (*sempiternitas*) and eternity (*aeternitas*). He is clearly influenced by Augustine and Neoplatonism. Like Augustine, Boethius appears to hold to a dynamic view of time, but also, perhaps inconsistently, to hold that all of time—past, present and future—is 'present' to God.

In recent years philosophers have appealed to Boethius as the origin of a durational rather than a timeless view of God's eternity. I shall first examine the two foremost durational interpretations, and argue that neither offers a satisfactory notion of what is meant by timeless duration. I shall then attempt to place Boethius in the Neoplatonic context in which he stood and show that it is highly unlikely that Boethius himself had a urational rather than a timeless view of God's temporal mode of being. Finally I shall examine two seemingly contradictory streams within Boethius' doctrine of time, and suggest that perhaps Boethius departed from Augustine's psychological theory of time and held to a realist dynamic theory, but at the same time grasped tacitly that his concept of divine timelessness logically implied a static theory of time, thus resulting in somewhat of an inconsistency in his theory.

Boethius' Views on God and Time

In Boethius' famous statement that 'Eternity is the whole, perfect, and simultaneous possession of endless life,' some have seen a doctrine of timeless (successionless) duration. This is, however, an unlikely interpretation. Boethius clearly held the following:
- Time is dynamic—yet all events (including future events) somehow exist timelessly in eternity.
- God's foreknowledge is of future events themselves. All times—past, present and future—are equally 'present' to God.

Eternity as Timeless Duration

The best-known passage from Boethius is found in *The Consolation of Philosophy*, in a context in which Boethius is attempting to reconcile divine foreknowledge and human freedom:

> The common judgment of all rational creatures holds that God is eternal. Therefore let us consider what eternity is, for this will reveal both the divine nature and the divine knowledge.

> Eternity is the whole, perfect, and simultaneous possession of endless life. The meaning of this can be made clearer by comparison to temporal things. For whatever lies in time lives in the present, proceeding from past to future, and nothing is so constituted in time that it can embrace the whole span of its life at once. It has not yet arrived at tomorrow, and it has already lost yesterday; even the life of this day is lived only in each moving, passing moment. Therefore, whatever is subject to the condition of time, even that which—as Aristotle conceived the world to be—has no beginning and will have no end in a life coextensive with the infinity of time, is such that it cannot rightly be thought eternal. For it does not comprehend and include the whole of infinite life all at once, since it does not embrace the future which is yet to come. Therefore, only that which comprehends and possesses the whole plenitude of endless life together, from which no future thing nor any past thing is absent, can justly be called eternal. Moreover, it is necessary that such a being be in full possession of itself, always present to itself, and hold the infinity of moving time present before itself.[77]

Here is the celebrated definition of eternity as 'simultaneous possession of endless life.'[78] But the celebrity of the definition does not mean it is either clear or correct. On first reading, there appears to be a contradiction between 'simultaneous possession' or 'possession all at once,' in a non-extended moment, and 'endless, illimitable life,' which seems to entail endless temporal extension. At least intuitively it would seem that life must involve some duration, that a life lived instantaneously is not anything like what we would mean by 'life.'

Recently at least two sophisticated attempts have been made to explain the notion of a duration that is not a temporal extension as an interpretation of the definition of Boethius. Neither, I shall argue, ultimately succeeds.

In a remarkable and influential article, Eleonore Stump and Norman Kretzmann develop a 'durational' view of God's eternity that they base on Boethius.[79] They see four ingredients in the passage quoted above: First, anything eternal has life. Thus numbers, propositions or the universe are not eternal, although the first two might be atemporal and the third sempiternal. Second, the life of an eternal being cannot be limited—not in the sense of unlimited extension, but rather in the sense that an unextended point or an instant cannot be limited in extent. Third, illimitable life entails a duration of a special sort. And fourth, an eternal being possesses its life completely, all at once. Of particular interest here

---

[77] Boethius, *The Consolation of Philosophy*, tr. Richard Green (Indianapolis: Bobbs-Merrill, 1962), V, prose 6
[78] '*Aeternitas . . . est interminabilis vitae tota simul et perfecta possessio.*' Stump and Kretzmann translate the phrase 'complete possession all at once of illimitable life': Eleonore Stump and Norman Kretzmann, 'Eternity,' *Journal of Philosophy* 78 (1981), p. 430. I would suggest, although it is a minor point, that 'endless' captures the notion of the Latin *interminabilis* better than 'illimitable.'
[79] Ibid., pp. 429-58. In this chapter I am concerned only with Stump and Kretzmann's interpretation of Boethius; I shall deal with their own proposal of 'ET [eternal-temporal] simultaneity' and its place in their atemporal theory of God's being in Chapter 6.

are the last two claims, which together say that Boethius conceives of eternity as having 'infinite, timeless duration.'

The fourth claim is unproblematic, if we take 'life' as a temporally extended event and not a biological notion: Any temporal being, considered at t_2, has a stage of its life at the earlier time t_1 and another stage of its life at the later time t_3. Thus it cannot possess its complete life all at once. So if an eternal being truly possesses its life all at once (*tota simul*), it must be outside time.

But what is to be said of the third claim that this possession of complete life all at once is a duration? Clearly Boethius does not mean temporal extension, but temporal extension is what is meant by duration. Or so it would seem. Yet it is clear that Boethius does not think of eternity as point-like, for following Plato in the *Timaeus*, he says, 'For the infinite motion of temporal things imitates the immediate presence of His changeless life,' and explains that 'it seems to imitate to some extent that which it cannot completely express, and it does this somehow by never ceasing to be.'[80] The everlastingness of the universe certainly is a duration, and if it in any way imitates eternity, then eternity must too possess something of the quality of a duration. Otherwise, eternity would be more like an instant than like everlastingness.

Just how eternity can be timeless duration might be suggested in a passage from Boethius' *De Trinitate*:

> But the expression 'God is ever' denotes a single Present, summing up His continual presence in all the past, in all the present—however that term be used—and in all the future. Philosophers say that 'ever' may be applied to the life of the heavens and other immortal bodies. But as applied to God it has a different meaning. He is ever, because 'ever' is with Him a term of present time, and there is this difference between 'now,' which is our present, and the divine present. Our present connotes changing time and sempiternity; God's present, abiding, unmoved, and immovable, connotes eternity. Add *semper* to *eternity* and you get the constant, incessant and thereby perpetual course of our present time, that is to say, sempiternity.[81]

Our 'changing time' in its 'perpetual course' involves succession. But in God's eternity there is no succession. Stump and Kretzmann interpret this to mean that eternity contains no earlier or later points:

> Because an eternal entity is atemporal, there is no past or future, no earlier or later, *within* its life; that is, the events constituting its life cannot be ordered sequentially from the standpoint of eternity. But in addition, no temporal entity or event can be earlier or later than, or past or future with respect to the whole life of an eternal

---

[80] Boethius, *Consolation*, V, prose 6.
[81] Boethius, *De Trinitate* IV, 64-77, in *Boethius: The Theological Tractates*, tr. H. F. Stewart and E. K. Rand, ed. H. F. Stewart (Cambridge, MA: Harvard University Press, 1946). Although Boethius was mistaken in the etymology of 'sempiternity,' the meaning he attaches to the concept is clear.

entity, because otherwise such an eternal entity would itself be part of a temporal series.[82]

Atemporal duration is duration none of which is not—none of which is absent (and hence future) or flowed away (and hence past). Eternity, not time, is the mode of existence that admits of fully realized duration.[83]

This is an explicit denial not only that there is a sequence of events in eternity, but further that A-properties or B-relations of temporal events are applicable to those events in eternity. Stump and Kretzmann seem to believe that there are no distinct points at all in eternity. But this leads to serious problems— problems sufficient to convince us to reject the Stump/Kretzmann interpretation of Boethius.

If, on one hand, there are literally no distinct points in eternity, how then can God be said to know individual temporal events within creation? Consider two qualitatively identical events of photon emission: an atom emits a photon at t_1, reabsorbs a photon at t_2, and re-emits a photon at t_3. Now the only means by which these events are differentiated is by their temporal reference points t_1 and t_3. But if there are no distinct points in the duration of timeless eternity, then these two events cannot be numerically distinct.[84]

If, on the other hand, eternity contains points but these are indistinct or unordered, the situation is no better. Stump and Kretzmann propose what they call 'ET simultaneity' (to be discussed in detail in Chapter 6) by which each point in the temporal order is simultaneous with the 'whole' of eternity. But if A-properties and B-relations of the temporal order cannot be applied to that order from the standpoint of eternity, how can God know what came before what?

Thus it seems that the Stump/Kretzmann interpretation of atemporal duration faces two insurmountable difficulties and should be rejected. Similar difficulties (together with difficulties associated with the concept of ET simultaneity) lead Brian Leftow to consider a second interpretation of Boethius. According to Leftow, in the passage from *De Trinitate* quoted above, Boethius contrasts the eternal 'now' with the 'moving now' of tensed time: 'Perhaps, then, Boethian eternity is like an extension in tenseless time. Perhaps, that is, it involves earlier and later, yet none of it "passes away" or is "yet to come," as tensed theories say that phases of time do.'[85] On this interpretation, all moments of the life of an eternal being are lived at once but still are ordered as earlier and later. Leftow calls this 'Quasi-Temporal Eternity,' or QTE. He says, 'The life of a being with QTE is an extension in which positions are ordered as earlier and later. Yet none of it "passes away" or is "yet to come," as we think happens with temporal

---

[82] Stump and Kretzmann, 'Eternity,' p. 434.
[83] Ibid., p. 445.
[84] This objection is similar to one raised by Fitzgerald in his response to Stump and Kretzmann. Paul Fitzgerald, 'Stump and Kretzmann on Time and Eternity,' *Journal of Philosophy* 82 (1985), p. 264.
[85] Leftow, *T&E*, p. 120.

lives. This likens extension in QTE to an extension in tenseless time.'[86] Clearly this view avoids the problem with the individuation of temporal events and the divine knowledge of temporal order which plague Stump and Kretzmann, but leads Leftow into other difficulties, such as a tacit commitment to static time. Leftow does not seem to realize this. In comparing QTE to tenseless time, he rejects the charge that tenseless theories of time 'spatialize' time:

> The tenseless theorist can, for instance, note that space and tenseless time have different relations to spatiotemporal particulars. The same concrete particular thing can be wholly present at many points along the extensive time dimension of tenseless time, as I am wholly present at 1 P.M. and at 1:01 P.M. Concrete particular things cannot be wholly present at any point along a spatial extension . . . . That tenseless time does and space does not permit concrete particulars full location at a point is a large difference between the two. Again, I can be wholly present at many points in tenseless time at the same place, but I cannot be wholly present at many points at the same time.[87]

It has been argued that tenseless theories of time are committed to perduring, four-dimensional objects, just as dynamic theories of time are committed to enduring, three-dimensional objects.[88] If that argument is sound, then Leftow's 'non-spatialized' explanation of QTE fails, for perduring objects do not wholly exist at a point in tenseless time. Leftow would doubtless reject attributing temporal parts to a being with QTE, but it seems that, on his construal of QTE, such an inference is forced upon him. Boethius is committed to divine simplicity just as firmly as Augustine (or Leftow himself), and he too would reject the conclusion that God has temporal parts.

A little later Leftow relies on his 'non-spatialized' concept of tenseless time when he contrasts the way in which QTE-beings and temporal beings occupy their durations. He recalls Boethius' claims that 'whatever lives in time lives in the present, proceeding from past to future, and nothing is so constituted in time that it can embrace the whole span of its life at once,' and that an eternal being would 'comprehend and include the whole of infinite life all at once.' So, for a temporal being, its different temporal stages are not simultaneous while, for a QTE-being, all the temporal stages of its life are simultaneous:

> Boethius' contrasting point about QTE is that despite the fact that eternal point e+1 is later than eternal point e, since an eternal being exists '*tota simul*,' what

---

[86] Ibid.
[87] Ibid.
[88] Trenton Merricks, 'On the Incompatibility of Enduring and Perduring Entities,' *Mind* 104 (1995), pp. 523-31; Peter van Inwagen, 'Four-Dimensional Objects,' *Noûs* 24 (1990), pp. 245-55; William Lane Craig, 'God and Real Time,' *Religious Studies* 26 (1990), pp. 335-38.

occupies e does so simultaneously (in some non-temporal sense of simultaneity) with occupying e+1.[89]

What can be meant by a 'non-temporal sense of simultaneity'? Initially, it seems that simultaneity has nothing but a temporal sense. Leftow is uncharacteristically vague here. We might think of physical copresence as a spatial analog and say that distinct points on a continuous line were 'non-temporally simultaneous.' But Leftow does not want us to 'spatialize' time. I must confess an inability to grasp the concept of non-temporal simultaneity if spatial construals or analogs are ruled out. How can points e and e+1 be distinct points if they are not temporally distinct, nor distinct in a static four-dimensional space-time? I conclude that QTE falters as an explanation of Boethius' concept of eternity.[90]

If the two best proposals for explaining Boethius' concept of eternity as duration—Stump/Kretzmann's duration without succession and Leftow's QTE—both have inconsistencies, then perhaps the concept of timeless duration itself is incoherent, as a number of authors have claimed.[91] But perhaps these modern interpretations of Boethius are attempting to remove from his Neoplatonic metaphysics a concept that cannot stand up outside it.

Neoplatonic Influences on Boethius

What was said above about Neoplatonic influences on Augustine applies equally to Boethius. In other words, Boethius shared a metaphysical commitment to God's simplicity, strong immutability, and timelessness. I wish now to point to certain Neoplatonic ideas, which, I believe, show that Boethius did not, in fact, conceive of eternity as timeless duration but rather as atemporality.

The concept of eternity that attracted the attention of later thinkers of all schools comes from Parmenides. In a crucial sentence Parmenides says, 'Nor was it ever, nor will it be, since it now is, all together, one, continuous.'[92] Sorabji notes that at least eight different interpretations of these lines have been offered;[93] what is important here, however, is how subsequent philosophers in the Platonic tradition construed Parmenides.

---

[89] Leftow, *T&E*, p. 121.

[90] Leftow advances additional arguments intended to motivate a concept of eternal atemporal duration, but since they all rest on the coherence of QTE, if it falters so do they (*T&E*, pp. 125-46). It should be noted that Leftow does not himself endorse QTE; he merely claims that it is a coherent reading of Boethius. He himself opts for an 'Anselmian' understanding of eternity.

[91] Among those who take Boethius' definition as inconsistent or incoherent are William Kneale, 'Time and Eternity in Theology,' *Proceedings of the Aristotelian Society* 61 (1960-61), p. 99; Richard Swinburne, *The Coherence of Theism*, rev. ed. (Oxford: Clarendon Press, 1993), pp. 218-21; Sorabji, *TC&C*, pp. 108-13.

[92] Parmenides, *The Way of Truth*, 8.1.5-6, tr. by Sorabji, *TC&C*, p. 99.

[93] Sorabji, *TC&C*, pp. 99-108. In what follows I am indebted to Sorabji's treatment in chapters 2, 3, 5 and 8.

Plato, in the *Parmenides* (141A-D), represents Parmenides as saying that the One is not in time since whatever is in time is becoming older, and then its former self would be younger than it, so it could not be One. As discussed briefly above, Plato applied this thinking to the Forms, conceiving of them as timeless. In the *Timaeus*, Plato claims that the Forms cannot grow older. In *Timaeus* 37C6-38C3 he suggests that the Forms are eternal in the sense of timeless; hence 'was' and 'shall be' are not applicable to them. And in a *locus classicus* he says this about the Forms:

> What is it that always is, and has no coming into being, and what is it that always is coming into being, and never is? The one is to be grasped by the mind with reason, and is always in the same state. The other is opined by opinion combined with irrational sense perception, and keeps coming into being and going out of existence, but never has real being. Again, everything which comes into being necessarily does so through some cause, for nothing can have its genesis without a cause. Now a creator necessarily makes everything good, whenever he produces the form and the features of his work by looking towards what is always in the same state and using something of that kind as a model.[94]

Sorabji argues that Plato's use of 'always' was ambiguous between temporal everlasting duration and timelessness, but in this text the timeless sense can be seen.[95]

Plotinus (205-270) offered a resolution to the ambiguity in Plato by stipulating that, when speaking of the archetypal ideas, 'always' has a non-temporal sense, denoting true being as opposed to that which comes to be.[96] Plotinus developed an ontology in which the soul is midway between the intelligible world of *Noûs* and the phenomenal world (although reality is actually a continuum between particulars and the One). The transcendent phase of the soul is timeless, while its immanent phase shares time with the body.[97]

Iamblichus (d. *ca.* 325) differentiated between static and flowing time, although Sorabji argues that the distinction drawn by Iamblichus does not correspond readily to the modern one between the A-series and the B-series. Following Plato's distinctions between largeness itself, the largeness in Simmias, and Simmias who is large,[98] Iamblichus distinguished an unparticipated universal, the participated universal that it generates, and the particular that participates in the participated (generated) universal. So also for time: there is generative time, generated time, and the events and particulars that participate in generated time. Generated time is 'lower' time; it flows; it resides in its participants. The

---

[94] Plato, *Timaeus*, 27D6-28B1.
[95] Sorabji, *TC&C*, pp. 110-12.
[96] Ibid., p. 112.
[97] Paul C. Plass, 'Timeless Time in Neoplatonism,' *The Modern Schoolman* 55 (1977), pp. 1-3.
[98] Plato, *Phaedo*, in *Readings in Ancient Greek Philosophy: From Thales to Aristotle*, ed. S. Marc Cohen, Patricia Curd, and C. D. C. Reeve, tr. G. M. A. Grube (Indianapolis, IN: Hackett, 1995), 102A-D.

'superior' time is static, always the same in form.⁹⁹ Thus Iamblichus introduced into Neoplatonism the distinction between static and flowing time, but the static time is the time in which the Forms reside, and, rather than being conceived as atemporal duration, should rather be conceived as timelessness.

Proclus (410?-485) elaborated on Plotinus' and Iamblichus' treatment of the threefold distinction between timelessness/time/things-in-time. Further, he distinguished succession from time itself, which gives structure to change without changing itself. Thus he introduced a transcendent time, distinct from the flow of time that is measured by change:

> Proclus' reduction of time to an eventless flow sounds in some respects like Newton's conception of absolute time: 'absolute, true and mathematical time of itself and from its own nature flows equally without relation to anything external .... Relative, apparent and common time is some sensible and external measure of duration by means of motion.' Newtonian time is entirely without content, while Proclus' transcendent time contains the patterns or causes of temporal events, but it does so in such a way that they constitute a completed whole.¹⁰⁰

Again, the important point to note is that static or transcendent time is the realm of the Forms—the patterns of temporal events and particulars.

Finally, Ammonius (435-517) supported the timeless, rather than the durational, view. Ammonius was an older contemporary of Boethius, although whether Boethius in Italy knew of Ammonius in Alexandria, or whether they both were influenced by Proclus (who was Ammonius' teacher) is far from clear. In a commentary on Aristotle's *De interpretatione*, Ammonius notes the sense which must be given to 'is':

> For we shall not allow anyone to say that the knowledge of the gods runs along with the flow of things, nor that anything with them is past or future. Nor shall we allow that 'was' and 'will be' are used among them, words which, as we have heard from Plato's *Timaeus*, signify some change. Only 'is' is used, and that not the 'is' which is counted in with 'was' and 'will be' and contrasted with them, but the 'is' which is conceived before the level of temporal images, and which signifies the gods' undeviating unchangeability. This is the 'is' which the great Parmenides also says belongs to all that is thinkable, when he says 'for it was not, nor will it be. All together, it *is* alone.'¹⁰¹

With Ammonius' reference to Parmenides we have come full circle in this section. What I believe this survey shows is that the Neoplatonic tradition, in which both Augustine and Boethius stand, did not develop a concept of atemporal duration but rather located the eternal forms or archetypal ideas in a realm of timelessness which transcends the realm of changing particulars and the moving now.

Sorabji summarizes in this way:

---

[99] Sorabji, *TC&C*, pp. 33-45.
[100] Plass, 'Timeless Time,' p. 9.
[101] Ammonius, *in Int.* 136.17-25, tr. by Sorabji, *TC&C*, p. 102.

> Parmenides was **groping** for the concept of timelessness. Plato clouded the issue with his talk of 'always'. Plotinus restored the idea of timelessness, thereby influencing the Neoplatonists and some prominent Christians. Philoponus dissented, but Boethius paid no attention, and transmitted the traditional concept to the Latin middle ages. In the sub-plot, Plotinus distinguished non-temporal senses of the words involved, but some of his successors preferred to use two distinct words for temporal and non-temporal eternity.[102]

Craig agrees that Sorabji is correct in this assessment,[103] and I think it unlikely that Boethius represents as radical a departure from the philosophical tradition he knew as Stump and Kretzmann assert, or as would be necessary if Leftow's QTE were the correct interpretation.

The point I am making is this: the novel concept of atemporal duration should probably not be attributed to Boethius. He, like Augustine before him and (as we shall see) Anselm and Aquinas after him, applied the concept of timelessness to God's eternity—a move that is quite natural given the Christian acceptance of Neoplatonism in the early centuries of the Christian era. It should be said, however, that in applying the idea of God's timelessness to the problem of foreknowledge and freedom in *The Consolation of Philosophy*, V, Boethius does make a significant advance on previous attempts to resolve the fatalist dilemma.

The Nature of Time

If Boethius did not introduce a new concept of eternity with his famous definition, he does begin to move away from Augustine's psychological/dynamic view of time in certain passages.

First, it is clear that Boethius regarded (at least some) future actions or events as contingent, and thus propositions about them as lacking truth value. In posing the dilemma of divine foreknowledge and human freedom, he assumes the contingency of the future:

> If this is so, how does God foreknow future possibilities whose existence is uncertain? If He thinks that things will inevitably happen which possibly will not happen, He is deceived. But it is wrong to say that, or even to think it. And if he merely knows that they may or may not happen, that is, if He knows only their contingent possibilities, what is such knowledge worth, since it does not know with certainty?[104]

And in his commentary on Aristotle's *De interpretatione*, Boethius writes that:

---

[102] Ibid., pp. 120-21 (emphasis in original). Sorabji does acknowledge that the picture is not entirely as neat as this summary would indicate; for example, he ignores the Gnostics with their quite different conception of eternity, but justifies doing so because they represent a quite different tradition from the mainstream of Neoplatonist-early Christian-medieval theology tradition he is considering.

[103] Craig, *PDF*, pp. 92-6.

[104] Boethius, *Consolation*, V, prose 3.

> ... the syllogism is of this mode: if every affirmation is definitely true or false, negation will come out in the same way, so that everything happens by an inevitable reason or necessity; and if this is so, free choice perishes. But this is impossible; therefore it is not true that every affirmation or negation is definitely true or false.[105]

Here Boethius is following Ammonius and the Peripatetics in a traditional Aristotelian logic in which truth-value gaps exist for future contingent propositions. Since he accepts a correspondence theory of truth, he seems to believe that future states of affairs do not exist.[106]

Throughout this discussion Boethius has assumed a dynamic theory of time that is clearly more realist than Augustine's. But of course Boethius goes on to develop a theory of God's eternity in which all events—past and therefore accidentally necessary, or future and contingent—are equally present to God. However, God's knowledge of future contingents does not confer on them any kind of necessity:

> He knows, as if present, which events are occurring contingently and which necessarily. His knowledge imposes no absolute necessity on the things He knows, but only a conditional necessity: *if* He knows them, then they must exist—but there is no necessity that He know them. Therefore, events which for us lie in the future are known by God as present and as occurring contingently, in sofar [*sic*] as they are the product of our free decisions.[107]

With this conclusion, Boethius may or may not have solved the dilemma of foreknowledge and freedom; that is not my concern. But in reaching the conclusion, it seems to me that Boethius departs from his earlier dynamic view of time. Perhaps he senses that his theory of time cannot sustain his conclusion. At any rate, he begins using concepts that have the consequence of moving from a dynamic to a static theory of time. In doing so, Boethius introduces two metaphors that will prove crucial in understanding Aquinas' theory of time and eternity:

> Consider the example of a number of spheres in orbit around the same central point: the innermost moves towards the simplicity of the center. ... Therefore, the changing course of Fate is to the simple stability of Providence as ... that which is generated is to that which *is*, as time is to eternity, as a circle is to its center.[108]

---

[105] Boethius, *In librum Aristotles de interpretatione*, unpublished tr. by Marilyn McCord Adams, cited in Craig, *PDF*, p. 85.
[106] Craig, *PDF*, pp. 85-7.
[107] Ibid., p. 98.
[108] Boethius, Consolation, IV, prose 6. The last phrases of this passage are '*ad quod est id quod gignitur, ad aeternitatem tempus, ad punctum medium circulus, ita est fati series mobilis ad providentiae stabilem simplicitatem.*' V. E. Watts translates, 'like that between that which is coming into being and that which is, between time and eternity, or between the moving circle and the still point in the middle.' Cited by Leftow, *T&E*, p. 135.

Here Boethius compares two triples of concepts: (i) that which is generated/time/circle, and (ii) that which is/eternity/the center of a circle. What this comparison points to is that true being, the temporal mode of which is eternity, is like the point at a circle's center, while generated being, the temporal mode of which is time, is like the circumference. It is difficult to read any concept of duration into this comparison, but a timeless conception of eternity does fit. And it seems to be in just such a way that Aquinas will use the image.[109]

A more significant feature of the comparison, however, is the move towards static time adumbrated in comparing time to the circumference of a circle. This would seem to indicate that all of time is 'equally present' to the center point (God's eternity). Or, since all points on the circumference of a circle coexist with its center, all temporal events must really be simultaneously present to God.

The second example, in which we hear an echo of Augustine, is this:

> Since God lives in the eternal present, His knowledge transcends all movement of time and abides in the simplicity of its immediate present. It encompasses the infinite sweep of past and future, and regards all things in its simple comprehension as if they were now taking place. Thus, if you will think about the foreknowledge by which God distinguishes all things, you will rightly consider it to be not a foreknowledge of future events, but knowledge of a never changing present. For this reason, divine knowledge is called providence, rather than prevision, because it resides above all inferior things and looks out on all things from their summit.[110]

Zagzebski translates the concluding phrases, 'for it [divine knowledge] is far removed from matters below and looks forth at all things as though from a lofty peak above them.'[111] The metaphor of God seeing all things from a summit is developed by Aquinas, as we shall see. But here again are indications of a static view of time. God's knowledge, like God himself, is timeless, yet it encompasses the infinite sweep of past and future. How could this be unless the past and the future existed in some way to be so encompassed? The summit metaphor thus is directly parallel to the center/circumference metaphor. The 'matters below' comprise past and future events, all of which must be simultaneously present to the view from the lofty peak above.

To be fair, it would be too much to say that Boethius introduces a static theory of time, for he does not actually say that past and future are temporally present to God. In speaking of God's knowledge he says that past and future are 'before the mind of God.' Yet he uses the same visual model of cognition as

---

[109] The analogy of the circle was used frequently by Plotinus: *Enneads* I, 7, 1; V, 1, 12; VI, 5, 4-5, 11; VI, 8, 18; VI, 9, 8. See Katherin A. Rogers, 'Eternity Has No Duraration,' *Religious Studies* 30 (1994), pp. 5f.

[110] Boethius, *Consolation*, V, prose 6.

[111] Linda Trinkaus Zagzebski, *The Dilemma of Freedom and Foreknowledge* (New York: Oxford University Press, 1991), p. 38.

Augustine, saying, for example, in the paragraph immediately following the one just quoted:

> 'Why then do you imagine that things are necessary which are illuminated by this divine light, since even men do not impose necessity on the things they see? Does your vision impose any necessity upon the things which you see present before you?' 'Not at all,' I answered.
>
> 'Then,' Philosophy went on, 'if we may aptly compare God's present vision with man's, He sees all things in his eternal present as you see some things in your temporal present. Therefore, this divine foreknowledge does not change the nature and properties of things; it simply sees things present before it as they will later turn out to be in what we regard as the future.'[112]

Perhaps we could say, to borrow from what Sorabji said of Parmenides, that Boethius is groping towards a static theory of time.

Summary

Brian Leftow compares the history of philosophical theology to an hourglass, flourishing in the academies of late Antiquity, then dwindling and dribbling down to medieval times through a relatively few sources, only to be reborn from those sources and flourish anew. As regards the concepts of time and eternity, Augustine and Boethius were the neck of the hourglass.[113]

While some recent philosophers have found in Boethius a doctrine of eternity as atemporal duration, I believe I have shown not only that the explanations of atemporal duration which have so far been offered have serious if not fatal flaws, but also that the doctrine owes more to the Neoplatonic tradition in which Boethius stood than to viable metaphysical considerations. More likely, Boethius held to a timeless eternity little different from Augustine's concept. However, Boethius does not relate God's knowledge to the timeless Neoplatonic ideas (a move which allowed Augustine to retain dynamic time) but rather to the events in time themselves. As a consequence Boethius seems to exhibit a movement towards static time—a movement that will become pronounced in Aquinas.

St Anselm

Anselm of Canterbury (1033-1109) shows the marked influence of Augustine as well as familiarity with Boethius. Like them, Anselm emphasized the atemporality of God, but in introducing the idea of eternity as a dimension containing time he gives a different sense to the notion of atemporal duration than did Boethius, yet one that seems to move him closer to a static view of time. Indications are found in the *Monologium* and the *Proslogium*, as well as in his late work, the *Concordia*.

---

[112] Boethius, *Consolation* V, prose 6.
[113] Leftow, *T&E*, p. 112.

God's Simplicity

Anselm, like Boethius and Augustine before him, held that God is simple—a doctrine rooted in Neoplatonism, as discussed above. Nevertheless, Anselm's treatment of God's simplicity seems to go a step beyond Augustine and Boethius in terms of what it brings to the discussion of God's temporality.

St Anselm on God and Time

- God is simple, therefore immutable and timeless.
- Time is dynamic (in a Boethian sense).
- The future is present to eternity.
- Eternity is a successionless duration. But the single 'now' of eternity contains all temporal intervals.

In *Monologium* 16, Anselm presents this argument:

> It seems, then that by *participation* in this quality, that is, justice, the supremely good Substance is called just. But, if this is so, it is just through another, and not through itself. But this is contrary to the truth already established, that it is good, or great, or whatever it is at all, through itself and not through another. So, if it is not just, except through justness, and cannot be just, except through itself, what can be more clear than that this Nature is itself justness? . . . Hence, if it is inquired what the supreme Nature, which is in question, is in itself, what truer answer can be given, than *Justice*?
>
> . . . Moreover, what we see to have been proved in the case of justness, the intellect is compelled to acknowledge as true of all attributes which are similarly predicated of this supreme Nature. Whatever such attribute is predicated of it, then, it is shown, not of what character, or how great, but what it is.[114]

Generalized (as the last paragraph authorizes), Anselm's argument is as follows:

27. Whatever is F is F through participation in F-ness.
28. So God is F through participation in F-ness.
29. But God is F through himself.
30. Hence, God = F-ness.

In the same context Anselm says, of the supreme Nature (or God), 'when it is called just it is properly conceived of as being justness, but not as possessing justness.' That is, Anselm maintains that the identity statement 'God = justice' is more proper than the predication 'God is just.'

In endorsing the identity statement containing the abstract term as preferable, Anselm seems to lay the foundation for ascribing to God certain

---

[114] Anselm, *Monologium*, in *St Anselm: Basic Writings*, tr. S. W. Deane, 2nd ed. (La Salle, IL: Open Court, 1962).

features of universals. A universal can be wholly present at several spatial and temporal locations at once without being composite.[115]

In *Monologium* 17 Anselm denies that God is composite—another statement of simplicity. He has already asserted that God exists *a se*—that is, independently. From this it follows that God is not compounded of various goods:

> For, everything which is composite requires for its subsistence the things of which it is compounded, and, indeed, owes to them the fact of its existence, because, whatever it is, it is through these things; and they are not what they are through it, and therefore it is not at all supreme.

Having laid his foundation in terms of God's simplicity, Anselm proceeds in chapters 18-24 to consider the relation of God to time. He will return to the conclusions he has established from consideration of divine simplicity in order to resolve a paradox he develops.

Temporal Omnipresence and 'Omniabsence'

In *Monologium* 18-19, Anselm argues that God must be eternal—that is, without beginning or end. There are no surprises here. But he introduces a somewhat surprising line of argument in chapters 20 and 21, showing both that God exists at every time and at no time.

In support of the conclusion that God exists in every place and at every time, Anselm takes as his first premise the doctrine that God is causally sustaining creation at all times, and as his second premise the principle that a cause must exist when and where its effect exists. The first premise is part of classical theism, but the second may be open to objection. Anselm does not discuss action at a spatial or temporal distance; he takes it as given that a cause must be spatially and temporally contiguous with its effect. How likely is it that this premise is true?

Now, no matter what we make of possible objections to the second premise based on such things as field theory or Quantum Mechanics, Anselm has, I think, a plausible line of argument open to him. Let us assume that God acts causally at a given time and location, and the effects of God's action are at some later time and distant location. But by the first premise, God exists at every point in time and space between the cause and the effect, sustaining not only the points of space and time, but also the causal chain itself. And he is present at the terminus of the causal chain; he is present with the effect.[116] As Anselm says,

---

[115] Anselm assumes an ontology that is realist about universals. If nominalism or conceptualism is true, and if abstract terms refer instead to tropes or scattered particulars, his argument does not go through.

[116] Even if we agree that simultaneous causation is ruled out within the 3+1 dimensions of physical space-time, if God's existence is in a supertemporal dimension as Anselm suggests (see below), then what would appear as simultaneous causation within 3+1 dimensions might be possible.

'[God] must exist everywhere and always, that is, in every place and at every time.'[117] Thus, Anselm has established the conclusion of temporal omnipresence:

> TOP.   God is present at every time.

In chapter 21, Anselm argues for what seems the opposite conclusion. He begins by asserting a trilemma: necessarily, (i) God is present at every time, or (ii) at some time, or (iii) at no time. For the hypothesis of a *reductio*, Anselm accepts (i), which is TOP, the conclusion of his previous chapter. This forces a dilemma: either God is wholly present at every time, or only a part of him is present at every time. But the latter horn is rejected: God does not exist in part at one time, and in part at another. Being simple, God cannot have temporal parts. Consequently, Anselm considers the former horn of the dilemma: God must be wholly present at every time. But this, too, leads to unacceptable results. For God cannot be wholly present in different times all at once, nor can he be wholly present in different times successively. Thus TOP results in a contradiction, and must be rejected. But (ii) of the trilemma is unacceptable, given Anselm's understanding of God's causal necessity to the existence of creation. Thus, (iii) must be true, and Anselm has established the conclusion of temporal 'omniabsence.'

> TOA.   God is present at no time.

Anselm regards his arguments for TOP and TOA as valid, and begins *Monologium* 22 by asking, 'How, then, shall these propositions, that are so necessary according to our exposition, and so necessary according to our proof, be reconciled?' At this point he returns to his suggestion of chapter 17 that, in certain aspects, God is relevantly like universals.

> Perhaps the supreme Nature exists in place and time in some such way, that it is not prevented from so existing simultaneously, as a whole, in different places or times, that there are not more wholes than one; and that its age, which does not exist, except as true eternity, is not distributed among past, present, and future.[118]

The Dimension of Eternity

With this foundation, Anselm proceeds to develop his concept of eternity. He brings to the discussion great originality, which we would expect of the inventor of the Ontological Argument, suggesting the concept of eternity as a higher dimension transcending space and time. He grants that 'the law of space and time,' by which he means the principle that a single entity cannot be wholly present in different places or at different times simultaneously, applies to a certain class of entities. But he maintains that there are entities of a different class to which this law does not apply. Specifically, he maintains that:

---

[117] Anselm, *Monologium*, 20.
[118] Ibid., 22.

> . . . place is only predicable of objects whose magnitude place contains by including it, and includes by containing it; and that time is predicable only of objects whose duration time ends by measuring it, and measures by ending it. Hence, to any being, to whose spatial extent or duration no bound can be set, either by space or time, no place or time is properly attributed.[119]

The thought is this: an entity is located in a place only if its spatial extent is bounded by the spatial limits of that place. An entity can be in a place but can 'spill over' beyond the boundaries of that place, and thus not be located in (Anselm's phrase is 'contained by') that place. The Rocky Mountains are in Colorado, for example, but because the Rockies 'spill over' into other states, it cannot properly be said that the Rockies (as an entity) are located in (within the boundaries of) Colorado. Similarly, an entity is located in time only if its duration is bounded by the temporal boundaries of that time. If it 'spills over' into adjacent temporal regions, it cannot be said to be located at—contained by—that time. In both cases, the respective entity is in a place and at a time, but is not contained by the place or time. Anselm concludes:

> But in the case of the supreme Being, the first sense only is intended, namely, that it is present; not that it is also contained. If the usage of language permitted, it would, therefore, seem to be more fittingly said, that it exists *with* place or time, than that it exists *in* place or time. For the statement that a thing exists *in* another implies that it is *contained*, more than the statement that it exists *with* another.[120]

So Anselm now has in hand the solution to the paradox he developed. God 'exists in every place and time because [he] is absent from none; and [he] exists in none, because [he] has no place or time.'

> We have sufficient evidence, then, to dispel the contradiction that threatened us; as to how the highest Being of all exists, everywhere and always, and nowhere and never, that is, in every place and time, and in no place or time, according to the consistent truth of different senses of the terms employed.[121]

Anselm's concept seems to be that of God existing in a 'superdimension' that contains space and time but transcends them. This is born out in the *Proslogium*, chapter 21, where Anselm states explicitly, 'For, as an age of time contains all temporal things, so thy eternity contains even the ages of time themselves.' Just how the 'higher' dimension of eternity can contain space and time, without itself involving sequence, Anselm does not say.

But there is more to eternity. That God is without beginning or end was Anselm's starting point. But, based on his doctrine of divine simplicity, he concludes that eternity is not endless succession, since God cannot have temporal parts. Here he draws on Boethius' definition of eternity:

---

[119] Ibid., 22.
[120] Ibid.
[121] Ibid.

> Hence, if this Being is said to exist always; since, for it, it is the same to exist and to live, no better sense can be attached to this statement, than that it exists or lives eternally, that is, it possesses interminable life, as a perfect whole at once. For its eternity apparently is an interminable life, existing at once as a perfect whole.[122]

Anselm, at this point, has the following concepts of time and eternity. Time is dynamic, more in a Boethian than an Augustinian (psychological) sense. Eternity is a successionless duration. But the single 'now' of eternity contains all temporal intervals. Since God is eternal, and since by the doctrine of simplicity God = eternity, God himself contains all temporal intervals. In a way analogous to universals, God can then be 'with' every time and place, but is not located 'in' time or place.

In his late work *De concordia*, Anselm gives indications as to how time and eternity may be related in the sense he desires them to be:

> In eternity there is only a present, which is not a temporal present like ours, but an eternal present, in which all of time is contained. As the present time contains every place and the things which are in any place, so the eternal present contains at once the whole of time and whatever exists at any time. . . . Eternity has its own simultaneity, in which exist all things which exist at the same place or time and all things which are diverse in place or time.[123]

This passage suggests how time and eternity may be interpreted in terms of dimensionality. Objects that have discrete coordinates in n dimensions may all have the same coordinate in the n+1st dimension. For example, all the objects present on my desk have discrete spatial locations in three-dimensional space, but they all have the same coordinate in the fourth dimension when time is incorporated, yielding 3+1 dimensions. Similarly, Anselm thinks, anything in space and time may have discrete locations in 3+1 dimensions, but may all have the same coordinate in the dimension of eternity. In this way God can be present at all places and in all times—or with all places and times, to use Anselm's preferred locution—without being located in space and time.

Leftow notes another implication of the dimension model.[124] What has no extension in a lower dimension may turn out to be extended in a higher dimension. For example, a point in a two-dimensional plane may be the intersection of that plane with a line perpendicular to it in three dimensions. So while in two-dimensional space the point was unextended, in the higher dimensional world of three-dimensional space the point was in fact extended. Similarly, if eternity is a higher dimensional world containing our 3+1 space, the conclusion that God has no extension in time does not entail he has no extension (duration) in eternity.

---

[122] Ibid., 24.
[123] Tr. Leftow, *T&E*, p. 212.
[124] Ibid., p. 214.

Summary

While standing firmly in the tradition of Augustine and Boethius, Anselm suggests creative alternatives to their views on God and time. And, like them, Anselm proves to be a fruitful seedbed for contemporary reflection. Just as Stump and Kretzmann build their theory of God's relation to time on Boethius, so Leftow builds his on Anselm. I shall reserve comment on Leftow's theory for the next chapter. However, I must note here one salient respect in which I find Anselm's analysis flawed, and another in which I find promise.

Although Anselm speaks throughout as if time is dynamic, it is not at all clear that he can reconcile his dimensional concept of eternity with dynamic time. Like both Augustine and Boethius, Anselm believes both that God had knowledge of future events and that at least some future events were contingent. Unlike Boethius, Anselm does not apply his theory of time and eternity to the solution of the problem of freedom and foreknowledge. Nevertheless, there is evidence that Anselm believes that the dimension of eternity is 'present' to all future times as well as present and past. But this entails the existence of the future; even God cannot be present to a non-existing place or time! And if the future already exists, then time cannot be truly dynamic after all. Here, then, is a conceptual inconsistency in Anselm's account.

On the positive side, Anselm's theory of time and eternity, if stripped of the constraints imposed by the doctrine of divine simplicity, is very fruitful. I have already suggested that God's existence is temporal; Anselm provides the outlines of a theory in which God's temporality in some sense contains the physical time of the universe that we know. Details will be left for chapter 9.

St Thomas Aquinas

Thomas Aquinas (1225-1274), to a greater extent than any medieval philosopher since Augustine, labors over issues of God and time. Like those before him, Aquinas based much of his discussion on the simplicity of God, and, like Boethius, developed his views primarily in considering the dilemma of divine foreknowledge and human freedom. Based on constraints imposed by divine simplicity and immutability, Aquinas regarded God as atemporal. In elucidating his view of eternity Aquinas draws on both the point-like and the timeless duration models. Unlike Augustine, Boethius and Anselm, Aquinas did not accept that God could have foreknowledge of future contingents as future. As a consequence, he seems to have been led to a view of time as static. Aquinas' views are found in *Summa Theologiae* and *Summa Contra Gentiles*, as well as in other works.

> **St Thomas Aquinas on God and Time**
>
> - Time: Driven by the need to respond to fatalism, Aquinas' final model of time seems to be B-theoretic, although he talks throughout as if time is dynamic.
> - Divine simplicity entails divine timelessness.
> - Eternity: Aquinas had at least three different models of timelessness available to him: Augustine's timelessness *simpliciter*, Boethius' point-like model, and Anselm's timelessness as a supertemporal dimension. Aquinas makes use of them all in different contexts.

The Challenge of Fatalism

Unlike Augustine, who in *De libero arbitrio* does not seek a solution to the fatalist challenge in his theory of time, it is precisely there where Aquinas locates his solution.

The fatalist problem, simply stated, is this. If at time t_1 God knows that E will occur at t_2 (where $t_1<t_2$), then since God's knowledge is infallible, it supposedly follows that E necessarily will occur. The fatalist problem already had a long history of discussion on which Aquinas could draw, going back at least to Aristotle.[125] But he seems to have been unsatisfied with previous solutions to a particular version of the problem. Based upon the inalterability of the past, and so of the inalterability (i.e. necessity) of God's knowledge in the past of all that was yet to come, there would seem to be no respect in which any possible future event could be contingent. Therefore, Aquinas was forced to strike out in new directions to develop his own solution.

Aquinas asks how God can have knowledge of nonexistent things. He draws a distinction between the class of nonexistents that are no longer or are not yet, and those nonexistents that are not *simpliciter*. God's knowledge of these two classes is different:

> But not all non-beings have the same relation to His knowledge. For those things that are not, nor will be, nor ever were, are known by God as possible to His power. Hence, God does not know them as in some way existing in themselves, but as existing only in the divine power. These are said by some to be known by God according to a knowledge of simple understanding. The things that are present, past, or future to us God knows in His power, in their proper causes, and in themselves. The knowledge of such things is said to be a knowledge of vision.[126]

In the case of God's knowledge of future contingents, that knowledge would be the 'knowledge of vision.' Using the metaphor of sight, Aquinas recalls Aristotle's discussion of a man seeing Socrates sitting. Given that he sees Socrates sitting at

---

[125] See Craig, *PDF*, for a thorough treatment of the history of the problem.
[126] St Thomas Aquinas, *Summa Contra Gentiles*, tr. Anton C. Pegis (Notre Dame, IN: University of Notre Dame Press, 1975), I, 66,8, hereafter *SCG*.

this moment, Socrates' sitting is temporally necessary. But it remains a contingent fact that Socrates should sit at this moment. Once a contingent event has come about, it has become certain—the man seeing it cannot be mistaken (barring, of course, cognitive malfunction at that moment)—but that does not change the nature of the event from contingent to necessary. As Aquinas puts it,

> The contingent is opposed to the certitude of knowledge only so far as it is future, not so far as it is present. For when the contingent is future, it can not-be.... But in so far as the contingent is present, in that time it cannot not-be. It can not-be in the future, but this affects the contingent not so far as it is present but so far as it is future. Thus, nothing is lost to the certitude of sense when someone sees a man running, even though the judgment is contingent. All knowledge, therefore, that bears on something contingent as present can be certain.[127]

So if God's foreknowledge is of the character of 'knowledge of vision,' then that foreknowledge, though certain, does not make the foreknown event necessary, and the fatalist challenge is averted.

Aquinas thus must show how God has 'knowledge of vision' of future contingent events. Clearly, God cannot know a future contingent event simply by knowing the cause of the event, for that would mean both that the cause determined the event and it would no longer be contingent, and also that God's knowledge would not be by vision but by simple understanding. But how can God 'see' the future, including future contingents?

The answer lies in Aquinas' understanding of God's eternity and its relation to time. Here he recalls the illustration used by Boethius:

> We may see an example of sorts in the case of a circle. Let us consider a determined point on the circumference of a circle. Although it is indivisible, it does not co-exist simultaneously with any other point as to position, since it is the order of position that produces the continuity of the circumference. On the other hand, the center of the circle, which is no part of the circumference, is directly opposed to any given determinate point on the circumference. Hence, whatever is found in any part of time coexists with what is eternal as being present to it, although with respect to some other time it [the part of time] be past or future. Something can be present to what is eternal only by being present to the whole of it, since the eternal does not have the duration of succession. The divine intellect, therefore, sees in the whole of its eternity, as being present to it, whatever takes place in the whole course of time. And yet what takes place in a certain part of time was not always existent. It remains, therefore, that God has a knowledge of those things that according to the march of time do not yet exist.[128]

This example helps us understand how Aquinas believes all temporal events are present to God. It does not entail, as Kenny seems to think, that God sees all

---

[127] Ibid., I, 67, 2.
[128] Ibid., I, 67, 7.

events as simultaneous with each other.[129] It is God's seeings of the events, not the events themselves, which are simultaneous. Aquinas explicitly declares that a point on the circumference of a circle stands in an ordered relation to other points so as to produce the continuous circumference. Similarly, each point in time stands in a relation of earlier or later to other points of time, so as to produce the continuous temporal series. Those ordered relations would be part of what God would see as he sees each event simultaneously.

In several places Aquinas uses another of Boethius' examples—that of the view from a lofty peak. Looking down on a line of travelers, a person on the peak sees all of them in a single moment, while a person along side the road would see each traveler pass successively. 'God is wholly outside the order of time, stationed as it were at the summit of eternity, which is wholly simultaneous, and to Him the whole course of time is subjected in one simple intuition.'[130] (Here again we can see the consequences of the doctrine of divine simplicity, according to which, since God = God's knowledge, any divisions in God's knowledge are ruled out.)
In another text the import is even clearer:

> Were someone to see many travelers along a road successively, over a certain period of time, in each part of that time he would see some passers by as present, so that over the whole time of his vision he would see every traveler as present. He would not see all as present at once because the time of his seeing is not all-at-once. If his seeing were able to exist all at once, he would see at once all as present, although they do not all pass by as present at once. Whence because the vision of God's knowledge is measured by eternity, which is all at once and yet includes all of time. . . God sees what happens in time not as future but as present.[131]

Strictly speaking, and as Augustine claimed, God does not foreknow anything, if foreknowledge means to know an event before it occurs. For God see simultaneously all events in the temporal series—past, present and future. To put it a bit more technically, God timelessly knows all events, and he timelessly knows which propositions are true at any point in the temporal series. But this knowledge is like our knowledge of the present in that it does not confer any necessity upon the events or propositions known. So the fatalist challenge seems to be defeated.

But more must be said about the nature of eternity and of time.

Eternity

Not surprisingly, Aquinas, like those before him, insists that God, being simple and immutable, must be timeless. He had at least three different models of

---

[129] Anthony Kenny, 'Divine Foreknowledge and Human Freedom,' in *Aquinas: A Collection of Critical Essays*, ed. Anthony Kenny (Garden City, NY: Anchor Books, 1969), p. 264.
[130] Aquinas, *In Perihermeneias* 14, 20, tr. Craig, *PDF* p. 107. Thomas uses the same image in *Summa Theologica* Ia, 14, 13 ad 3; and in *De Veritate* II, 12, and *Compendium Theologiae* 133.
[131] Aquinas, *De Veritate* 2, 12, tr. Leftow, *T&E*, pp. 181f.

timelessness with which to work: Augustine's timelessness *simpliciter*, Boethius' point-like model, and Anselm's timelessness as a supertemporal dimension. Apparently Aquinas felt that none of the models available to him was adequate, for he makes use of them all in different contexts.

Aquinas begins with the doctrine of divine simplicity and argues in an Augustinian fashion that simplicity entails immutability, and then that immutability entails eternity. But he adds, necessarily, only God is simple, and consequently, necessarily, only God is eternal.[132] Perhaps this is the thinking behind the comparison of eternity to a point. Thus he says,

> Moreover, God's understanding has no succession, as neither does His being. He is therefore an ever-abiding simultaneous whole—which belongs to the nature of eternity. On the other hand, the duration of time is stretched out through the succession of before and after. Hence, the proportion of eternity to the total duration of time is as the proportion of the indivisible to something continuous; not, indeed, of that indivisible that is the terminus of a continuum, which is not present to every part of a continuum (the instant of time bears a likeness to such an indivisible), but of that indivisible which is outside a continuum and which nevertheless co-exists with any given point in the continuum.[133]

On this model, eternity is unextended. But there are other indications that Aquinas, like Anselm, understood eternity as some sort of timeless duration. This is because God is a living being, and 'that which is truly eternal not only exists but lives. Now living includes in a way activity. . . and flow of duration is more apparent in activity than in existence.'[134] That God is alive, then, entails that eternity be some sort of duration, albeit timeless duration.

Aquinas says more than Boethius, however, in allowing that we may affirm different propositions of God at different times, not because God changed in himself through time, but because we (or other temporal things) did. 'Thus there is nothing to prevent these names that import relation to the creature from being applied to God temporally, not by reason of any change in Him, but by reason of the change of the creature.'[135] A parallel would be our proclamation that the sun is coming out, when there is no change in the sun itself, but rather in our relation to the sun when a cloud has blown away. (In this way Aquinas makes room for our normal ways of speaking about God—and indeed ways of speaking about God found in Scripture—without concluding that such locutions are completely equivocal.) As Leftow says, 'Aquinas allowed a simple, timeless being to have an extrinsic history [but not an intrinsic history] and so to be extrinsically temporal.'[136] That is, the life God lives is extended in duration, but points in that

---

[132] St Thomas Aquinas, *Summa Theologica*, tr. the Fathers of the English Dominican Province (New York: Benzinger Brothers, 1948), Ia, 9, 1; 10, 2, hereafter *ST*.
[133] Aquinas, *SCG*, I, 66, 7.
[134] Aquinas, *ST*, Ia, 10, 1 *ad* 2, tr. Leftow, *T&E*, p. 150.
[135] Aquinas, *ST*, Ia, 13, 7.
[136] Leftow, *T&E*, p. 157.

life cannot be ordered by the earlier/later relations. All are simultaneous, or as Boethius had said, *tota simul*.

We may well wonder whether such duration really can be called a life, and whether it is really coherent to speak of eternity as both point-like and extended. My purpose here is not to answer these questions, but only to describe Aquinas' theory of eternity.

The Nature of Time

Aquinas frequently speaks of time's passage in terms that imply a dynamic theory of time, just as Augustine, Boethius and Anselm did.[137] But at the same time he often uses language that implies a static theory of time in which the future has the same ontological status as the present. For example, he discusses the being which future contingent singulars possess:

> The contingent is in its cause in such a way that it can both not-be and be from it; but the necessary can only be from its cause. But according to the way both of them are in themselves, they do not differ as to being, upon which the true is founded. . . . Now, the divine intellect from all eternity knows things not only according to the being that they have in their causes, but also according to the being that they have in themselves. Therefore, nothing prevents the divine intellect from having an eternal and infallible knowledge of contingents.
>
> Again, when it is said that God *knows* or *knew this future thing*, . . . this is not future with reference to the divine knowledge, which, abiding in the moment of eternity, is related to all things as present to them. . . . we cannot say that this is known by God as non-existent, so as to leave room for the question of whether it can not-be; rather, it will be said to be known by God in such a way that it is seen by Him already in its own existence.[138]

The contingent and the necessary do not differ in their being, says Aquinas. So in seeing a future contingent, God sees it already in its own existence. Now we might interpret this to mean that future contingents exist in their forms or essences in the mind of God, but Aquinas denies this:

> God knows all contingent events not only as they are in their causes but also as each of them is in actual existence in itself. . . . Hence, all that takes place in time is eternally present to God, not merely, as some hold, in the sense that he has the intelligible natures of things present in himself, but because he eternally surveys all things as they are in their presence to himself.[139]

Aquinas clearly believes that God knows future contingents as actual existents and not merely as essences or intelligible natures, as Augustine apparently held. But if future contingents actually exist for God, externally to himself, then they must

---

[137] 'As the apprehension of time is caused in us by the fact that we apprehend the flow of the *now*': Aquinas, *ST*, Ia, 10, 2, to cite just one example of many.

[138] Aquinas, *SCG*, I, 67, 3 and 9.

[139] Aquinas, *ST*, Ia, 14, 13.

exist on an ontological par with the present. That is, the entire temporal series must exist as a whole.

That this is Aquinas' meaning is undeniable in this passage from the *Compendium theologiae*:

> Even before they [future contingents] come into being, He sees them as they actually exist, and not merely as they will be in the future. . . . Contingent things, regarded as virtually present in their causes with a claim to future existence, are not sufficiently determinate as to admit of certain knowledge about them, but, regarded as actually possessing existence, they are determinate, and hence certain knowledge is possible. . . . For His eternity is in present contact with the whole course of time, and even passes beyond time. We may fancy that God knows the flight of time in His eternity, in the way that a person standing on top of a watchtower embraces in a single glance a whole caravan of passing travelers.[140]

Summary

Aquinas used the thinking of the great philosopher-theologians before him, and built on their views. But the most difficult aspect of his theory is the inference that, probably unconsciously, Aquinas held to a static theory of time, more strongly even than his predecessors. Storrs McCall concludes, 'We cannot accept Aquinas's view of the universe *sub specie aeternitatis* without implying that time as human beings perceive it is unreal.'[141] And Craig arrives at the same evaluation, concluding:

> . . .what Aquinas' doctrine of God's eternity and knowledge of future contingents was seen to imply seems to be positively affirmed by Aquinas, namely, that the past, present, and future are all ontologically on a par with each other. Accordingly, Thomas held to a B-theory of time. Nevertheless, I find it inconceivable that he consciously adhered to such a theory of time. For him becoming was not mind-dependent, but real. . . . Aquinas seemed to hold both to a dynamical view of time and to the actual existence of all temporal things for God in eternity. Despite this, however, I must admit that I can only make sense of Aquinas' position on God's foreknowledge and future contingents by interpreting him as proponent of the B-theory of time.[142]

Conclusion

Clearly, the consensus that emerged during the Middle Ages was that eternity, the temporal mode of God's being, was timeless. Further, aside from Augustine's psychological theory, there was a strong tendency to affirm the intuitive idea that the passage of time is real. But I also noted that the attempt to make the two

---

[140] Tr. William Lane Craig, in 'Was Thomas Aquinas a B-Theorist of Time?' *New Scholasticism* 59 (1985), p. 482.
[141] Storrs McCall, *A Model of the Universe*, p. 46.
[142] Craig, 'Was Thomas Aquinas a B-Theorist of Time?' p. 483.

concepts cohere led more and more clearly to a static, B-theoretic concept of time. As I shall argue below, a timeless view of God's existence entails the B-theory of time, so such a move was perhaps forced upon these thinkers.

Even a brief survey such as this one shows clearly the influence of Neoplatonic metaphysics in the work of Augustine, Boethius, Anselm and Aquinas. Most notably in the concept of divine simplicity, immutability and eternity, the Neoplatonist metaphysics also affected the medieval philosophers' views of the nature of time. In the next chapter I shall examine contemporary theories of God's timelessness that seek to avoid the difficulties associated with Neoplatonism. But, as we have reasons to prefer a dynamic theory of time (Chapter 2), if the contemporary theories are all shown to entail the B-theory, then we will have reasons to reject the medieval consensus of an atemporal understanding of God's being.

Chapter 6

Atemporality: Contemporary Statements

The traditional view of God as atemporal, which emerged from the Middle Ages, holds considerable attraction for contemporary philosophers as well. The attraction stems from two sources. The first is respect for the philosophical theological tradition, including commitments to strong immutability and divine simplicity. The second is commitment to a static theory of time.

In this chapter I shall consider contemporary proponents of the atemporal view, including Eleonore Stump and Norman Kretzmann, Brian Leftow, and Paul Helm. The arguments these philosophers offer in favor of atemporality include both theological and philosophical strands. I shall not attempt a complete representation of their work, but will summarize and critique what I take to be the crux of their arguments.

First, though, we need to set the stage. A ubiquitous complaint against the atemporal view is that it cannot make sense of God's causal activity in the actual temporal order, if time is dynamic. While not unknown in the Middle Ages (St Thomas, for example, attempts to deal with the problem), the issue has been felt more keenly in contemporary treatments. The problem may be framed in two ways. First, since causes precede their effects in time, if God brings about any effects in time, he must somehow be temporally present at the time of the cause. But how can something that is atemporal, and thus isolated from the A-series of events that constitute time, have any interaction with that series without itself becoming 'infected' with temporality? Moreover, it is not at all clear how God can, in one timeless act, bring about effects that are spread out in time (since on a dynamic theory of time, the future does not exist). A second way to frame the problem is to ask what atemporal causation means. Our grasp of the concept of causation may not be complete, but it surely seems to involve temporal notions (even if simultaneous causation is possible).

The first two atemporalist arguments we will consider attempt to give an account of what it is for an atemporal being to be 'present' to temporal entities, for, if successful, the objections just mentioned will dissolve. The third atemporalist cheerfully accepts a B-theoretic account of time, so the problems do not arise for him: presumably, an atemporal God could act timelessly on the timelessly existing four-dimensional space-time manifold.

Eternal-Temporal Simultaneity: Eleonore Stump and Norman Kretzmann

Co-authors of numerous articles, Eleonore Stump and Norman Kretzmann combine contemporary analytical philosophy with exegetical skill in ancient and medieval material. In a well-known and widely influential article, they offer an ingenious construal of divine timelessness.[1] In the previous chapter I was critical of one aspect of their theory—namely, the conclusion that eternity was undifferentiated, containing no distinct points and no ordered sequence of events. I argued that such a conception of eternity would entail that two qualitatively identical events could not be numerically distinct, and that on such a view even God could not know what came before what. Be that as it may, I want here to turn attention to another concept introduced by Stump and Kretzmann—the notion of ET-simultaneity.

> **ET-Simultaneity (Stump and Kretzmann)**
>
> ET-simultaneity defines a simultaneity relation between eternal and temporal entities so that an eternal entity can be 'present' (causally, relationally) to a temporal entity. It marks a notable contribution to the contemporary discussion of God's temporal mode of being.

ET-Simultaneity

Stump and Kretzmann recognize that if God's temporal mode of being is timeless, then some relation must be defined by which God can somehow be present to the 'now' of temporal creatures. 'What we want now is a species of simultaneity—call it ET-simultaneity (for eternal-temporal simultaneity)—that can obtain between what is eternal and what is temporal.'[2] ET-simultaneity proves difficult to define, however, for the relata—eternity and time—are distinct modes of existence and irreducible, either to each other or to a third mode of existence common to both time and eternity.

To get past this difficulty, Stump and Kretzmann consider the relativity of simultaneity under the Special Theory of Relativity (STR). A relativized definition of simultaneity must specify a given observer and that observer's reference frame. Analogously, a definition of ET-simultaneity must specify two observers and two reference frames. Accordingly, Stump and Kretzmann characterize ET-simultaneity in this way:

> Let 'x' and 'y' range over entities and events. Then:
> (ET) For every x and for every y, x and y are ET-simultaneous iff
> (i) either x is eternal and y is temporal, or vice versa; and

---

[1] Eleonore Stump and Norman Kretzmann, 'Eternity,' *Journal of Philosophy* 78 (1981), pp. 429-58.
[2] Ibid., p. 436.

(ii) for some observer, A, in the unique eternal reference frame, x and y are both present—i.e., either x is eternally present and y is observed as temporally present, or vice versa; and
(iii) for some observer, B, in one of the infinitely many temporal reference frames, x and y are both present—i.e., either x is observed as eternally present and y is temporally present, or vice versa.³

Several features of the definition are worth noting. If x and y are ET-simultaneous, they cannot be temporally simultaneous, since at least one of the relata must be eternal. Further, the relation of ET-simultaneity is symmetric, but non-reflexive and non-transitive. No event can be ET-simultaneous with itself; and the fact that x is ET-simultaneous with y and y is ET-simultaneous with z does not entail that x is ET-simultaneous with z.

Stump and Kretzmann believe that their definition immediately deflects at least one criticism of divine atemporality posed by Anthony Kenny:

> ... on St. Thomas' view, my typing of this paper is simultaneous with the whole of eternity. Again, on this view, the great fire of Rome is simultaneous with the whole of eternity. Therefore, while I type these very words, Nero fiddles heartlessly on.⁴

In response, Stump and Kretzmann note that the notion of ET-simultaneity means that the following three propositions form a consistent set (at least at the time of their writing):

(A) Richard Nixon is alive in the temporal present.
(B) Richard Nixon is alive in the eternal present.
(C) Richard Nixon is dead in the eternal present.

This is not to be understood to mean that Nixon is both alive and dead, or, like Schrödinger's cat, in a superposition of alive- and dead-states, but rather that

> One and the same eternal present is ET-simultaneous with Nixon's being alive and is also ET-simultaneous with Nixon's dying; so Nixon's life is ET-simultaneous with and hence present to an eternal entity, and Nixon's death is ET-simultaneous with and hence present to an eternal entity, although Nixon's life and Nixon's death are themselves neither eternal nor simultaneous.⁵

I confess that I find Stump's and Kretzmann's explication of ET-simultaneity quite puzzling on several counts. In criticism of the notion of ET-simultaneity I shall suggest that the notion is seriously deficient in at least four respects: individuation

---

³ Ibid., p. 439.
⁴ Ibid., p. 447. The quotation is from Anthony Kenny, 'Divine Foreknowledge and Human Freedom,' in *Aquinas: A Collection of Critical Essays*, ed. Anthony Kenny (Garden City, NY: Anchor Books, 1969), p. 264.
⁵ Ibid., p. 443.

of eternal events, the role of the stipulated observer, the implicit assumption of the B-theory of time, and the role of definition in the overall structure of the Stump and Kretzmann argument.

Difficulties with the Concept of ET-Simultaneity

1. Eternal events cannot be individuated..
2. The notion of a temporal (atemporal) observer 'observing' eternity (time) is unclear.
3. ET-simultaneity entails static time, but to do the required work it must serve to explain problems that arise from dynamic time.
4. A stipulative definition is simply inadequate to meeting the dialectical explanatory burden.

Individuation of Eternal Events

Consider the domain of the variables x and y. Stump and Kretzmann state that x and y range over entities and events. But what is an eternal event? Presumably events are particulars, and so it should be possible to specify identity conditions for events and thus to individuate them. Several proposals for the identity conditions of events have been offered, but I will show that none will serve to elucidate what is meant by an eternal event.

Donald Davidson maintains that two events are identical iff they have the same causes and effects, so that:

$$x = y \leftrightarrow [(z) (z \text{ caused } x \leftrightarrow z \text{ caused } y) \& (z)(x \text{ caused } z \leftrightarrow y \text{ caused } z)].^6$$

But, for two reasons, this will not do for the purposes of individuating eternal events, even ignoring the problem of explaining what eternal or timeless causation means. First, on the causal theory of time argued in Chapter 2, causes must temporally (not just logically) precede their effects. Consequently any causal identity conditions for an event would seem to render that event temporal. Perhaps Stump and Kretzmann would reply that eternal events are individuated by their temporal causes and effects; indeed, they discuss how petitionary prayer might legitimately be said to 'cause' a divine response. My response would be that this is not a standard understanding of causation, and indeed this difficulty alone—accounting for the causal efficacy of petitionary prayer—is one of the major motivations behind the growing acceptance of a temporal understanding of God's being.[7] Second, be that as it may, there still remains the problem that Davidson's definition is circular as an individuation. As Carol Cleland points out, on Davidson's view, causes and effects are themselves events, and so 'the circularity

---

[6] Donald Davidson, 'The Individuation of Events,' in *Actions and Events*, ed. Ernest LePore and Brian P. McLaughlin (New York: Blackwell, 1985), p. 179.
[7] See David Basinger, *The Case for Freewill Theism: A Philosophical Assessment* (Downers Grove, IL: InterVarsity Press, 1996), pp. 105-22.

in Davidson's proposal has to do with the fact that it individuates events only if they are already individuated.'[8]

John Lemmon offers a second proposal, arguing that two events are identical iff they occupy the same space-time region.[9] The standard criticism of Lemmon is that it certainly seems possible that different events occupy the same space-time region. But, regardless of the merits of this criticism, Lemmon's definition will not serve Stump and Kretzmann simply because of its dependence on space and time.

Cleland herself has offered a third proposal.[10] She suggests that events are concrete changes—that is, 'the time-ordered exemplification of differing states by the same concrete phase.'[11] But of course there can be no time ordering in the timeless duration of eternity, so Cleland's proposal will not help Stump and Kretzmann.

Jaegwon Kim has offered perhaps the most widely accepted proposal. He maintains that an event is the possession of a property by a physical object at a time, and thus the identity of events is determined by the identity of properties, physical objects and times.[12] The problem here is the requirement that events are constituted by physical objects. If Strawson's 'No-Space World' is possible,[13] then Kim's conditions fail in any case, but even if a world of disembodied sounds is not logically possible, certainly the eternal entities comprising eternal events need not be—and indeed, most likely will not be—physical entities. Perhaps Stump and Kretzmann would want to adopt a reformulation of Kim's definition, and leave out the requirement that the objects possessing properties be physical. In this case, they would also need to amend Kim's 'times' to 'the unique eternal present.' But since the eternal present is unique, and since necessarily (they argue) there can be only one eternal entity, the somewhat weak result is that there is only one eternal event—namely, that unique eternal present in which God possesses all at once his illimitable life. But this just is eternity, according to the Boethian definition adopted by Stump and Kretzmann. Substituting this result back into their definition, we are left with vacuous statements to the effect that eternity and temporal events or entities are ET-simultaneous iff eternity and temporal entities are observed to be simultaneous. Unless Stump and Kretzmann can offer a more cogent elucidation of an eternal event, their definition is in difficulty.

---

[8] Carol Cleland, 'On the Individuation of Events,' *Synthese* 86 (1991), p. 230.
[9] John Lemmon, 'Comments on D. Davidson's "The Logical Form of Action Sentences",' in *The Logic of Decision and Action*, ed. Nicholas Rescher (Pittsburgh: University of Pittsburgh Press, 1966), pp. 98-9.
[10] Cleland, 'Individuation,' pp. 231-51.
[11] Ibid., p. 245.
[12] Jaegwon Kim, 'Causation, Nomic Subsumption, and the Concept of Event,' in *Supervenience and Mind: Selected Philosophical Essays* (New York: Cambridge University Press, 1993), originally published in the *Journal of Philosophy* 70 (1973), pp. 217-36.
[13] Peter F. Strawson, *Individuals* (London: University Paperbacks, 1979), pp. 59-86.

'Observing' Eternity

A second deficiency in the Stump/Kretzmann account is the role played by the observer A or B. Take the temporal observer B. What would it mean to say that B observes (or perceives) an event or an entity as eternally present? It is clear, assuming a causal theory of perception, that temporal beings can only perceive events or entities within their past light-cones—that is, events or entities which occupy a location in a temporal series. But eternal events and entities, by definition, do not exist in the temporal series. Given that time and eternity are different modes of existence, how can a temporal being observe an event or entity in a timeless mode of being? If it were as straightforward as Stump and Kretzmann seem to assume, then there would be no difficulty at all with the concept of ET-simultaneity in the first place. The 'observation' in question certainly cannot be perceptual, for perceptual observation by human beings is, at least contingently, a temporal process. On the other hand, if by 'observes' Stump and Kretzmann mean that B becomes convinced by rational argument that x is eternal, then why use the language of observation at all? And it is just this very rational argument that is at issue here.

Or take the eternal observer A. It is no clearer what Stump and Kretzmann mean by an observer in the eternal reference frame.[14] The notion of an observer in a particular reference frame entered the discussion via a consideration of the relativity of simultaneity under STR. But under STR, the notion of an observer is linked, at least in theory, to the speed-of-light propagation of causal connections or signals. In eternity, however, this notion is meaningless. Now, in a footnote Stump and Kretzmann state: 'It is important to understand that by "observer" we mean only that thing, animate or inanimate, with respect to which the reference frame is picked out and with respect to which the simultaneity of events within the reference frame is determined.'[15] But their discussion shows that there is—indeed, of necessity can only be—one eternal being, God. So it seems to follow that God observes Nixon alive and observes Nixon dead in the same timeless duration of eternity, undifferentiated, with no sequential points. Therefore, Kenny's objection returns.

Static Time

A third deficiency hides in condition (iii) in the definition, which 'provides that an eternal entity or event observed as eternally present (or simply as eternal) by some temporal observer B is ET-simultaneous with every temporal entity or event.'[16] But regardless of the temporal location of the observer, only a static theory of time could render this condition coherent. Later, Stump and Kretzmann maintain, 'From the standpoint of eternity, every time is present, co-occurrent with the whole

---

[14] William Hasker raises similar objections in *God, Time, and Knowledge* (Ithaca, NY: Cornell University Press, 1989), pp. 165-6.
[15] Stump and Kretzmann, 'Eternity,' p. 438 n. 15.
[16] Ibid., p. 439.

of infinite atemporal duration.'[17] And in their discussion of Nixon's death at some time in the future (with respect to the time of their writing), they say:

> We can show the implications of this account of ET-simultaneity by considering the relationship between an eternal entity and a future contingent event. . . . Nixon's death some years from now *will be* present to those who will be at his deathbed, but *is* present to an eternal entity. It cannot be that an eternal entity has a vision of Nixon's death before it occurs; in that case an eternal event would be earlier than a temporal event. Instead, the actual occasion of Nixon's dying is present to an eternal entity.[18]

On a dynamic theory of time, future events do not exist, and so could not be ET-simultaneous (or simultaneous under any other possible definition of simultaneity) with anything. So ET-simultaneity would certainly seem to require a B-theory of time. However, Stump and Kretzmann confuse the issue with their very next sentence:

> It is not that the future pre-exists somehow, so that it can be inspected by an entity that is outside time, but rather that an eternal entity that is wholly ET-simultaneous with August 9, 1974, and with today, is wholly ET-simultaneous with August 9, 1990 [the illustrative date of Nixon's death], as well.[19]

This strikes me as a case of wanting to have it both ways. Not even an eternal entity can be simultaneous with a nonexistent event (any more than an omnipresent entity can be 'present to' a nonexistent spatial location), and no stipulative definition of ET-simultaneity can make it so. If Stump and Kretzmann had simply bit the bullet and adopted a B-theory of time, then their definition would be more plausible. (Indeed, when they assert that 'the past is solely a feature of the experience of temporal entities,'[20] they surely sound like B-theorists.) Of course, if my conclusions above about the nature of time are correct, the whole concept of ET-simultaneity would then be invalidated.

The Role of Definition

The fourth criticism I shall offer has to do not so much with the conceptual details of the definition as with role played by the definition in the structure of Stump's and Kretzmann's overall argument. They clearly expect the reader to see that the definition of ET-simultaneity makes comprehensible the notion that an entity having eternity as its temporal mode of being can be related to entities and events in time. But can their definition do this work?

Consider an analogy. In the debate between nominalism and realism about universals, the nominalist often charges that the realist's concept of a

---

[17] Ibid., p. 441; my emphasis.
[18] Ibid., p. 442.
[19] Ibid.
[20] Ibid., p. 454.

universal as an immaterial entity that can be wholly present in numerically distinct material objects at the same time is incoherent. Suppose, following Stump and Kretzmann, that the realist were to propose a definition of 'IM (for immaterial-material)-copresence' as follows:

> Let x and y range over properties and objects. Then:
> (IM) For every x and for every y, x and y are IM-copresent iff
> (i) either x is immaterial and y is material, or vice versa; and
> (ii) for some observer, A, in the unique immaterial reference frame, x and y are both copresent—i.e., either x is immaterially present and y is observed as materially present, or vice versa; and
> (iii) for some observer, B, in one of the infinitely many material reference frames, x and y are both copresent—i.e., either x is observed as immaterially present and y is observed as materially present, or vice versa.

Such a definition is clearly possible, but it probably would not convince the nominalist, who could plausibly maintain that an incoherent idea cannot merely be defined into coherence. Definitions simply can't do that work. Similarly, if ET-simultaneity is an incoherent concept, the incoherence cannot be masked behind a definition.

Summary

Despite the celebrity of Stump's and Kretzmann's article, their account of eternity has not been widely accepted. In particular, many philosophers have found the definition of ET-simultaneity, which is essential to their explication of how a timelessly eternal being could stand in relation to temporal entities, to be incoherent.[21] I have offered criticisms of the concept above, but ultimately I must concur with Stephen Davis: 'But is Stump and Kretzmann's notion of eternity coherent? I do not believe so, although I am not able to prove that it is incoherent.'[22] So while I cannot show in any rigorous manner that the concept is incoherent, I believe there are difficulties of a sufficiently significant nature to warrant the conclusion that Stump's and Kretzmann's attempt to defend God's timelessness is untenable.

---

[21] Among the critical discussions of Stump and Kretzmann's paper, the following are significant: Stephen T. Davis, *Logic and the Nature of God* (Grand Rapids, MI: Eerdmans, 1983), pp. 16-22; Paul Fitzgerald, 'Stump and Kretzmann on Time and Eternity,' *Journal of Philosophy* 82 (1985), pp. 260-9; Hasker, *God, Time, and Knowledge*, pp. 163-6; Brian Leftow, *Time and Eternity* (Ithaca, NY: Cornell University Press, 1991), pp. 167-80, hereafter *T&E*; Linda Trinkaus Zagzebski, *The Dilemma of Freedom and Foreknowledge* (New York: Oxford University Press, 1991), pp. 41-2; William Lane Craig, *God, Time, and Eternity: The Coherence of Theism II: Eternity* (Dordrecht: Kluwer Academic Publishers, 2001), pp. 79-96.
[22] Davis, *Logic and the Nature of God*, p. 19.

Eternity as a Reference Frame: Brian Leftow

In his *Time and Eternity* Brian Leftow offers a very thorough defense of God's atemporality.[23] He claims that his view of God's atemporality is neutral with respect to tensed or tenseless time (although I shall argue below that his view, like all atemporal views, is not neutral after all). A careful scholar of the medieval philosophers and theologians, and at home with the logical and analytical tools of contemporary philosophy, Leftow is committed to divine simplicity, and feels that divine atemporality is entailed thereby. While he may well be correct regarding the entailment, I believe that I have already shown that divine simplicity rests more heavily on a Neoplatonic metaphysics than on biblical exegesis or arguments of philosophical theology, and need not be considered an essential part of the Judeo-Christian tradition.

> **The Reference Frame of Eternity (Brian Leftow)**
>
> The idea behind defining the reference frame of eternity is this: eternity may be viewed as a 'superdimension' within which the temporal dimension exists, just as three-dimensional space is a higher dimension within which a two-dimensional surface lies.

Even though it may have been the commitment to simplicity that drove him to a timeless view of God's being, Leftow lists 16 other arguments in favor of timelessness. In addition, he applies his theory to a range of theological and philosophical problems, making a prima facie case for the greater explanatory power of his view.

Interestingly, Leftow argues for a very strong position—the modal claim that God is necessarily timeless:

> But even if God must be timeless, the question still stands: *can* he be so? For if God must be timeless and cannot, what follows is not that God is nonetheless timeless but that the concept of God is inconsistent. By arguing then that God must be timeless, then, I put myself at risk of being forced to grant that God necessarily does not exist.[24]

I would reject this strong line, for I do not believe that it can be shown either that God necessarily must be timeless or that he necessarily cannot be. Rather, I wish to challenge Leftow's tacit claim that he has offered a satisfactory theory of God's atemporality.

---

[23] Brian Leftow, *T&E*. Aspects of his view are found in 'Time, Actuality and Omniscience,' *Religious Studies* 26 (1990), pp. 303-21; 'Why Didn't God Create the World Sooner?' *Religious Studies* 27 (1991), pp. 157-72; 'Timelessness and Foreknowledge,' *Philosophical Studies* 63 (1991), pp. 309-25; and 'Eternity and Simultaneity,' *Faith and Philosophy* 8 (1991), pp. 148-79. See also Ian Leftow, 'Timelessness and Divine Experience,' *Sophia* (1992), pp. 43-53.
[24] Leftow, *T&E*, p. 282.

The Eternal Reference Frame

Leftow's construction of a theory of time and eternity is most interesting, drawing heavily from sources as wide-ranging as St Anselm, STR, and tense and modal logic. The resulting theory is complex and richly textured, and a full explication of the theory is beyond the scope of this work. However, I shall give a succinct overview of the theory before directing criticism at several crucial points. Leftow locates his theory in the Anselmian tradition that construes eternity as a 'superdimension' containing space and time (see the exposition of Anselm's views in the previous chapter). He offers the following as an 'outline' of his position:

> I hold that God is timeless. I take it that the relations between God and temporal beings are as Chapter 10 sets forth [to be examined below] and that God is temporally omnipresent and omnicontiguous, as if eternity were a higher dimension in which He and temporal things coexist.[25]

Leftow's original contribution, as he suggests, lies in his definitions of the relations between God and temporal beings—relations that spell out the dimensional notion of eternity in terms of a reference frame. These relations form the heart of his theory and offer, he believes, a superior alternative to Stump's and Kretzmann's ET-simultaneity.

First, Leftow offers the 'Zero Thesis':[26]

> Z. The distance between God and every spatial creature is zero.

Next is the thesis he calls 'M':[27]

> M. There is no change of any sort involving spatial, material entities unless there is also a change of place, i.e. a motion involving some material entity.

These two theses allow Leftow to claim the following:

> My argument, then, is that given the Zero Thesis, (M), and one very general property of time (its being a fourth dimension of an extensive continuum), it follows that in the actual world there is no motion or change relative to God. So if a frame of reference is a system of objects at rest relative to one another, then it appears that God and all spatial objects share a frame of reference, one in which nothing changes... relative to God, the whole span of temporal events is actually there all at once. Thus in God's frame of reference, the correct judgment of local simultaneity is that all events are simultaneous. But all events are simultaneous in

---

[25] Ibid., p. 267.
[26] Ibid., p. 222.
[27] Ibid., p. 227.

Atemporality: Contemporary Statements 169

no temporal reference frame. Therefore the frame of reference God shares with all events is atemporal.[28]

Leftow proposes that the relations between God and spatio-temporal entities can be regarded as those defined by a reference frame. But the concepts of a temporal and an eternal reference frame are not at all intuitive. Leftow offers a lengthy and involved series of definitions in which he attempts cash out the concept of a temporal and an eternal reference frame:[29]

N. 'Now' is a primitive term such that occurring now does not entail having a position in a B-series.

AO. An event E A-occurs iff E occurs now.

BO. An event B-occurs iff E's location in a B-series is t, and it is now t.

BS. Two events are B-simultaneous iff they have the same location in a B-series in the same reference frame.

AS. Two events are A-simultaneous iff they are B-simultaneous and they B-occur.

ERF. R is an eternal reference frame iff R is such that necessarily, all events that A-occur in R A-occur A-simultaneously-in-R. Alternatively, R is an eternal reference frame iff within R, the relations 'earlier' and 'later' can hold only between locations in the atemporal analogue of a B-series and not between B-occurrences.

TRF. R is a temporal reference frame iff it is not the case that R is such that necessarily, all events that A-occur in R A-occur simultaneously. Alternatively, R is a temporal reference frame iff within R, the relations 'earlier' and 'later' can apply not only between locations in a B-series but also between A- and B-occurrences.

EE. K is a timeless entity iff K can A-occur/exist but cannot B-occur/exist: that is, iff K can exist now but cannot be located in a B-series. Alternatively, K is eternal iff K can be A-simultaneous but cannot be B-simultaneous with other entities.

---

[28] Ibid., pp. 227-8.
[29] Ibid., pp. 238-41. The labels of the definitions are mine, added to assist in the discussion following.

> TE. K is a temporal entity iff K can B-occur/exist: that is, iff K can have a location in a B-series.

These definitions are dense and elaborate, but are simpler than Stump's and Kretzmann's theory in that the only kinds of simultaneity involved are temporal; they do not invoke a *sui generis* species of ET-simultaneity. Leftow concludes:

> If these definitions hide no nasty surprises, they let us say that a temporal thing can occur within the atemporal reference frame without compromising the absolute distinction between temporal and timeless things or reference frames. They let us say that an eternal frame of reference includes an A-simultaneous array of events located sequentially in temporal B-series. They also let us say that events A-occur in both eternity and time, but B-occur only in time. Yet temporal events that A-occur in eternity also B-occur in time, and so occur in eternity as ordered in timeless analogues of their B-relations.[30]

If Leftow has succeeded, his theory does solve the problem of how a timeless God could be related to temporal entities. But are there, after all, 'nasty surprises' awaiting investigation?

Evaluation

Since Leftow's development of his theory begins with the Zero Thesis, let us begin our evaluation there as well. Once again, the Zero Thesis is:

> Z. The distance between God and every spatial creature is zero.

This thesis does not seem to me to be prima facie intuitive. Leftow explains his derivation of (Z). He asserts:

> D1. There is no distance between God and any spatial entity.

Any defender of traditional theism would be tempted to accept (D1) as a statement of God's 'omnipresence.' But there is a hidden ambiguity in D1. Leftow wants to read it as

> D2. There is a distance-relation between God and any spatial entity, and the distance between them is zero.

He acknowledges, however, that many philosophers would want to read (D1) as

> D3. It is not the case that there is a distance-relation between God and any spatial entity.

---

[30] Ibid., p. 241.

So before accepting (Z), we must decide between (D2) and (D3).

Leftow allows that a defender of (D3) might claim that the doctrine of God's omnipresence does not entail that God is located at every spatial position, and indeed other attributes of God (for example, that he is immaterial, that he is omnipotent and omniscient, and so on) entail that God does not have any spatial location. But distance is just a relation between two spatial locations. Hence (D2) cannot be correct, since it cannot be meaningful to speak of a distance relation between one spatial and one non-spatial entity. In other words, (D2) violates a type restriction on the relata. But, remarks Leftow, (D3) is a straightforward denial of the first conjunct of (D2). If that conjunct is not meaningful, neither is its denial.

This line of argument is flawed in two respects. First, a linguistic objection may be offered. For, 'It is not the case that there is a distance relation' is not strictly equivalent to 'There is no distance relation' if certain use/mention distinctions are observed. 'There is no distance relation' is how Leftow wants to construe (D3) as a denial of (D2), in which case 'distance relation' is used, and if this notion were meaningless in (D2) it is also meaningless in this reading of (D3). But we can read (D3) as making mention of 'distance relation.' Read in this way, (D3) asserts that 'It is not the case that there is a "distance relation",' presumably because the term mentioned is strictly meaningless. Consider this parallel. A person enamored of Lewis Carroll might assert, 'There are slithy toves.' The denial, 'There are no slithy toves,' might be guilty of using a meaningless term and hence of being meaningless itself. But to state, 'It is not the case that there are "slithy toves"' is simply and straightforwardly to deny that a referent exists of the term 'slithy toves,' and this sentence surely is meaningful.

Difficulties with the 'Eternal Reference Frame' Concept

1. Confusion over God's 'spatial location.'
2. The temporal indexical 'now' cannot entail a B-property, but this seems implicit in any concept of 'now.'
3. Inability to explain what temporal extension means in a timeless reference frame.
4. Implication that each relativistic reference frame would have its own independent analogue in the eternal reference frame.

Second, Leftow's defense has the unwelcome consequence of making it impossible to deny that a particular relation exists based on a violation of a type restriction on the relata. For example, 'Hamlet was an ancestor of Queen Margrethe of Denmark' is simply false, not because Hamlet is a meaningless term, but because the fictional prince is just not the sort of thing that can enter into the 'ancestor of' relation with a real queen. It is a violation of a tacit type restriction. But, on Leftow's account, a violation of a type restriction on the relata renders the assertion or the denial of relation meaningless. Consequently one could not meaningfully state that 'It is not the case that Hamlet was an ancestor of Queen Margrethe of Denmark.' Since this consequence is absurd, Leftow's account must be flawed. So Leftow's defense against the first objection to (D2) fails. It is indeed meaningful to maintain a type restriction on the relata of distance relations, and to deny (D2) based on a violation of that restriction.

A second objection to (D2) considered by Leftow goes as follows. If God has no spatial location, then he has no spatial relations. That is, generalized and by contraposition, for any x, if x has spatial relations, then x has spatial location. But, says Leftow, this claim is false. 'Is not located in' expresses a spatial relation. And surely God and sets, to name but two examples, stand in this relation to every spatial region. So perhaps the objector means that for any x, if x has positive spatial relations, then x has spatial location. Leftow responds that 'is external to' expresses an apparently positive spatial relation which is equivalent to 'is not located in.' Thus, Leftow concludes, 'the objector to (D2), then, must give us a suitable non-question-begging account of what makes a spatial relation positive. This task looks difficult.'[31]

I would agree that the task is difficult, and that Leftow's response to the second objection to (D2) is reasonable. So perhaps (D3) is not the better reading of (Z). But we are left without any compelling argument that establishes that (D2) is a better reading either. What rides on the distinction?

Leftow wants (Z), conjoined with (M), to entail that nothing changes with respect to God. (M) says that there is no change of any sort involving spatial, material entities unless there is also a change of place—that is, a motion involving some material entity. So if there is a zero spatial distance between God and every material entity, then there can be no motion of any spatial entity with respect to God, and so no change with respect to God. But of course, if (Z) is read as (D3), the argument will not go through. If it doesn't, then Leftow is not warranted in asserting that God shares an atemporal reference frame with every material object.

But even if (Z) should be read as (D2), it still is not at all clear that the notion of an atemporal reference frame is coherent. Several additional difficulties remain to be uncovered here. The first is with the definition of 'now':

> N. 'Now' is a primitive term such that occurring now does not entail having a position in a B-series.

It is apparent why Leftow phrases this definition as he does: the phrases 'eternal now' or 'eternal present' cannot entail having B-relations. Earlier, he argued that a single durationless event that did not stand in earlier or later relations with any other event would not count as a temporal event.[32] And on (N), eternity, the 'eternal now,' is not a temporal entity, which is what Leftow wants.

But it is hard to resist the conclusion that Leftow's definition is ad hoc and question-begging, for 'now' seems to be the epitome of a temporal term. On the A-theory, what else could 'now' mean? And even on the B-theory, as we saw in Chapter 2, 'now' could be given a token-reflexive translation using 'simultaneous with,' or understood as picking out a time. But both options involve

---

[31] Ibid., p. 234.
[32] Ibid., pp. 31, 237. I argue against this claim below in Chapter 9.

a B-determination. Leftow needs an instantaneous 'now' that possibly is not preceded nor succeeded by any other instants. On the causal theory of time, 'now' marks the separation between a present cause and a future effect. Michael Tooley argues that the concept of the future is not analytically basic, but is to be analyzed in terms of the concept of the present, the concept of the earlier-than relation, and the tenseless existential quantifier.[33] If these considerations are on target, then the indexical 'now' implicitly entails a position in a B-series, and Leftow's analysis of other features of his theory will be undercut.

Further difficulties arise in Leftow's definition of an eternal reference frame:

> ERF. R is an eternal reference frame iff R is such that necessarily, all events that A-occur in R A-occur A-simultaneously-in-R. Alternatively, R is an eternal reference frame iff within R, the relations 'earlier' and 'later' can hold only between locations in the atemporal analogue of a B-series and not between B-occurrences.

What Leftow wants this definition to accomplish is to allow temporal events to retain something like their temporal ordering in eternity, but still to occur simultaneously in eternity. The retention of something analogous to temporal ordering is necessary, of course, if God is to know what we know about the temporal order of events. Yet temporal events cannot be temporally ordered in eternity, or else eternity would have temporal parts—a conclusion Leftow would strenuously reject. Does ERF do what Leftow needs it to do and, at a more basic level, is it even coherent?

Let us take first Leftow's primary definition, and cash it out in terms of his other definitions AO, AS, BS and BO:

> ERF'. R is an eternal reference frame iff R is such that, necessarily, all events that occur now (AO) in R occur now (AO) at the same location t in a B-series in R (AS, BS), and it is now t (BO).

Since 'it is now t' can be phrased as 't is present,' and since there is only one time location in eternity, the eternal present, this can be rephrased as follows:

> ERF''. R is an eternal reference frame iff R is such that, necessarily, all events that occur now (AO) in R occur now (AO) at the same location, the eternal present, in a B-series in R (AS, BS), and eternity is present (BO).

---

[33] Michael Tooley, *Time, Tense and Causation* (Oxford: Oxford University Press, 1997), chapter 5, hereafter *TT&C*.

But this reading is confused, since the present is not part of a B-series. Perhaps instead of 'B-series' we should read 'atemporal analogue of a B-series,' as Leftow indicates in his alternative definition. Thus:

> ERF'''. R is an eternal reference frame iff R is such that, necessarily, all events that occur now (AO) in R occur now (AO) at the same location, the eternal present, in the atemporal analogue of a B-series in R (AS, BS), and eternity is present.

But this doesn't remove the confusion, for what is meant by the location of the eternal present in the atemporal analogue of a B-series? Since eternity cannot be earlier than or later than any temporal moment, it would seem that the location of the eternal present would be just the entire atemporal analogue of a B-series. If that is the case, then the definition reduces simply to the trivial claim that an ERF is one in which all events occur simultaneously in eternity. Something has gone wrong here. Leftow has not given us a coherent sense of an ERF.

There are additional problems with the concept of a reference frame itself. Leftow nowhere offers a general definition of a reference frame. A Cartesian reference frame is a system of orthogonal axes with a specified origin.[34] Now, Leftow endorses the conclusions of STR, so his TRF must be a relativistic reference frame. A relativistic reference frame is one in which four axes are orthogonal in hyperspace but in which an arbitrary clock must be included, since time is relative to the reference frame. Immediate worries arise as to what could serve as the atemporal analogue of this clock in the ERF. One answer would be to maintain that there are many ERFs, one for each relativistic reference frame that exists. Aside from the bloated ontology this approach yields, it would seem clearly to violate the spirit of Leftow's definitions, the point of which seems to be to offer a unique relation between the temporal order and eternity. And Leftow rejects this response, claiming that, by applying the relativity of simultaneity, one can see that:

> ... an event occurs in eternity simultaneously with all other events, but this does not entail that the event occurs at the same time as all other events in any other reference frame. ... Events are present and actual all at once in eternity, but present and actual in sequence in other reference frames.[35]

Nevertheless, the problem remains. How can very many—perhaps infinitely many—relativistic reference frames, each with its own clock, be represented in a single ERF in which there is no clock but merely an atemporal analogue of the ordering determined in each individual reference frame by B-determinations? Since in the Einsteinian interpretation of STR adopted by Leftow there is no distinguished temporal reference frame by which to order events occurring in all relativistic reference frames, how can these all be correlated and ordered in

---

[34] In GTR other reference frames are more appropriate, but the following point still holds *mutatis mutandis*.

[35] Leftow, *T&E*, pp. 234-5.

eternity? I find this inconceivable, but this might be due to my limited powers of imagination rather than any incoherence in Leftow's model itself.

But there are still further problems. Whether reference frames are depicted graphically or described verbally or in mathematical notation, they all necessarily involve extension along axes. So to speak of an 'atemporal analogue of a B-series' in an eternal reference frame is to do nothing more than to specify that within that reference frame some axis represents a B-series. The points on such an axis represent dates; the directions along the axis represent the earlier-than or later-than relations. Now while it is all well and good to speak of such an axis as being present all at once, it is nevertheless extended. And what is extended necessarily has proper parts. Divine simplicity maintains that God's being is equal to his eternity. Hence if the atemporal analogue of the B-series possesses the atemporal analogue of proper temporal parts, then God's being possesses the atemporal analogue of proper temporal parts. Once this is realized, we are faced with a choice. We may cling to the standard conception of divine simplicity and reject the conclusion that God's being has any kind of proper parts, or we may accept that it does. In the former case, we must reject Leftow's attempt at defining eternity as a reference frame, and either retreat to a Boethian concept of eternity, with all its difficulties, or develop some new concept that will learn from and avoid the problems of Stump and Kretzmann and Leftow. In the latter case, we might as well accept that God's being has proper temporal parts instead of some 'atemporal analogue' of proper temporal parts, and develop a concept of God as a temporal being.

The 'Classical View' Revisited: Paul Helm

It will be instructive to look briefly at one additional advocate of the atemporalist conception of God's being, Paul Helm. I shall examine his arguments for divine atemporality and show that they can be expected to lead to difficulties similar to those which affect Stump and Kretzmann and Leftow. This will set the stage for a general argument against all theories of divine atemporality that I shall present in the final section of this chapter.

The 'Classical View' (Paul Helm)

1. A Perfect Being would be absolutely immutable, and hence must be atemporal.
2. A Perfect Being would not be subject to the vicissitudes of temporal passage; he could not lose part of his perfect life.

Helm offers a number of arguments supporting God's atemporality, of which I shall note what I take to be the three strongest.[36] First is an interesting argument, a

---

[36] Paul Helm, *Eternal God* (New York: Oxford University Press, 1988); 'Timelessness and Foreknowledge,' *Mind* 84 (1975), pp. 516-27; *Faith and Understanding* (Grand Rapids, MI:

version of which often arises in response to the Kalām version of the Cosmological Argument:

> The idea that God exists in an infinitely backward extending time . . . requires that an infinite number of events must have elapsed before the present moment could arrive. And since it is impossible for an infinite number of events to have elapsed, and yet the present moment has arrived, the series of events cannot be infinite. Therefore either there was a time when God began to exist, which is impossible, or God exists timelessly.[37]

The argument is this:

1. Either time is eternal or it had a first moment.
2. If time is eternal then an infinite series of events has elapsed before the present moment.
3. But it is impossible that an infinite series of events has elapsed.
4. Hence time is not eternal.
5. Hence time had a first moment.
6. But if time had a first moment, then if God is temporal, then God's existence had a first moment.
7. But it is impossible that God's existence had a first moment.
8. Therefore God is not temporal.

There are a number of weaknesses in this argument. First, (2) is possibly false. For there is another possibility: a single event of infinite duration elapsed prior to the present, or prior to the first event of the sequence of events which we count as finite time. But even if (2) stands, (3) may be false. For while the impossibility of an actually instantiated infinite temporal series has been argued by many, most notably Craig,[38] it is possible to respond that it is an actually instantiated infinite series of points in physical time which is impossible, and God's infinite time is not physical time.[39] Further, one could argue against (7) that, although time had a first moment and God's existence is temporally coextensive with time, that does not entail that God came into existence with the first moment of time, since it is possible that God existed timelessly 'before' the creation of time ('before' here being taken in a logical rather than a temporal sense, since it is impossible that there be any time before time).[40] Consequently, (2) and (3) will not do the work needed to support (7). So this argument fails.

Eerdmans, 1997), chapter 4; 'Divine Timeless Eternity,' in *God and Time: Four Views*, ed. Gregory E. Ganssle, (Downers Grove, IL: InterVarsity Press, 2001).

[37] Helm, *Eternal God*, pp. 37-8.

[38] William Lane Craig and Quentin Smith, *Theism, Atheism, and Big Bang Cosmology* (Oxford: Clarendon Press, 1993).

[39] Alan G. Padgett, *God, Eternity and the Nature of Time* (New York: St Martin's Press, 1992).

[40] See Craig, *God, Time and Eternity*, pp. 256-80.

In a second argument Helm claims that, assuming God is immutable, God's timelessness is entailed. 'A God who acts but is immutable . . . must be timelessly eternal, since any action in time (as opposed to an action the effect of which is in time) presupposes a time before the act, and a time when the act is completed, and thus presupposes real change.'[41] Of course, God might not act, but the God of traditional theism does act. So the argument proceeds:

9. Every temporal act has a finite duration.
10. Hence every temporal agent undergoes change
    a. either from not acting to acting, or
    b. from acting to not acting.
11. Therefore if God is temporal, he undergoes change.
12. But if God is immutable, then he does not undergo change.
13. Therefore God is not temporal.

But (9) and (10) are false. For it is possible that God is temporal, and the universe is eternal, and that God acts providentially at every moment of time to sustain the universe in its physical existence.[42] Then although God's providential act is temporal, it had no beginning and no end, and so no finite duration (contra [9]). Moreover, in this case, even though God is temporal and thus a temporal agent, he never undergoes a change from not acting to acting, or vice versa (contra [10]). So to make the argument go through, Helm would need to modify (9) and (10) significantly. But regardless of whether such modification is possible, the argument still turns on what was earlier characterized as strong immutability, which claims that it is impossible that God undergo any kind of change. But if this interpretation of immutability is rejected in favor of weak immutability, according to which God necessarily cannot change in his essential nature, then Helm's argument will not go through no matter how (9) and (10) are patched up.

Difficulties with the 'Classical' View

1. The absolute changelessness (and hence timelessness) of a Perfect Being rests upon two questionable assumptions:
   a. divine simplicity; and
   b. strong immutability.
2. While it is true that a temporal being suffers loss with time's passage, the force of the objection is weakened by two observations:
   a. the richness of some experiences necessarily involves temporal passage;
   b. if time is dynamic, timeless life is not possible.

A third argument urged by Helm is quite interesting in its own right. From the standpoint of Anselmian Perfect Being theology, Helm claims that divine temporality:

---

[41] Helm, *Eternal God*, p. 90.
[42] This counter to Helm's argument is suggested by Leftow, *T&E*, pp. 79-80.

> ... flouts the basic theistic intuition that God's fullness is such that he possesses the whole of his life *together*. To many, the idea that God is subject to the vicissitudes of temporal passage, with more and more of his life irretrievably over and done with, is incompatible with divine sovereignty, with divine perfection and with the fullness of being that is essential to God.[43]

Now, I confess to feeling the pull of this line of argument. There is a sadness about life in which the most sublime moments pass irrecoverably, never again to be present. No memory, however good, is a perfect substitute for the experience. (Of course, graciously, the most agonizing and painful moments also pass irrecoverably and, graciously, the memory is never as painful either.)

But is such a life defective in any clear sense? An affirmative answer seems to miss an important point. What we are granting as lamentable is a result of the passage of time. But many experiences necessarily depend upon temporal passage—for example, the experience of hearing a symphony. What would that be like if it were not extended in time? How would a timeless person enjoy a symphony? And if God is temporal, certainly his perfect memory and his perfect foreknowledge (on the traditional understanding of foreknowledge) would attenuate the 'loss' to some degree.

Finally, to anticipate somewhat, I shall argue in the next section that if God is timeless, then time is tenseless. But if we have good reason to believe that time is tensed, then we have to reject a timeless God. So perhaps the alleged 'limits' on the fullness of God's life are metaphysically necessary limits: no genuine life could avoid them.

In the end, I do not find Helm's arguments compelling. But even if they were, we could still ask for some conception of the relation between eternity and temporal events. Helm deflects this request, asking us to imagine the relation between God and the world as analogous to that between an author and the characters of a novel:

> The author is the author of the whole, but it does not make sense to ... ask whether his writing ... the novel is simultaneous with any of the events in the work. His writing, being in time, is simultaneous with what he writes, but the act of writing is not simultaneous with any of the events that occur in the work, not even if the author writes himself into the novel.[44]

This analogy has a certain prima facie attractiveness. But the attractiveness masks several problems. For while the events of the temporal series actually occur, the events of a novel do not occur at all. And while Helm is trying to argue that God is atemporal, the novelist, whose writing is clearly successive, is necessarily temporal. These disanalogies show that Helm's attempt to avoid offering a

---

[43] Helm, 'Divine Timeless Eternity,' pp. 30-31. Leftow offers the same line of argument as one of his 16 arguments for timelessness. See *T&E*, pp. 278ff.
[44] Helm, *Eternal God*, pp. 30f.

proposal for the relation between eternity and time, however nicely constructed, must fail.

In the absence of compelling arguments in favor of atemporality, and in the absence of accounts of the eternity/time relation, Helm's statement of divine atemporality is ultimately no more successful than that of Stump and Kretzmann or of Leftow, which is to say, not successful at all.

Atemporality and Static Time[45]

In the discussions of divine atemporality in this and the previous chapter, one common element may be seen to be lurking in the background of each. That element is a commitment, tacit or overt, to static time. An interesting feature of the debate over God's temporal mode of being is that those who argue for divine temporality generally do so by beginning with an argument for tensed, A-theoretic, time.[46] On the other hand, those who argue for divine atemporality generally claim to be neutral with respect to tenseless, B-theoretic, or tensed, A-theoretic time.[47]

> **Timeless God, Tenseless Time**
>
> This section offers a general argument to the effect that if God is timeless, then time must be tenseless. But if we have good reasons to deny the consequent of this conditional and affirm dynamic time, then by *modus tollens* we have good reasons to believe that God is temporal.

I shall argue that theories of divine atemporality are not, in fact, neutral with respect to theories of time, but rather entail a commitment to tenseless time. This is an important result, since, if the reasons given in Chapter 2 to reject tenseless time are good ones, then we will have good reasons to reject all theories of divine atemporality.

A number of different authors have offered arguments that attempt to show that the assumption of God's timelessness is inconsistent with certain other truths about God that are central to Christian theism, together with facts entailed by dynamic time. Typically, the arguments focus on God's omniscience. These arguments generally are of two forms. The first form is semantic:[48]

---

[45] This section is adapted from Garrett DeWeese, 'Timeless God, Tenseless Time,' *Philosophia Christi* 2:2:1 (2000), pp. 53-9.
[46] See references in footnotes 48 and 49 below.
[47] For a recent example, see Brian Leftow, *T&E*, p. 17.
[48] Versions of the semantic argument are found in A. N. Prior, 'The Formalities of Omniscience,' reprinted in *Papers on Time and Tense* (Oxford: Oxford University Press, 1968), pp. 26-42; Norman Kretzmann, 'Omniscience and Immutability,' *Journal of Philosophy* 63 (1966), pp. 409-21; Nicholas Wolterstorff, 'God Everlasting,' in *Contemporary Philosophy of Religion*, ed. Steven M. Cahn and David Shatz (Oxford: Oxford University Press, 1982), pp. 77-98; Richard M. Gale, 'Omniscience-Immutability Arguments,' *American Philosophical Quarterly* 23 (1986), pp. 319-35. Sorabji traces the

A. There are essentially tensed truths (for example, those expressed by propositions containing the temporal indexical 'now').
B. A timeless being cannot know essentially tensed propositions.
C. Therefore, if God is timeless, there are truths he cannot know.

The second form of the argument is metaphysical:[49]

D. Time is dynamic (that is, temporal becoming is real).
E. Hence, what is real is constantly changing.
F. A timeless being cannot experience change in its knowledge.
G. Therefore, if God is timeless, he cannot know what is real.

In both cases, the conclusion is apparently inconsistent with the assumption of God's omniscience. But since a commitment to divine omniscience is much more firmly entrenched in (Judeo-Christian) theism than is a commitment to divine atemporality, it is the latter that should be given up.

Parenthetically, we should note that these are in fact separate arguments, for (A) and (D) are not equivalent. It might be the case, for instance, that time is tenseless and (A) still be true because the subjective experience of temporal passage makes tense and temporal indexicals necessary features of linguistic expression. Or it might be the case that (D) is true and also that all tensed sentences or sentences containing temporal indexicals could be translated into, or analyzed in terms of, tenseless sentences.

Both arguments, it can be seen, depend for their punch upon the claim that a timeless being cannot know certain facts or propositions. It may be, however, that this claim is misguided. William Alston, for example, claims that God's knowledge is direct intuitive knowledge and thus non-propositional or non-discursive.[50] Alston is neutral with respect to God's temporal mode of being. But Gregory Ganssle argues that, if God's mode of knowing is direct intuition, then God must be atemporal. This is because, argues Ganssle, direct intuition is not a time-spanning relation.[51] It is possible that other versions of this temporalist argument strategy may be susceptible to defeat in an analogous way given appropriate reconceptualizations of the divine attribute in question. But it is not clear that a different conception of God's attributes will avoid the problem I am

---

history of these arguments in *Time, Creation and the Continuum: Theories ion Antiquity and the Early Middle Ages* (London: Duckworth, 1983), pp. 260-61.

[49] Versions of the metaphysical argument are found in Stephen T. Davis, *Logic and the Nature of God*, pp. 13-24; William Lane Craig, 'Was Thomas Aquinas a B-Theorist of Time?' *New Scholasticism* 59 (1985), pp. 475-83; Delmas Lewis, 'Eternity Again,' *International Journal for Philosophy of Religion* 15 (1984), pp. 73-9; William Hasker, 'Yes, God Has Beliefs!' *Religious Studies* 24 (1988), pp. 385-94.

[50] William P. Alston, 'Does God Have Beliefs?' in *Divine Nature and Human Language: Essays in Philosophical Theology* (Ithaca, NY: Cornell University Press, 1989).

[51] Gregory Ganssle, 'Atemporality and the Mode of Divine Knowledge,' *International Journal for Philosophy of Religion* 34 (1993), pp. 171-80.

considering—namely, that a timeless conception of God entails tenseless time. For the argument I shall now offer is general; it does not depend upon any assumptions about God's attributes.

I shall make two initial assumptions: first, that God is timeless (atemporal); and, second, that time is dynamic (tensed, or A-theoretic).

Here, then, is the argument:

14. The ordered sequence of temporal events T exists in eternity in an atemporal analogue A of the temporal order.
15. There is a function $\mathscr{F}$ (T,A) which maps points (moments, events) in the temporal order T onto points in the atemporal analogue A of the temporal order.
16. The future does not exist.
17. No function can map nonexistent values onto real values.
18. $\mathscr{F}$ (T,A) cannot map future points onto the atemporal analogue A.
19. Hence, A cannot contain atemporal analogues of future points of T.
20. Either A grows as the temporal order grows, or A is incomplete.
21. A cannot grow, since there can be no change in an atemporal entity.
22. A cannot be incomplete, for then the actual time sequence T would, at any time after the first event, be larger than A.
23. Therefore, if God is timeless, then time is tenseless.

The first premise in this argument is the one which does most of the work:

14. The ordered sequence of temporal events T exists in eternity in an atemporal analogue A of the temporal order.

This premise might seem to be the most easily questioned. It may fairly be asked how T can 'exist' in A. The term 'atemporal analogue' is from Brian Leftow, but the idea behind the term is surely discernible in Boethius' image of a circle, representing the temporal order, and its center, representing a timeless God, to whom all points on the circumference—the temporal order—are present. It is also discernible in the image used by both Boethius and Aquinas of a line of travelers observed from a tower or a hilltop, where the travelers represent the temporal order and the lofty observer represents God. I confess that I am not clear myself about this notion. It might mean either that temporal events themselves actually do exist in eternity, but ordered by something other than past, present and future (this seems to be Leftow's meaning), or it might mean that the atemporal analogue A is a representation in eternity of the temporal sequence T (which seems to be more in line with the meaning of Stump and Kretzmann and Paul Helm).

Either way, though, it might be objected that the ordered sequence of temporal events does not exist in eternity at all, either as events or as representations of events. The argument might run like this: By the doctrine of divine simplicity, God = his eternity, so there can be nothing else that is 'in eternity' or is eternal. God's relations to the temporal order T are just that—to T and not to any analogue or representation of that order in eternity. But this objection fails, for there remains the problem that if time is dynamic, then the relations in which God stands to temporal entities must change. But this is impossible *ex hypothesi*. Aquinas recognized this, and developed his doctrine, regarded as untenable by most contemporary philosophers, that God has no real relations at all with the temporal world. Consequently we must assume that there exists an A of some sort—an atemporal analogue of T.

So, given (14), it follows that

15. There is a function $\mathscr{F}$ (T,A) which maps points (moments, events) in the temporal order T onto points in the atemporal analogue A of the temporal order.

This can be seen as follows. A must be as well-ordered as T, or it will not be a representation of T. Given our assumption that time is dynamic, T will be ordered by A-theoretic determinations. Let us grant that all A-determinations can either be translated *salva veritate* into or analyzed in terms of B-determinations, so that the atemporal analogue A will be ordered according to B-determinations. But there must be some relation that does the translating or analyzing. That relation will be a function that takes as its input sequential points (moments, events) of T and gives as its output points in A ordered by B-determinations. Let that function be $\mathscr{F}$ (T,A). We need not specify further just how that function does its work, but the existence of $\mathscr{F}$ cannot reasonably be doubted.

The next step in the argument is the recognition, given the assumption of dynamic time, that

16. The future does not exist.

That is, there are no real points (moments, events) of the future. The future does not exist *at all*, in any sense.

Next, I claim that it follows from the nature of a function that

17. No function can map nonexistent values onto real values.

It might be objected that this assumption is false. Why cannot a person define a function that maps her future grandchildren onto, say, the natural numbers? To see why not, we may represent the function as a set of ordered pairs. The first member of each pair will be a grandchild, and the second will be a natural number. Now, if the grandchild is actual—that is, if she has been born—then there will indeed be an ordered pair, for example, [Sarah,1]. But if the grandchild is not actual, then either there will be a variable placeholder or an actually existing ersatz entity playing the role of a non-actual grandchild, or there will not be an ordered pair at all. If the former, the pair will be something like, [n,7]. But then once the seventh grandchild is born the variable placeholder will be replaced by a reference to the grandchild. Now, given the axiom of set theory that two sets are identical iff they contain exactly the same elements, it is obvious that the function as defined by the set of ordered pairs it generates will be different before and after the arrival of the grandchild, since clearly [n,7] is not the same element as [Sammy,7]. If, on the other hand, it is the case that there are not ordered pairs representing unborn grandchildren, then with the birth of each successive grandchild a new ordered pair will be produced. But by the principle that, necessarily, a set exists only if its elements exist, clearly in this case the sets of ordered pairs, and hence the function defined by the set, will be different with each new blessed event. Consequently I take (17) as true.

Then, from (16) and (17), we have:

18. $\mathscr{F}$ (T,A) cannot map future points of T onto the atemporal analogue A.

In other words, $\mathscr{F}$ (T,A) cannot take points in the nonexistent future as arguments.

19. Hence, A cannot contain atemporal analogues of future points of T.

But if A cannot contain future points, we are faced with the following dilemma:

20. Either A grows as the temporal order grows, or A is incomplete.

But neither horn of the dilemma is acceptable:

21. A cannot grow, since there can be no change in an atemporal entity.
22. A cannot be incomplete, for then the actual time sequence T would, at any time after the first event, be larger than A.

Either A grows as future points (moments, events) are added to T, or else A is only complete with respect to T at a single time t and incomplete with respect to T at any later time t + n. But both of these alternatives are unacceptable. Hence if A is to be complete, then all future points must be contained in T, which means that time cannot be dynamic. So the only way to escape the dilemma is to reject the assumption that led to the dilemma—namely, that time is dynamic. Consequently, our conclusion is this:

23. If God is timeless, then time is tenseless.

So it seems that any theory that posits a timeless God is committed, at least tacitly, to tenseless time.

Brian Leftow has tried to head off this line of attack. He asks, 'Does occurring in eternity entail tenseless time?' He wants to remain neutral between tensed and tenseless theories of time:

Now, I am not going to enter the lists for or against tenseless theories of time. Rather, whatever the merits or demerits of a tenseless view, I hope to show now that the existence of a timeless being and of the eternal simultaneity relation I suggest is compatible with a tensed theory of time, i.e. with the claim that only present (and perhaps past) events exist in time, so that there is a genuine and radical ontological distinction between present (and perhaps past) events and future events.[52]

Leftow's argument involves invoking the relativity of simultaneity, a firmly entrenched assumption of Einsteinian interpretations of STR. Of course, the denial of absolute simultaneity poses a problem for all theories of dynamic time. But I argued in Chapter 3 that there are versions of STR that do not entail a denial absolute simultaneity, and so offer adequate responses to this difficulty. What these responses show is that the versions of STR which entail the relativity of simultaneity are in fact just those versions which are committed to static or tenseless time. And, of course, this poses no threat to my argument! Thus I conclude that Leftow's attempt to maintain neutrality between tensed and tenseless time is unsuccessful.

If my argument succeeds, then I have shown that any atemporalist construal of God's being entails the B-theory of time.[53] But since I claim that we have good reasons to reject tenseless time, then by *modus tollens*, any concept of a timeless God must also be rejected.

Conclusion

At the end of the discussion, the atemporalist view, while perhaps not incoherent, suffers from serious, perhaps insurmountable, difficulties. The two best contemporary attempts to make sense of divine atemporality—that of Stump and Kretzmann and of Leftow—have been shown to contain apparent contradictions and conceptual flaws, and it has also been demonstrated that a third contemporary attempt—that of Helm—can be expected to suffer from analogous problems.

Further, a significant positive result of this chapter is the conclusion that all atemporal conceptions of God's being must at least tacitly assume static time. Although similar to other arguments that have been advanced to this effect, my argument differs in that it does not rely on any conclusions drawn from God's knowledge, but is instead a general argument. If my argument succeeds, then I have shown that all atemporalist construals of God's being entail a B-theory of time. But if the argument of Part I is sound, then atemporalist construals of God's being must be rejected.

---

[52] Leftow, *T&E*, p. 231.

[53] Note that this argument, formed in terms of God, can be extended to *any* atemporal concrete entity—that is, any timelessly existing entity that possibly stands in a causal relation. Whether there might in fact be any such entities other than God, I do not know.

Chapter 7

A Medieval Dissent: God is Temporally Everlasting

As we saw in Chapter 5, the consensus among medieval philosophers and theologians was that God existed atemporally. I attributed the consensus primarily to views of God's simplicity and immutability, which had their origins in Neoplatonism. Further, the threat of theological fatalism motivated discussions of the nature of time and God's foreknowledge, and common models of the relationship between God and time had the implication that time is static. While I have argued that an atemporal view of God's being entails static time, and have referred to passages in Boethius, Anselm, and Aquinas which are most naturally interpreted as entailing static time, it is not so clear that these thinkers consciously adopted a static concept of time. Indeed, I think William Lane Craig is correct to maintain that it is almost inconceivable that Aquinas—and, I would add, the others we have considered—consciously held to a B-theory of time.[1]

But the medievals do not speak with one voice. John Duns Scotus presents the first unequivocal affirmation of dynamic time and a satisfying account of temporal contingency. William of Ockham generally accepted Duns Scotus' view of time and added some interesting features of his own to the discussion. Both, because of a clear commitment to dynamic time, offer views of God's relation to time that can be seen as precursors to the modern temporalist views of God's being. It must be noted that, as with the earlier thinkers we have looked at and many lesser figures in the medieval period as well, the presenting issue was not the nature of time, but rather the issue of God's infallible knowledge of future contingent propositions. In dealing with this issue, the nature of time and God's temporal mode of being naturally came into play, but they were not the central focus of the discussions. Consequently, the discussions are not as clear for our purposes as we might wish. Nevertheless, in exploring the views of these two thinkers, we will see how they opened the door to a departure from the consensus of earlier centuries.

The development of thought on these issues did not advance much after Duns Scotus and Ockham until the sixteenth-century Spanish Jesuit Luis de Molina. Although he himself held to divine atemporality, Molina's views of dynamic time and his notable contribution of the doctrine of Middle Knowledge make him a suitable figure with which to end this chapter.

---

[1] William Lane Craig, 'Was Thomas Aquinas a B-Theorist of Time?' *New Scholasticism* 59 (1985), p. 482.

John Duns Scotus

John Duns Scotus (1266-1308) invites careful attention because of his clear view of dynamic time and his rejection of the reality of the future. This view leads him to suggestions that God's relation to time is of a different sort than that suggested by his predecessors, notably Augustine and Aquinas. The historian of philosophy Calvin Normore observes:

> When considered against the background of Aquinas' position, Scotus' discussions of God's knowledge signal the clash of two fundamentally different ways of conceiving the nature of time. The first, which seems to have been Boethius' and may have been Aquinas', conceives the difference between past and future as perspectival rather than ontological. . . . The second view, the one for which Scotus argues and the one which seems to be taken more or less for granted in the first quarter of the fourteenth century, sees the difference between past and future as an objective difference, one that exists for God as well as for us.[2]

Duns Scotus' Views of God and Time

- Synchronic contingency: the obtaining of a contingent state of affairs C at t does not entail that C's complement is impossible (i.e. that C is necessary) at t.
- Synchronic contingency entails dynamic time.
- If time is dynamic, the future does not exist in any sense.
- Therefore, accounts of God's foreknowledge as foresight, developed by Boethius, Anselm and Aquinas, are flawed.

The salient discussion is found in Duns Scotus' commentary on the *Sentences* of Peter Lombard.[3] In his *Lectura* I 39 Duns Scotus comments on Lombard's

---

[2] Calvin Normore, 'Future Contingents,' in *The Cambridge History of Later Medieval Philosophy*, ed. Norman Kretzmann, Anthony Kenny and Jan Pinborg (Cambridge: Cambridge University Press, 1982), p. 367.

[3] By the time of Duns Scotus, the academic path of a prospective theologian was a well-defined 13-year course of study. After nine years in the Faculty of Theology at Oxford, Scotus became a *baccalaureus sententiarius*. In the first year of this two-year stage Duns Scotus prepared his *Lectura*, notes for lectures on Lombard's *Sentences*, the obligatory theological text of the day, and in the second year he delivered these lectures, in the academic year 1298-99. Following another two years of study, he became a *magister designatus* and in 1302 was chosen to teach at the University of Paris. In Paris he began, as was customary, preparing his *Ordinatio*, an elaboration of his lectures on the *Sentences*. However Duns Scotus died in November of 1308 before his extensive and elaborately detailed *Ordinatio* could be completed. Since both the *Lectura* and the *Ordinatio* are commentaries on Lombard's *Sentences*, they both follow the same divisions and subdivisions. Duns Scotus' views that are relevant to this study are found in the section concerning God's knowledge in Book I, Distinctions 38-39. However, in the *Lectura* Distinction 38 is missing, and in the *Ordinatio* Distinction 39 is missing. For additional details concerning Duns Scotus' life and career, and the textual tradition of his writings, see A. Vos Jaczn., H. Veldhuis, A.H. Looman-Graaskamp, E. Dekker, and N.W. den Bok,

doctrine of God's knowledge. There he interacts with the theological tradition from Augustine (not merely Lombard), and with the great non-Christian philosophers such as Aristotle, Avicenna and Averroës. He refutes the common view that God's immutable foreknowledge makes the future necessary, and in so doing develops (or discovers) what has been called synchronic contingency. This is the key to understanding Duns Scotus' views on time and God's relation to it.

Synchronic Contingency

The debate over Aristotle's understanding of future contingents (propositions, states-of-affairs, and events), set forth in *De interpretatione* 9, is well known. Duns Scotus took Aristotle, along with Avicenna, whom he regarded as the most important Islamic philosopher, as sharing a necessitarian viewpoint.[4] Against necessitarianism, Duns Scotus desired to defend a Christian position in which not all matters concerning human existence are necessary. For Duns Scotus, God and his essential attributes are necessary. And so are certain relations between contingent states of affairs (called, since Aristotle, 'the necessity of the consequent'). But he wants to deny that all future events or states of affairs are necessary, and to do so he develops his theory of synchronic contingency.

Duns Scotus' theory of synchronic contingency can best be understood in contrast to (his understanding of) the modal ontologies of Parmenides and Aristotle.[5] As noted in Chapter 5, Parmenides regarded Being as necessary and immutable. There was in essence only one state of affairs; what appears to us to be successive states of affairs, or change, is a matter of deceptive sensory data. Parmenides' modal ontology may be represented graphically as in Figure 7.1 (a), in which S is the only possible state of affairs.

Formally, a Parmenidean modal system would hold the following:

1. $\sim\Diamond$ (S1 & $\neg$S1)
2. $\Diamond$ S1$\rightarrow\sim\Diamond\neg$S1 $\equiv$ $\Diamond$ S1$\rightarrow\Box$ S1 (where $\rightarrow$ is strict implication)
3. S1$\rightarrow\sim\Diamond\neg$S1 $\equiv$ S1$\rightarrow\Box$ S1
4. $\sim\Diamond$ (S1 & $\neg$S2)

'Introduction,' *John Duns Scotus: Contingency and Freedom: Lectura I 39* (Dordrecht: Kluwer, 1994), pp. 1-15. Hereafter, unless otherwise noted, quotations of Scotus will be from this translation of the *Lectura*; references to the introduction and commentary of Vos Jaczn. *et al.* will be cited as *C&F*.

[4] See Scotus, *Lectura* I 8.235, 237; Vos Jaczn., *C&F*, p. 21. It is beside the point here whether Scotus properly interpreted Aristotle or not. For a thorough discussion of differing interpretations of Aristotle, see William Lane Craig, *The Problem of Divine Foreknowledge and Future Contingents from Aristotle to Suarez* (Leiden: E. J. Brill, 1988), chapter 1; hereafter *PDF*.

[5] My discussion in this section relies upon Vos Jaczn., *C&F*, pp. 23-38.

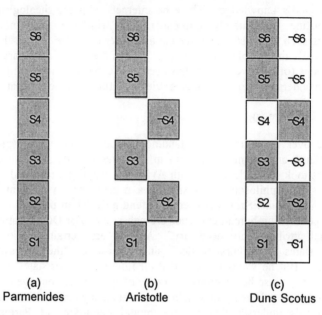

(a) Parmenidean necessitarianism; (b) Aristotelian diachronic contingency; (c) Scotist synchronic contingency.

Figure 7.1 Differing views of contingency

Aristotle departed from Parmenides' radical necessitarianism, offering an ontology in which change and contingency were real features of the world. However, the Aristotelian theory of contingency held that a state of affairs S is contingent only if a different state of affairs ¬S is possible at a different moment. In one of Aristotle's more famous pronouncements, he declares, 'What is, necessarily is, when it is; and what is not, necessarily is not, when it is not.'[6] Thus since ¬S is not possible at the same moment at which S is the case, S is necessary at that moment.[7] The Aristotelian conception is depicted graphically in Figure 7.1 (b), where S is a mutable state of affairs. It can be seen that this model is nothing more than an ontology that permits change through time; it is not a model that allows contingency in the modern sense of the word. Aristotelian contingency, in contrast to Scotan (or modern) contingency, may be termed diachronic contingency.

Formally, an Aristotelian modal system would hold the following, where (1) through (3) are identical to the Parmenidean system:

---

[6] Aristotle, *De interpretatione*, 9, 19a23-24.
[7] See (3) below. Duns Scotus explicitly considers this Aristotelian aphorism as an objection to his view in *Lectura* I 39.55 and refutes it in §58.

1. $\sim\Diamond$ (S1 & $\neg$S1)
2. $\Diamond$ S1$\rightarrow$$\sim$$\Diamond$$\neg$S1 $\equiv$ $\Diamond$ S1$\rightarrow$$\Box$ S1
3. S1$\rightarrow$$\sim$$\Diamond$$\neg$S1 $\equiv$ S1$\rightarrow$ $\Box$ S1
5. $\Diamond$ (S1 & $\neg$S2)

Duns Scotus was the first clearly to develop a theory of genuine contingency in which a state of affairs S is contingent only if $\neg$S is possible at the same time, hence, synchronic contingency. This is represented graphically in Figure 7.1 (c), where S is a mutable state of affairs. Formally, a Scotist modal system would hold the following, where only (1) is held in common with Parmenides, and (1) and (5) in common with Aristotle:

1. $\sim\Diamond$ (S1 & $\neg$S1)
5. $\Diamond$ (S1 & $\neg$S2)
6. $\Diamond$ S1$\rightarrow$ $\Diamond$ $\neg$S1 $\equiv$ $\Diamond$ S1$\rightarrow$$\sim$$\Box$ S1
7. S1$\rightarrow$ $\Diamond$$\neg$S1 $\equiv$ S1$\rightarrow$$\sim$$\Box$ S1

But textual support must be furnished for these conclusions. Duns Scotus introduces a useful distinction in discussing diachronic contingency:

> One kind of contingency and possibility is that the will is successively related to opposite objects, and this possibility and contingency results from its mutability. And according to this possibility a distinction is mare regarding possible propositions which are composed of contrary and opposite terms, such as *Something white can be black*. And according to the divided sense the proposition is true, as far as terms are understood to have a possibility at different times, such as *Something white at a can be black at b*.[8]

That Duns Scotus is discussing diachronic contingency is clear; it is also clear that by 'divided sense' he means a proposition understood as a conjunction of two further propositions. In the 'composite sense' his example formalized would be

8. (x) $\Diamond$ (Wx$_{t1}$ & $\sim$Wx$_{t1}$)

where Wx means 'x is white' and t_1 and t_2 are temporal indexicals corresponding to his a and b. This is clearly false within an Aristotelian system of diachronic contingency in which (3) holds. But in a 'divided sense' the possibility operator is in a different place and modifies a different expression:

9. (x) (Wx$_{t1}$ & $\Diamond$$\sim$Wx$_{t2}$)

This distinction enables Duns Scotus to attack the Aristotelian axiom expressed by (3):

---

[8] Duns Scotus, *Lectura*, I 39.48.

We must say that with respect to this proposition:
Everything which is, when it is, is necessary [Aristotle, *De int.* 9]
we must make the distinction between the composite and the divided sense. . . . In the present case in the composite sense the proposition is true, denoting the necessity of concomitance; then the meaning is:
Everything which is when it is, is necessary
denoting that
Everything is when it is
is necessary. In the divided sense the proposition is false and the necessity of the concomitant is denoted as follows:
Everything which is, when it is, is necessary.
Then the meaning is:
Everything which is, is necessary, when it is,
and this is false, because something contingent is not necessary when it is.[9]

By 'necessity of concomitance' Duns Scotus means both the logical and the temporal necessity of the consequence.[10] Thus a temporal indexical is needed, so his interpretation of Aristotle would be

10. $(x)(x_{t1} \rightarrow \Box\, x_{t1}) \equiv (x)(x_{t1} \rightarrow \sim\Diamond\sim x_{t1})$

But Duns Scotus has argued that S_{t1} & $\Diamond\sim S_{t1}$ (see (7) above). Hence (10) is false. Thus:

11. $(x)(x_{t1} \rightarrow \Diamond\sim x_{t1}) \equiv (x)(x_{t1} \rightarrow x_{t1})$

The Nature of Time

What of significance follows from the theory of synchronic contingency? Given that Duns Scotus' interpretation of contingency must be temporally indexed, we may conclude the following. Aristotle's modal system is not incompatible with a dynamic view of time, but only if bivalence is denied for future contingent propositions (that is, the indeterminacy of future contingent states of affairs creates a truth-value gap). Aristotle's system is, however, equally compatible with static time in which the future states of affairs already exist, thus making (10) true.

Duns Scotus' system, on the other hand, is incompatible with static time, for (11), the time-indexed version of (7), will be false on a static theory of time. By insisting that what obtains in the present is not necessary, Duns Scotus does not, of course, mean that both a state of affairs and its complement could simultaneously obtain. Rather, he replaces the received doctrine of the necessity of the present with the claim that there is no *repugnantia terminorum*, or semantic inconsistency, in the possibility that the complement of the actual state of affairs should obtain.[11] At the same time, Duns Scotus affirmed the necessity *per*

---

[9] Ibid., I 39.58.
[10] Vos Jaczn., *C&F*, p. 135.
[11] Normore, 'Future Contingents,' pp. 368-9.

accidens of the past.[12] So his discovery of synchronic contingency entails the first unequivocal historical expression of dynamic time and temporal contingency, and thus of true temporal becoming.

God and Time

Anyone sympathetic to a temporalist construal of God's being might hope that Duns Scotus' clear articulation of dynamic time and temporal contingency would lead him to rethink the traditional, timeless view of God. But he does not do so. Perhaps this is because he, like his predecessors in the Scholastic tradition, believed firmly in the doctrine of divine simplicity (and, as a corollary, strong immutability).[13] But this entails that there are no divisions in God's being, so a fortiori there are no temporal divisions. Thus he speaks of the one single moment of eternity in which there is no succession.[14]

The entire argument of *Lectura* 39 is to reject necessitarianism, or the view that God's immutable foreknowledge necessitates the future. Duns Scotus' doctrine of synchronic contingency is a major part of that argument, as is his rejection of the classical assumption that immutability entails necessity. Duns Scotus holds that necessity entails immutability, but argues that God's knowledge of contingents, although immutable, is itself contingent.[15] The logical apparatus and technical distinctions that he deploys in *Lectura* 39 are impressive indeed, but cannot concern us here. Rather, we will conclude our consideration of Duns Scotus by looking at his explication of God's eternal knowledge of temporally contingent events as illustrative of how he conceives of God's relation to time.

Duns Scotus cites, and rejects, two theories of how God can know future contingents. The first theory, that of Bonaventure and of Augustine, is that ideas in God's mind represent contingent things along with all their essential and accidental

---

[12] The concept of necessity *per accidens*, or accidental necessity, had been articulated in the work of the thirteenth-century logician William of Sherwood; it will play a large role in William Ockham's solution to the problem of foreknowledge and freedom. See Linda Trinkaus Zagzebski, *The Dilemma of Freedom and Foreknowledge* (New York: Oxford University Press, 1991), pp. 15-17, hereafter cited as *DFF*.

One way of characterizing accidental necessity is as follows: A proposition p is accidentally necessary at time t in possible world w iff
   (i) p is metaphysically contingent, and
   (ii) p is true at t and at every moment after t in every possible world w^* such that w^* shares the same history with w at t.

A proposition p is metaphysically necessary iff p is true at every time in every possible world; metaphysically impossible iff it is true at no time in no possible world; and metaphysically contingent iff it is neither metaphysically necessary nor metaphysically impossible. See Alfred J. Freddoso, 'The Necessity of Nature,' *Midwest Studies in Philosophy* 11 (1986), p. 219.

[13] John Duns Scotus, *A Treatise on God as First Principle*, 2nd ed., tr. and ed. Allan B. Wolter (Chicago: Franciscan Herald Press, 1966), 4.1-11; 4.75-86.

[14] For example, Duns Scotus, *Lectura*, I.39.54.

[15] Ibid., I 39.77, which explicitly refers to Aristotle, *Metaphysics* Δ (1015a20-b15).

properties and relations, including causal relations.[16] But Duns Scotus rejects this theory, claiming that simple ideas cannot represent all of reality, particularly the aspect of (synchronic) contingency. So an idea of a future contingent that will become actual is no different from an idea of a future contingent that is merely possible and will never become actual.

The second theory Duns Scotus cites is that of Aquinas, according to which everything is present to God in eternity.[17] In his response, Duns Scotus' dynamic view of time is clear as he rejects two essential commitments of the Thomist solution. First, he says that 'the current of time' is 'infinitely flowing,'[18] with the consequence that the future cannot be present to God in the way in which the present is. In support of this claim he offers an analogy with space. Despite the affirmation of God's 'immensity' or omnipresence, it is still true that God cannot be present to a spatial location that exists merely potentially. In the same way God cannot be present to a time that exists merely potentially. 'I grant that the immensity is present to every place, but not to every actual and potential place. . . . Thus, eternity will not, by reason of its infinity, be present to any non-existent time either.'[19]

Duns Scotus accepts the principle that

12. If x is present to y, then both x and y exist.

In other words, copresence is a real dyadic relation that can only hold between relata that exist simultaneously. Specifically, in terms of the spatial analogy, Duns Scotus holds that

13. If spatial location x is present to God, then both x and God exist.

But of course God exists, and (in the great Anselmian tradition) God exists necessarily. So the conjunct involving God in the consequent will always be true. Thus, taking the contrapositive, Duns Scotus maintains that

14. If spatial location x does not exist, then x is not present to God.

This argument Duns Scotus takes as unassailable, given the truth of (12). Applying the analogy, then, to time, he claims that from (12) it follows that

15. If temporal location x is present to God, then both x and God exist.
16. If temporal location x does not exist, then x is not present to God.

But the future does not exist, as Duns Scotus has already argued. Hence

---

[16] Ibid., I 39.18-22.
[17] Ibid., I 39.23-30.
[18] Ibid., I 39.27.
[19] Duns Scotus, *Ordinatio*, 1.38-39.8, cited in Craig, *PDF*, p. 130.

17. The future is not present to God.

Clearly the verbs in the argument must be taken as tensed and not tenseless. But in that case, the argument implies that God is temporal. Duns Scotus does not draw that conclusion in so many words, but it would seem that the inference is certainly valid.

At least two further points of Duns Scotus' rejection of Aquinas' solution deserve comment. He argues that things that cannot coexist with each other cannot coexist with a third thing either. Thus, since past times and future times cannot coexist with each other, they cannot coexist with eternity, and so cannot coexist with God from eternity. Further, if things that exist successively in time coexist in eternity, then absurdities follow. For it is possible for contradictories to exist successively in time, so that the proposition 'Adam exists' is true at t_1 and 'Adam does not exist' is true at t_2. But if the two coexist in eternity, then both would be true simultaneously, which is absurd. In these points, his commitment to 'flowing time' is clear, as is at least a hint of God's temporality.

Second, Duns Scotus objects to the Thomist view that though God's initial act of causation is necessary, the introduction of secondary causes brings in contingency. On this view there could be no true synchronic contingency, since given an initial necessary cause, all subsequent effects are determined.

Having rejected two well-known Scholastic theories of how God can have infallible knowledge of future contingents, Duns Scotus embarks on his own solution to the problem. Greatly simplified, it is this. A future contingent can be represented as a 'neutral proposition,' either p or not-p. Since nothing happens outside of God's will, it is the divine will that decided for one component, say, p. But the ground of God's knowledge of that decision cannot be p itself. This is because, from the doctrine of divine simplicity follows that of divine impassibility, according to which nothing created can causally affect God, and hence p cannot be the ground of God's knowledge of p. Nor can the ground of God's knowledge be the divine will itself, for then God's knowledge would be 'vilified' or 'devalued' by changing from the knowledge of the disjunction of p or not-p to the knowledge of one of the disjuncts, p. The ground of God's knowledge, therefore, must be God's essence: 'So the fact that God knows every true proposition has its ground in his essence, although the fact that this proposition received the truth-value 'true' and not its opposite—is determined by his will.'[20]

Whether or not this solution to the problem is successful is a question I will not pursue.[21] But I note that, in rejecting the two commitments of the Thomist solution, Duns Scotus has allowed for a conception of God's relation to time that is not so clearly committed to static time. Granted, he does not explicitly draw this conclusion. And from the fact that Distinction 39 is absent from his later

---

[20] Vos Jaczn., *C&F*, p. 147 (commentary on *Lectura* I 39.65).
[21] See the discussion in Craig, *PDF*, pp. 137-9.

Ordinatio,[22] Cross has inferred that he changed his mind about God's temporality, but did not yet have the details worked out.[23]

Summary

With his doctrine of synchronic contingency, Duns Scotus marks a major advance in the philosophy of time, offering a clear, unequivocal theory of dynamic time and temporal becoming. His views on time and contingency would have a profound impact on discussions of God's foreknowledge through the sixteenth century. However, in his philosophical theology, Duns Scotus did not—in fact, given the inertia of tradition, probably could not—move dramatically outside of the medieval consensus of God's timelessness.

William of Ockham

William of Ockham (1285-1347) offers a significant departure from the medieval consensus. Building on Duns Scotus' concept of contingency and dynamic view of time, Ockham proposes a solution to the problem of foreknowledge and free will that has attracted considerable interest in recent years.[24] More significant for the present study is Ockham's view that God's existence is temporally everlasting rather than timelessly eternal. In this, Ockham follows the logic of Duns Scotus' position beyond where even Duns Scotus himself went, and in so doing he apparently gives us the earliest sketch of a theory of divine temporality. Ockham's views emerge most clearly in his *Tractatus de praedestinatione et de praescientia Dei et de futuris contingentibus*,[25] as well as in his commentaries on Aristotle's writings.

---

[22] See above, footnote 3.

[23] Richard Cross, 'Duns Scotus on Eternity and Timelessness,' *Faith and Philosophy* 14 (1997), pp. 3-25. I find this suggestion plausible, although I'm not convinced by Cross's further claim that Scotus held throughout to a B-theory of time. Since he held firmly to his theory of synchronic contingency laid out in the *Lectura*, and since that theory is clearly linked to dynamic time, I believe that he remained an A-theorist. His inability to resolve the incompatibility of dynamic time and a timeless God could well have been the reason why he left Distinction 39 out of the *Ordinatio*.

[24] Marilyn McCord Adams, 'The Problem of God's Foreknowledge and Free Will in Boethius and William Ockham' (unpublished Ph.D. dissertation, Cornell University, 1967), and *William Ockham*, 2 vols (Notre Dame, IN: University of Notre Dame Press, 1987); Alfred J. Freddoso, 'Accidental Necessity and Logical Determinism,' *Journal of Philosophy* 80 (1983), pp. 257-78; Alvin Plantinga, 'On Ockham's Way Out,' *Faith and Philosophy* 3 (1986), pp. 235-69; William Lane Craig, 'William Ockham on Divine Foreknowledge and Future Contingency,' *Pacific Philosophical Quarterly* 69 (1988), pp. 117-35; Zagzebski, *DFF*, chapter 3: 'The Ockhamist Solution,' pp. 66-97.

[25] William of Ockham, *Predestination, God's Foreknowledge, and Future Contingents*, 2nd ed., tr. Marilyn Adams and Norman Kretzmann (Indianapolis: Hackett Publishing, 1983), hereafter *Treatise*.

The Nature of Time

William Ockham is known as the father of nominalism (although today his views might better be classified as a form of conceptualism), and much detail in his arguments is directed at attempts to reify qualities or posit universals existing independently of particulars.

Ockham on God and Time

- Adopted Scotus's view of synchronic contingency and dynamic time.
- A relational view of time follows from Ockham's nominalism.
- Ockham's solution to the fatalist dilemma has the consequence that God is temporal—probably the first departure from the medieval consensus.

From Aristotle Ockham took the definition that time is the 'number of motion with respect to before and after.'[26] With his nominalist understanding of numbers, where numbers must be the count or measure of some concrete thing, Ockham takes Aristotle to mean that time is the measure whereby we determine the duration of motion and, by extension, the duration also of rest and of the existence of any object subject to generation and corruption. Hence time is not a thing itself:

> Quantity, body, surface, line and place are not distinct really from substance and quality; because every body and surface and line and place is a substance or a quality... therefore, for the same reason, time is not really distinct from a substance or a quality; but no substance or quality is a thing distinct from permanent things; therefore, etc. Further, it has been shown that motion is not a thing distinct from permanent things; therefore, and for the same reasons, neither is time. And again, everything can be explained without such an absolute thing.[27]

Ockham accepts the principle, common among the medievals, that if there were no motion there would be no time. 'What Ockham is driving at, when he defines time as "the measure of all things whose duration can be certified by the intellect by means of something else better known to it," is that time is a certain non-spatial dimension of motion.'[28]

The thought goes like this: Ockham distinguishes two kinds of quantity, discrete and continuous. The former are numbered by a measure intrinsic to the things themselves. For example, a finite collection of numerically distinct individuals is numbered simply by counting the discrete individuals. On the other hand, the measure of a continuous quantity must be numbered using something distinct from the quantity itself. For example, a length of cloth is measured by an

---

[26] Aristotle, *The Physics*, Loeb Classical Library, tr. Philip H. Wicksteed and Francis M. Comfort (New York: G. P. Putnam's Sons, 1929), IV, 11.219b1.
[27] *Philosophia Naturalis*, IV, 2, tr. Herman Shapiro, in *Motion, Time and Place According to William Ockham* (New York: The Franciscan Institute, 1957), p. 95, hereafter *Motion*.
[28] Shapiro, *Motion*, p. 97; the quotation is from Ockham, *Phil. Nat.* IV, 3.

ell, or yardstick, which is a number extrinsic to the cloth. Since motion is continuous, no intrinsic measure is available, so an extrinsic measure—time—must be used.

What is this time, which is an extrinsic measure of motion? Ockham claims that there is one motion that is uniform, perfect, and most natural for the entire universe, the 'prime motion'—that of the heavenly sphere which moves circularly with maximum speed. (He acknowledges that, for practical purposes, the diurnal motion of the sun gives us a perfectly adequate measure.) But there seems to be circularity involved here. Motion is measured by time, but time is measured by a particular motion. So is time motion, and motion time? Ockham considers Aristotle's arguments against this equivalence, and finds them not generally valid when read literally, as employed against his own conclusions by his opponents. His argument hangs on a distinction he draws elsewhere between absolute and connotative terms. While the 'absolute' or real definition of time and motion are the same, their 'connotative' or nominal definitions differ; 'time' includes a mental aspect that motion does not:

> It is obvious from what has been said, that time is not something other than the prime motion, because through the prime motion. . . we know for how long temporal things endure, move, or rest. And still 'time,' and 'prime motion,' differ in definition, since 'time' imports a soul which is certified, and other things of which it is certified.[29]

Prime motion is one thing; but it takes a mind to recognize divisible parts in the prime motion that serves as the measure of motion, rest, or duration of physical objects. And this, Ockham asserts, is just what Aristotle himself really meant.

In a reply to one of the putative Aristotelian objections to his view, Ockham makes a significant point. The instant called 'now' which divides past from future refers to the part of prime motion that is 'here.' Since prime motion is uniform, successive parts of it will succeed one another uniformly, and so instants of time succeed one another uniformly. Thus it is incorrect to conceive of 'now' somehow as the generation of a new instant and the destruction of the past instant, as Ockham's opponents did. Temporal becoming (to use the modern phrase) is to be understood as the successive changes in position of prime motion.[30]

Much of Ockham's discussion of the nature of time strikes a contemporary philosopher of time as arcane and off the mark. Clearly the medieval discussion was shaped by concerns quite different from those that interest philosophers today. Nevertheless, as Ockham takes insights from both Aristotle and from Duns Scotus and develops his views on time, certain significant features may be seen. The first is a clearly dynamic theory of time. Temporal becoming is objectively real for Ockham, and there is an ontological asymmetry between past and future. A second feature is Ockham's refusal to hypostatize time. He would

---

[29] Ockham, *Phil. Nat.* IV, 7, tr. Shapiro, *Motion*, p. 99; see also Adams, *William Ockham*, vol. 2, pp. 876-80.
[30] Shapiro, *Motion*, p. 107.

have thus not only opposed Newton's concept of absolute time, but also, one presumes, the incorporation of time into a four-dimensional space-time. Third, in partially identifying time with motion, Ockham accomplishes two things. He underlines the mental aspect of the measurement of time, yet offers an objective ground for the flow of time. While no contemporary supporter of the dynamic theory of time would regard motion as the objective ground of time's flow, in his discussion of motion Ockham refers to his discussions of causation[31] and so may be read as anticipating, in some way, a causal theory of time.

God and Time

William Lane Craig claims:

> For Ockham the relationship between God's foreknowledge and future contingents was a literally conceived concern, for he held that God's eternity was not a state of timelessness, but that God, though immutable, endures throughout all past, present, and future time, which arises from the order of succession among changing things.[32]

Similarly, Zagzebski regards a solution to the foreknowledge/freedom dilemma as 'Ockhamist' if one of its features is that it 'assumes God exists in time.'[33] Unfortunately it is impossible to find as clear a statement in Ockham as these two authors imply. Nevertheless, I believe that their conclusions are justified inferences from his treatment of the relation between predestination, foreknowledge and future contingents in the *Treatise*.

Ockham brings his characteristic genius to the problem of foreknowledge and freedom. He rejects the solutions of Augustine, Aquinas and Duns Scotus before offering his solution, which involves denying the claim that all past-tense propositions are necessary *per accidens*. Some past tense propositions are really about the future, he claims, and consequently do not share the necessity of the past. Ockham believes that a proposition is true or false depending on whether the corresponding state of affairs obtains in the world. But because states of affairs are instantiated or, in Ockham's terms, 'posited in reality,' successively, a state of affairs will remain contingent until it is actually instantiated. Consequentially, any proposition about states of affairs in the world should contain a temporal indexical. If S is a state of affairs, one should say, 'S obtains at t_n.' Craig explains:

> So if it is now t_1 and S obtains at t_4, then the proposition *p* describing *S* is now determinately true, since its corresponding state of affairs obtains (tenselessly); but

---

[31] Ockham's views both of causation and of motion are highly complex and controversial. Shapiro, *Motion*, chapters 2 and 3, sees Ockham as the prototype of modern empiricists, whereas Adams, *William Ockham*, vol. 2, chapters 18, 'Efficient Causality,' and 19, 'Motion: Its Ontological Status and Its Causes,' argues that such readings of Ockham are mistaken.
[32] Craig, *PDF*, p. 146.
[33] Zagzebski, *DFF*, p. 66.

until $t_4 p$ is contingent. At t_4 and thereafter, once S has been instantiated, it is no longer contingent whether S obtains, and so p is necessary. . . . Prior to t_4 it is genuinely possible that S not be instantiated, even though it will be; thereafter, it is impossible. . . . [Ockham] means that prior to t_4, p is possibly true or possible false, but that at t_4 and afterward p is necessarily true or necessarily false. . . . Hence, for Ockham all future-tense propositions, whether causally necessary or causally contingent, are not necessary *per accidens*, though all are determinately true or false.[34]

Next, Ockham draws the distinction between genuine past and present tense propositions and those that are only verbally past or present tense:

> That proposition that is about the present in such a way that it is nevertheless equivalent to one about the future and its truth depends on the truth of the one about the future does not have [corresponding to it] a necessary proposition about the past. On the contrary, the one about the past is contingent, just as its [corresponding proposition] about the present. . . they are verbally (*vocaliter*) about the present or the past.[35]

While Ockham does not give clear criteria for distinguishing genuinely past- or present-tense propositions from merely verbally past or present ones, it seems clear enough that when the state of affairs to which a proposition refers is really a future state, then the proposition itself is only verbally past or present. Ockham's favorite example in the *Treatise* is 'Peter is predestinated.' Ockham does not believe that 'being predestinated' is a real property of an individual, but instead refers to the future state of affairs of God granting eternal blessedness to Peter. 'Therefore, "Peter was predestinate" is contingent just as is "Peter is predestinate".'[36] With this distinction, Ockham answers an objection posed by his necessitarian opponent:

> [Objection] 4. The proposition 'God predestinated Peter' was true from eternity. Therefore it cannot be false. Therefore it is necessary.
> [Reply.] I reject the inference, since many propositions were true from eternity that are false now. For example, 'the world does not exist' was true from eternity and nevertheless is false now. Thus I maintain that even if 'God predestinated Peter' will have been true from eternity, it can nevertheless be false and it can fail ever to have been true.[37]

All of this leads Ockham to his articulation of what has been called 'counterfactual power over the past,' or 'soft facts about the past.' He writes:

> [Objection 5a.] On the contrary: every proposition that is true now and can be false can change from truth to falsity. But suppose that the proposition 'Peter is predestinate' is true now and can be false (as is consistent); therefore, ['Peter is predestinate' can change from truth to falsity.]

---

[34] Craig, *PDF*, p. 150.
[35] Ockham, *Treatise,* 1.C. Bracketed phrases are the translators' clarifications.
[36] Ibid.
[37] Ibid., 1.E.

> [Reply.] I maintain that the major premise is false, since more is required—i.e., that the proposition that will be false or will be capable of being false was true at some time. Therefore, although the proposition 'Peter is predestinate' is true now and can be false, nevertheless when it will be false, it will be true to say that it never was true. Therefore it cannot change from truth to falsity.[38]

Ockham affirms that God's knowledge of a future contingent proposition is immutable, but that it is not necessary. Thus for any future contingent proposition p, 'God knows that p' is true only if p. But since p is temporally contingent up to the time of its instantiation, 'God knows that p' is also temporally contingent, even though immutable. Thus it is possible that, should ~p be true, it will have been the case that 'God knows that p' will always have been false. Freddoso writes:

> Ockham's view is precisely that even though the divine cognitive act remains intrinsically immutable, its objects change over time, with the (alleged) result that propositions ascribing knowledge to God derive their modal status from the modal status of the propositional objects of that knowledge.[39]

Since it is not of immediate concern here, I will not try to untangle the issue of just how God has knowledge of future contingents. Indeed, Ockham himself is rather unsettled:

> Assumption 6. It must be held beyond question that God knows with certainty all future contingents—i.e., He knows with certainty what part of the contradiction is true and which false. (Nevertheless all such propositions as 'God knows that this part—or that part—of the contradiction is true' are contingent, not necessary, as has been said before.)
> It is difficult, however, to see how He knows this [with certainty], since one part of the contradiction is no more determined to truth than the other.[40]

Regardless of the merits or demerits of 'Ockham's way out,' or of the hard facts/soft facts distinction, or of the principle of counterfactual power over the past,[41] the salient aspects of Ockham's underlying notion of God's relation to time

---

[38] Ibid., 1.F.
[39] Alfred J. Freddoso, 'St Thomas's Rejection of Ockham's Way Out,' unpublished paper, <http:// www.nd.edu/~afreddos/papers/futcon.htm>, p. 5.
[40] Ockham, *Treatise*, 1, Assumption 6.
[41] See Zagzebski, *DFF*, pp. 66-85 for a thorough discussion. I should note that some interpreters of Ockham take his notion of power over the past in a much stronger sense. Molina, for example, takes Ockham as arguing that the divided sense of a proposition, 'Peter's sin, which is foreknown by God, is able not to occur,' is true for two reasons: (i) If Peter were in fact not to sin, then his sin would never have been foreknown by God; and (ii) if Peter in fact does not sin, then God will *at that moment* bring it about that from eternity he foreknew that Peter would not sin. Molina accepts (i) but rejects (ii) on the grounds that (a) this would involve a change in God, and (b) there is no such power over the paSt Hence Molina takes the notion of power over the past as real and not merely counterfactual. It appears that what Molina really objected to was the idea of backwards causation. Luis de

can be seen in this discussion. Ockham believes that time is dynamic, that future events are genuinely contingent, and that he cannot specify how God knows future contingents. By rejecting the solutions of Augustine, Aquinas and Duns Scotus, which are possibly consistent with a timeless God, and by refusing to specify the way in which God does know future contingents, it may be concluded that Ockham is not attracted to solutions to the foreknowledge/freedom dilemma that involve an atemporal God. Before S is posited in reality, it is possible that ¬S will be posited. Therefore, either p or ~p is possibly true. But after S is posited in reality, then p is true and ~p is false. But Ockham believes that prior to S's being posited in reality, S is a contingent state of affairs and p is a contingent proposition, even for God! As Freddoso noted, the modal status of God's knowledge depends upon the modal status of the objects of his knowledge. This can only be the case if God himself experiences temporal becoming. There is a definite instant in God's being when S ceases to be potential and becomes actual, and when p ceases to be contingent and becomes necessary *per accidens*.

Summary

I believe it is clear enough that Ockham ventured further than his predecessors in articulating a doctrine of divine temporality. Whether or not he avoided making this conclusion explicit because it would be judged radical, perhaps even heretical, I cannot say. Nevertheless, it seems that those who regard divine temporality to be a critical part of Ockham's solution to the problem of foreknowledge and freedom are justified in that assessment.

Later Medieval Developments

The views articulated by Duns Scotus and Ockham apparently caused no small stir in Oxford, and even in the Church hierarchy. Ockham himself was forced to appear before the papal court at Avignon in 1324 to answer charges of heresy; in the end 51 propositions drawn from his work were found open to censure, and although Ockham was not formally condemned, he later found himself in the forefront of a dispute between his Franciscan order and the pope. This dispute led to Ockham's excommunication and flight from France to Italy and ultimately Germany, where he died. Just how much of Ockham's political troubles with the Church were due to his innovative theology, and just how much his theology was opposed because of his political troubles, is quite difficult to determine from the perspective of nearly seven centuries.

Thomas Bradwardine (1295-1349) reacted to both the Scotist and Ockhamist positions on temporal contingency. Essentially he held that if God was ever able to bring it about that Adam would not exist, then, since God's power was immutable, he must always have the power to bring it about that Adam never

---

Molina, *On Divine Foreknowledge (Part IV of the* Concordia*)*, tr. Alfred J. Freddoso (Ithaca, NY: Cornell University Press, 1988), IV.51.18, hereafter *Concordia*.

existed—in other words, God has the power to change the past as well as the future. Nevertheless, Bradwardine claimed, since past, present and future were equally dependent on the free will of God, then because God's will is prior to creation, the past, present and future were equally necessary. The whole history of the world is determined; God predestinates everything. In describing God's relation to time Bradwardine rehabilitates the Boethian image of a circle with God at the center.[42]

Bradwardine's reaction to Duns Scotus and Ockham seems to be a great step backwards in both the philosophy of time and in finding a solution to the problem of God's foreknowledge and human freedom. The dispute between the followers of Bradwardine and followers of Ockham spilled out from Oxford to Paris in the fourteenth century, with such notables as John Buridan on one side and Gregory of Rimini on the other. The fourteenth century closed and the fifteenth century passed with no notable advances in this dispute, and with one significant aspect of the dispute left virtually untouched—the question of how God could have knowledge of future contingents. As noted above, Ockham (and Gregory of Rimini as well) were content to say that we just do not know how God knows future contingents, but must simply accept by faith that knowledge of all truths is part of God's essential nature and consequently he does know future contingents, although this fact does not entail them becoming necessary in any sense.

The crux of the problem is the role causality plays in knowledge. The common view was that our knowledge is at least indirectly dependent upon being causally affected by the objects of our knowledge. But since future contingents do not exist, on the dynamic theory of time articulated by Duns Scotus and accepted by Ockham, Rimini, and their followers, they cannot cause any knowledge. And if it is claimed that future contingents can be known in virtue of their past or present causes, then their contingency is seriously undermined. A solution to this problem, in the form of a non-causal account of God's knowledge, did not come until the late sixteenth century.[43]

Luis de Molina

Luis de Molina (1535-1600) was a leading figure in a remarkable sixteenth century revival of Scholasticism in Spain, partly as a response to the Council of Trent (1545-63) and partly in reaction to the writings of Protestant theologians. Molina and his fellow Jesuit and contemporary Francisco Suarez (better known for his writings on the philosophy of law) developed and defended the doctrine of Middle Knowledge. Molina's doctrine of Middle Knowledge, and Alvin Plantinga's

---

[42] Normore, 'Future Contingents,' pp. 374-5.

[43] Interestingly, the great seventeenth century Rationalists all held to an atemporal view of God's existence, while the great British Empiricists took the opposing line. I shall not, however, trace the history of the debate through the period of Modern Philosophy, as very little in the way of new arguments was offered until the twentieth century.

independent rediscovery of it in the 1970s, have provoked a wide literature.[44] Although Molina's and Suarez's doctrine of Middle Knowledge promises to offer fascinating solutions to both the foreknowledge/freedom problem and the problem of evil, we shall be able to pay only superficial attention to the theory in this chapter.

Molina's Contribution: Middle Knowledge

Although Molina held to divine atemporality, he was a strong advocate of dynamic time and a libertarian account of human freedom. His remarkable contribution was his theory of Middle Knowledge (MK). Here is a rough outline of how MK is currently understood:
1. In the First Moment (Natural Knowledge), God knows all necessary truths (i.e. logical and metaphysical necessities), and thus what worlds are possible.
2. In the Second Moment (Middle Knowledge), God knows counterfactually what would be the case if he were to actualize any of the possible worlds.
3. Logically subsequent to his MK, God freely decides to actualize one world (creation)
4. In the Third Moment (Free Knowledge), God freely knows what his own causal contributions will be to the world he has actualized (providence).

What MK gains is a non-causal account of God's knowledge that retains his complete foreknowledge of future contingents. The salient difficulty with MK is explaining how the truth values of the items of MK, 'counterfactuals of freedom,' are grounded.

Molina's Theory of Time and Eternity

It is clear that Molina wants to uphold the dynamic theory of time developed by his predecessors such as Duns Scotus and Ockham. This is evident, for example, in his spirited defense of a strong libertarian or contra-causal concept of freedom, which partially grounds his discussion of the source of contingency in the world.[45] But it is in his discussion of Duns Scotus' refutation of Aquinas' theory of how temporal events are present to God in eternity that his insistence on dynamic time is most clear.

Disputations 48 and 49 of the *Concordia* are central to Molina's discussion. The title of Disputation 48 is, 'Whether All the Things That Exist, Have Existed, and Will Exist in Time Are Present to God from Eternity with Their

---

[44] Molina, *Concordia*; Alvin Plantinga, *The Nature of Necessity* (Oxford: Clarendon Press, 1974). Among the more significant discussions of the issue are Robert M. Adams, 'Middle Knowledge and the Problem of Evil,' *American Philosophical Quarterly* 14 (1977), pp. 109-17; Alfred J. Freddoso, 'Introduction' to *Concordia*, hereafter cited as 'Introduction'; Edward R. Wierenga, *The Nature of God: An Inquiry into Divine Attributes* (Ithaca, NY: Cornell University Press, 1989), chapter 5; William Hasker, *God, Time, and Knowledge* (Ithaca, NY: Cornell University Press, 1989), chapter 2; Zagzebski, *DFF*, chapter 5; Thomas Flint, *Divine Providence: The Molinist Account* (Ithaca, NY: Cornell University Press, 1998), chapters 3-7; William Lane Craig, 'Middle Knowledge, Truth-Makers, and the Grounding Objection,' *Faith and Philosophy* 18 (2001), pp. 337-52.
[45] Molina, *Concordia* IV.47.7-8.

Own Proper Existence.' In his discussion Molina first defends a Boethian conception of eternity:

> Eternity is in itself a certain indivisible duration, a simultaneous whole having as a unit an infinite durational latitude by virtue of which it coexists and corresponds as a whole with each interval and point of time—not unlike the way in which the human soul is wholly in the whole human body and wholly in each of its parts. . . . It follows that the whole of time and whatever exists or successively comes to exist in it coexists with and exists in the indivisible now of eternity, before which there was nothing and after which there is nothing, and in which there is found no before or after and no past or future, but only an indivisible, simultaneously whole duration.[46]

With this definition of eternity, it would seem that Molina must reject Duns Scotus' and Ockham's openness to a temporalist conception of God's being. What is interesting, though, is how Molina keeps insisting on dynamic time. For example, he summarizes Duns Scotus' first objection to Aquinas' theory:

> That which does not exist is not able to coexist with anything, since coexistence requires the existence of both terms. But future things do not exist, nor have they existed. Therefore, they do not coexist from eternity with either God or eternity, and hence they are not present to God from eternity with their actual existence.[47]

His response, in full, is instructive:

> As for the first argument by the opponents, given that the major premise is conceded, the minor premise may be conceded as well if its meaning is that future things do not yet exist in the present time and have not existed in the past. But then the consequence should be denied, if its consequent is understood as saying that future things do not coexist from eternity with either God or eternity *in the very now of eternity* which is consignified by the verb 'coexists.' For even though future things neither exist at the present time nor have existed in the past, they nonetheless exist in the indivisible now of eternity which embraces the future time in which they will exist. On the other hand, if in the consequent the verb 'coexist' signifies an *interval of time*, so that its real meaning is that future things coexist with neither God nor eternity in real time or in the imaginary time that has elapsed from eternity up until the present moment of time, then the consequence should be conceded. For since those things did not exist in that interval of time, they could not have coexisted in that interval with either God or eternity.[48]

In this passage Molina strains to hold a Boethian concept of eternity and dynamic time. For while he agrees with Aquinas that all temporal things are present to God in eternity in their real existence, he also holds that future things, which do not yet exist in time, do not now exist in their real existence in eternity. There is a

---

[46] Ibid., IV.48.2.
[47] Ibid., IV.48.14.
[48] Ibid., IV.48.14.

considerable tension here. In Disputation 49, titled 'Whether Future Contingent Things Are Known by God with Certainty Because They Are Present to Him with Their Own Existence, and Whether the Contingency of These Things Might Thereby Be Reconciled with Divine Foreknowledge,' he explores the tension. Desiring to have Aquinas as an ally, he asserts that St Thomas would, if asked, deny the view commonly attributed to him:

> First of all, despite the arguments just adduced, I would not dare to claim that St. Thomas, whom in all things I sincerely desire to have as a patron instead of an adversary, believed that God knows future contingents with certainty solely on the basis of the presence of things with actual existence. Rather, if he were asked about this issue, he would, I believe, affirm the contrary position.[49]

It is at this point that Molina introduces his idea of Middle Knowledge, which he will expand upon in Disputations 52 and 53. Subsequent passages illustrate his commitment to dynamic time:

> I do not believe, nor is it to be conceded, that (i) the things that come to be in time exist in eternity *before* they exist in time, or that (ii) they are present to God in eternity with their own existence *before* they are actually present in time, or that (iii) it is *because* things exist in eternity that God foreknows future contingents with certainty before they exist in time. This is why the proposition 'From eternity all things coexist with God or are present to God with their own existence outside their causes,' . . . contributes nothing, as I see it, either toward establishing the certitude of the divine foreknowledge concerning future contingents or toward reconciling the contingency of things with divine foreknowledge.[50]

It might seem at first that Molina does not intend 'before' to be taken in a temporal sense, but rather in a logical or conditional sense. But, from the three arguments he adduces for this claim, it is clear that it is in fact a temporal sense that he has in mind. He writes:

> It should not be thought that the things that come to be successively in time exist in eternity before they exist in time. . . . Yet this is what would have had to be true in order for it to be the case that it was because of the existence of things in eternity that God foreknew them with certainty before they came to be in time.
>
> But if this was the claim being made by Boethius, St. Thomas, and the others who affirm on this basis that God foreknows future contingents with certainty, then I frankly confess that I do not understand it, nor do I think that there is any way in which it can be true.[51]

It is indeed difficult to imagine that either Boethius or Aquinas would hold that future things exist now—at the present time—in eternity; perhaps Molina means that this is a claim to which they are logically committed. For if the sole ground of

---

[49] Ibid., IV.49.7.
[50] Ibid., IV.49.15.
[51] Ibid., IV.49.16.

God's knowledge of future contingents is their presence in eternity, then the proposition 'All future things exist in eternity' is now true. But Molina notes that the proposition has two readings. If 'exist' means the eternal present, then the proposition is true but unhelpful, for this merely asserts that nothing will exist in time without existing in reality. The defender of the Boethian/Thomist interpretation needs the stronger claim that 'exist' refers to the temporal present. In other words, all future things exist now in eternity even though they do not exist now in time. But that claim does in fact seem incoherent.

A bit later, Molina analyzes the model, used by Boethius, Aquinas and Duns Scotus, of the relation of the center of a circle to its circumference. His analysis underlines the difference dynamic time makes in the understanding of the model:

> While the circle is being drawn, the center point does not yet correspond to the part still to be drawn, but corresponds only to the part already drawn. And this is not because the center point is lacking something that is required in order for it to correspond to the part still to be drawn; rather, it is because what is lacking is that very part to which the center point, in itself already existing as a whole, would correspond.... In the same way, as long as the whole of time has not yet elapsed, indivisible eternity corresponds not to the whole of time, but to the part that has elapsed.[52]

While Molina's commitment to dynamic time is clear enough, in the spirit of Molina himself I must frankly confess that I cannot understand how dynamic time can correspond to eternity without there being a change in eternity as each moment succeeds its predecessor. Yet there can be little doubt that Molina conceives of the relation of time and eternity as being something close to this.[53] Perhaps it is the instability of this view that leads Molina to opt for his theory of God knowing future contingents in virtue of his ideas of them rather than in virtue of their actual coexistence in eternity. But as to a satisfactory explanation of the relation of time and eternity, Molina leaves the matter unresolved.

God, Time, and Middle Knowledge

Enough has already been said to indicate that Molina regarded time as dynamic and at the same time held that God is atemporal. Indeed, his concept of dynamic time is so robust that he will not allow that any future event be present in any real sense to God or in eternity. Yet Molina is more than willing to use tensed language to speak of God's knowledge and causal power. Freddoso thinks that Molina regards such tensed language as appropriately used of God 'so long as it is

---

[52] Ibid., IV.49.18.
[53] This is how Freddoso understands Molina, yet Freddoso himself offers no explanation of this seemingly confusing claim. See 'Introduction,' pp. 30-36.

carefully dissociated from the temporalist thesis that God's act of understanding and act of willing are properly and adequately measured by time.'[54]

Perhaps Freddoso's observation points a way to a better understanding of Molina's view of God's relation to time. Like all the Aristotelian Scholastics, Molina held that time was metaphysically contingent, dependent on motion. God, of course, is metaphysically necessary. The consequence was that no measure derived from contingent time could be applied to God's eternal duration. Eternity is a duration to which temporal determinations such as 'before' and 'after' cannot be applied.

Molina clearly thought that it was in virtue of his atemporality that God could be temporally omnipresent. Just as no one physical location is more accessible to God than another, so no one temporal location is more accessible to God than another. He is equally present to each. But as the future does not exist, it cannot be present to God in its actual existence. The conclusion, which Molina refrains from drawing explicitly, is that God experiences succession as successive moments of time come into being.

It seems that Molina could have given a good account of God's relation to the created order had he accepted the possibility that God experiences succession. But believing, as he did, that time is contingent and is determined by motion, such a step would have been difficult. He would have had to sever time from motion, and distinguish physical time from metaphysical time. Had he done so, he might have been open to a temporalist conception of God.

Nevertheless, Molina's theory of Middle Knowledge represents a remarkable advance in the understanding of the problem of God's foreknowledge and human freedom, and while discussion of the intricate details must be omitted here, a brief outline of the theory is important.

The earlier medieval discussions of God's knowledge in eternity centered on logical 'moments' of that knowledge. While the details varied somewhat,[55] they generally included, as the first moment, God's natural knowledge, by which God knows all metaphysically necessary states of affairs that would obtain should he create; and, as a second moment, God's free knowledge, by which God knows what he freely wills to do in the created world. God's natural knowledge is pre-volitional while his free knowledge is post-volitional, where the aspect of the divine volition in view is God's free decision to create the actual world.

The question to which Molina addresses himself in Disputations 52 and 53 is whether God's knowledge of future contingents is pre-volitional or post-volitional. He claims that it cannot be post-volitional, since in that case God would already know what causal contributions he would providentially make in the world. But that way of ordering the moments of God's knowledge cannot avoid fatalism, for it cannot retain genuine contingency. God's providence can be combined with indeterminacy in creation only if it is pre-volitional—that is, only if God knows what would result from conditional contingent states of affairs, and

---

[54] Ibid., p. 34.
[55] For an historical account see Craig, *PDF*.

how he would (or would not) act in such cases. But conditional contingents are counterfactual; they have to do with what would happen if. . .

The conditional character of God's knowledge of future contingents can only be maintained, then, if God's knowledge of them comes (conceptually) between his natural knowledge and his free knowledge—in other words, if God has Middle Knowledge:

> By (i) [natural knowledge] He knows which spatio-temporal arrangements of secondary causes are possible and which contingent effects *might* emanate from any such arrangement. By (ii) [Middle Knowledge] He knows which contingent effects *would in fact* emanate from any possible spatio-temporal arrangement of secondary causes. By (iii) [free knowledge] He knows which secondary causes He wills to create and conserve and how He wills to cooperate with them via His intrinsically neutral general concurrence. So given His natural knowledge, His middle knowledge, and His free knowledge of his own causal contribution to the created world, He has free knowledge of all absolute future contingents. That is, He has within Himself the means required for knowing with certainty which contingent effects *will in fact* emanate from the *actual* arrangements of secondary causes.[56]

Here, is Molina's ingenious solution to how God can know future contingents and still retain creature freedom. The Middle Knowledge solution, unlike any other solution, would allow both dynamic time and God's timeless knowledge of future contingents. But Middle Knowledge is also apparently fully compatible with a temporalist account of God's being and, if successful, could provide a rich account of God's knowledge of future contingents to either the temporalist or the atemporalist. It is not part of the present project to evaluate the strengths and weaknesses of Middle Knowledge, although I shall return briefly to the theory in the last chapter.

Summary

Luis de Molina's contribution to philosophical theology is secured by his theory of middle knowledge. It seems that it is this theory that enables him to retain a commitment to dynamic time and also the traditional atemporal conception of God's being. Yet in his free use of tensed language to speak of God's knowing and willing, and in his insistence that the future exists in no sense in eternity, Molina can be seen as pushing slightly beyond Duns Scotus and Ockham in the direction of a temporal God.

Conclusion

The medieval consensus, described in Chapter 5, was that God is timeless, with the corollary, generally unrecognized, that time is tenseless. But as the Middle Ages

---

[56] Freddoso, 'Introduction,' p. 24.

unfolded and progress was made in philosophical theology, a growing commitment to dynamic time may be seen, notably in the writings of Duns Scotus, Ockham and Molina. And in their writings we see, as well, the first adumbrations of a theory of divine temporality.

Chapter 8

Temporality: Contemporary Statements

Many contemporary philosophers have found a temporal conception of God's being, although historically the minority view, much more plausible than its competitor. As was the case with the atemporalist conceptions, differences in detail exist among various philosophers, but the common element is the view that God does experience a succession of states in his being and, subsequent to the creation of a (temporal) world external to himself, God stands in real temporal (A-theoretic) relations to that world. Thus God properly may be spoken of as temporal.

Motivation for adopting a temporalist view springs from two sources: first, several philosophers claim that a personal being such as God cannot be timeless because 'timeless person' is an incoherent concept. Second, many philosophers are convinced that the dynamic theory of time is correct, and offer arguments to the effect that if God stands in some real relations to the dynamic temporal order, then God must be temporal. Representative of the former are Nicholas Wolterstorff and Stephen Davis, while William Hasker, Richard Swinburne, Alan Padgett, and William Lane Craig are representatives of the latter.

In this chapter I shall briefly examine the crux of the arguments of certain of these philosophers. What will emerge are the general motivations for adopting a temporalist view, as well as the general contours of the relation a temporal God sustains to physical time.

The Argument from Personality

The Judeo-Christian tradition (as well as the other great monotheistic religion, Islam) holds that God is a person, albeit an infinite, eternal person. And while the question of how attributes of personhood can justifiably be predicated of an infinite, incorporeal, eternal being has a long history in philosophical theology, the dogmatic theologians and the religious creeds are unanimous in affirming that God is a person.[1] The history of theology has witnessed a continual struggle between the poles of God's transcendence and his immanence, between Tillich's 'Ground of Being' and popular religion's 'God is my friend.' In recent decades a number of theologians have emphasized the need to recover a proper balance between transcendence and immanence, and as a result have re-emphasized God's

---

[1] Christian theologians should say, rather, that God is *personal*, because they affirm that the three members of the Trinity individually are persons.

personhood. This effort can be seen in the work of Karl Barth. Theologian Donald Bloesch observes:

> Barth perceived that in their penchant for philosophical appellations for God, both medieval scholastic theology and Protestant orthodoxy rendered God impersonal, though this was not their intention. To describe God as the most perfect being and to identify his attributes primarily as aseity, immensity, simplicity, immutability, omnipotence and so on is to lose sight of the biblical witness that the true God is one who acts in history, who addresses humanity in its brokenness and despair, enters into human pain and suffering. The immobile and self-sufficient God of the Hellenistic philosophical tradition is a far cry from the God who loves and judges, who gives grace and withholds grace, who agonizes over human sin and seeks to rescue the human creatures from sin. We can and must avail ourselves of philosophical concepts, but we cannot let these concepts rule our thinking.... we should not use biblical images to illustrate and support a philosophical vision, but we may use philosophical concepts to clarify a biblical vision.[2]

In their articulation of God as a person, philosophical theologians have also revised and clarified concepts of God's being, and often these revisions have had the result that God is now perceived as being temporal rather than timeless. I shall not consider the broader question here of attributing personhood to an infinite spiritual being, but shall focus on the role temporality plays in understanding God's personhood.

Nicholas Wolterstorff

In an influential article,[3] Nicholas Wolterstorff argues that God is temporal. Two concepts support his argument: that time is dynamic, and that God is personal. In support of the former, Wolterstorff goes to some lengths to develop his own mechanisms to denote tensed and tenseless language, and argues that although a tenseless proposition expresses the truth conditions for a tensed proposition containing a temporal indexical, the two are not equivalent propositions. He believes that this is an argument for the reality of tense, and thus (although he does not use the terminology I have used) time is dynamic.

Now, I have shown that this argument is not sound (see my discussion in Chapter 2). Nevertheless, I am persuaded that the conclusion of dynamic time is correct. So I shall pass on to Wolterstorff's arguments from God's personhood.

---

[2] Donald G. Bloesch, *God the Almighty*, volume 3 of *Christian Foundations* (Downers Grove, IL: InterVarsity Press, 1995), p. 35.

[3] Nicholas Wolterstorff, 'God Everlasting,' in *Contemporary Philosophy of Religion*, ed. Steven M. Cahn and David Shatz (Oxford: Oxford University Press, 1982), pp. 77-98; originally published in *God and the Good: Essays in Honor of Henry Stob*, ed. Clifton J. Orlebeke and Louis B. Smedes (Grand Rapids, MI: Eerdmans, 1975). See also Wolterstorff's 'Unqualified Divine Temporality,' in *God and Time: Four Views*, ed. Gregory E. Ganssle (Downers Grove, IL: InterVarsity Press, 2001).

> **Divine Temporal Action (Nicholas Wolterstorff)**
>
> Wolterstorff's argument runs like this:
> 1. Time is A-theoretic.
> 2. The Bible describes God's actions in A-theoretic terms.
> 3. Some of those actions are such that the temporality of the event on which God acts infects the action with temporality.
> 4. Therefore, God is temporal.

Wolterstorff takes seriously the biblical claim that God is a God who redeems, who acts, who knows, who remembers, who anticipates. All of these, he argues, involve change, and if God changes, then he cannot be atemporal ('eternal' in Wolterstorff's terminology):

> If we are to accept this picture of God as acting for the renewal of human life, we must conceive of him as everlasting rather than eternal. God the Redeemer cannot be a God eternal. This is so because God the Redeemer is a God who *changes*. And any being which changes is a being among whose states there is temporal succession. . . . [O]*ntologically*, God cannot be a redeeming God without there being changeful variation among his states.[4]

Wolterstorff claims that the biblical writers, in describing God's actions in history, conceive of God as standing in what we have called a temporal A-series, and that God's temporal series is not distinct from that of God's creation. But Wolterstorff acknowledges that it does not immediately follow that God is temporal, since the use of such language by the biblical writers might be figurative. He then considers St Thomas' attempts to reconcile God's eternal acts with the reality of time. Thomas offers an explanation of how God acts in history by claiming that God eternally decides and wills to bring about some effect at a time, and that effect will come about at the time God willed it to come about without any further time-related action on God's part.

Wolterstorff does not find the Thomist theory satisfactory. To rebut it, he argues 'that in the case of certain of God's actions the temporality of the event that God acts on infects his own action with temporality.'[5] As an example, he offers God's acts of knowing: While an eternal (timeless) God could know when an event e occurred, he could not, argues Wolterstorff, know that it was occurring. For no one can know of event e that it is occurring except when it is occurring. So no one can know that e is occurring until e begins. And similarly for e's ending: no one can know of e that it has occurred until it has ended. As these things are infected by the temporality of e, so must be God's act of knowing them:

> He knows what is happening in our history, what has happened, and what will happen. Hence, some of God's actions are themselves temporal events. But

---

[4] Ibid., p. 78.
[5] Ibid., p. 93.

surely the nonoccurrence followed by the occurrence followed by the nonoccurrence of such knowledge constitutes a change on God's time-strand. Accordingly, God is fundamentally noneternal.[6]

The case of God's knowing which was analyzed can be extended to other acts of God, such as remembering and planning. 'But all of such actions are presupposed by, and essential to, the biblical presentation of God as a redeeming God. Hence God as presented by the biblical writers is fundamentally noneternal. He is fundamentally in time.'[7]

How strong is Wolterstorff's argument? Granting his assumption of dynamic time, I think his case has strong prima facie plausibility. However, let us consider his example of God's knowing that an event is occurring. Let p be 'Luther is nailing his theses to the church door' and let q be 'Nixon is announcing his resignation.' Clearly a timeless God could know the tenseless propositions 'p is true on October 31, 1517' and 'q is true on August 9, 1974.' But can an analysis be given whereby a timeless God could know that Luther was nailing when he was nailing or that Nixon was resigning when he was resigning—that is, the tensed propositions? Are these acts inescapably infected with temporality?

One way an atemporalist might respond would be to offer a translation of the tensed proposition into tenseless language, or to state the truth conditions of the tensed proposition in tenseless terms. But as I argued in Chapter 2, it is clear that while the translations so far proposed may retain the truth value of the tensed propositions, they do not contain the same information. The informational content in p differs from p´ in that p says the nailing is happening now. If this is correct, the translation proposal will not undercut Wolterstorff's argument. Whether any other translation proposal will retain all of the A-theoretic content in B-theoretic language is doubtful. But I am not prepared to claim it is impossible, and so this step in Wolterstorff's argument is weak.

But there may be a way to strengthen Wolterstorff's argument. In the example, God's knowing is construed as knowledge of propositions, and if tensed propositions can be translated without any loss of content into tenseless propositions, then there seems no barrier to a timeless God knowing them. Perhaps the same could be said for God's acts of remembering and planning. But if we take another example of God's action, this response may not be available. Wolterstorff frequently speaks of God's act of redeeming. Redemption, according to the biblical writers, is a multifaceted event which includes at least God's forgiving the sinner, changing from an attitude of wrath to one of love towards the sinner, adopting the sinner as his child, and imputing to the sinner the righteousness of Christ. These acts surely seem to be real and not 'mere Cambridge' changes in God as well as the redeemed sinner, so they must be infected with temporality.

---

[6] Ibid.
[7] Ibid., p. 95.

It might be possible as well to extend Wolterstorff's basic argument. In presenting God as a personal being, the biblical writers point to the possibility that human persons can have a personal relationship with God. This possibility is underlined in Jewish and Christian theology as well. It is plausible that the notion of a personal relationship must be cashed out in ways that would make it necessarily a dynamic process. Two neutrons can possibly stand in certain relations to each other, but could not be said to 'have a relationship.' Nor could a human being be said to 'have a relationship' with a neutron, or anything atemporal and strongly immutable. If such an argument from personality could be mounted, it would certainly be in the spirit of Wolterstorff's article.

Stephen Davis

A second philosopher who argues from the personal nature of God to the conclusion that God is temporal is Stephen Davis.[8] He concentrates on two attributes of God: 'the claim that God is the creator of the universe, and the claim that God is a personal being who acts in human history, speaking, punishing, warning, forgiving, etc.'[9]

Creation and Acts in History (Stephen Davis)

Davis's first argument—that an atemporal God could not create *ex nihilo*—may be questioned on the supposition that bringing about the first moment of time is itself a temporal act. His second argument is stronger, and is similar to Wolterstorff's.

Davis's argument from God as creator is clear.

1. God creates x.
2. x first exists at T.
3. Therefore, God creates x at T.

But (3) is ambiguous between

        3a. God, at T, creates x

and

        3b. God creates x, and x first exists at T.

---

[8] Stephen T. Davis, *Logic and the Nature of God* (Grand Rapids, MI: Eerdmans, 1983), pp. 11-14. A second argument Davis offers for regarding God as temporal is that the idea of a timeless being is probably incoherent. But since his argument is directed against views that we have already considered, such as Boethius, Anselm, Augustine, and Stump and Kretzmann, I shall not consider it here.

[9] Ibid., p. 11.

Clearly, (3a) entails that God is temporal, for a timeless being cannot perform an action at a time. So the defender of timelessness must opt for (3b). Since (3b) is the conjunction of (1) and (2), it is entailed by (1) and (2). But can a defender of timelessness adopt (3b)? No, says Davis, because 'we have on hand no acceptable concept of atemporal causation, i.e. of what it is for a timeless cause to produce a temporal effect.'[10]

In addition to the argument from God as creator, Davis cites several biblical texts that seem to locate God's 'personal, caring, involved' actions in time. He then concludes:

> But the obvious problem here is to understand how a timeless being can plan or anticipate or remember or respond or punish or warn or forgive. All such acts seem undeniably temporal. . . . On both counts, then, it is difficult to see how a timeless being can be the God in which Christians have traditionally believed. It does not seem that there is any clear sense in which a timeless being can be the creator of the universe or a being who acts in time.[11]

In this second stage, Davis's argument is quite similar to Wolterstorff's argument, and my suggested extension of that argument. Again, the argument needs to be formulated with more rigor, but I believe that there is a sound argument here.

I am also sympathetic to the first stage of Davis's argument, but I believe that it has a major weakness as stated. The problem is that Davis has not been careful enough in drawing out the ambiguity of (3). Grant that (3a) entails that God is temporal. But there is an additional ambiguity in the second conjunct of (3b) that arises from an ambiguity in (2).[12] This is ambiguous between

    2a. x first exists at T, and there is no time prior to T at which x does not exist

and

    2b. x first exists at T, and there is no time prior to T.

If (2a) is the correct reading of (2), then Davis's argument goes through. If, however, (2b) is the correct reading, then things are not so clear. For if (2b) is correct, (3b) would be read as

    3b'. God creates x, and x first exists at T, and there is no time prior to T.

In other words, God would bring about both the existence of x and of time at T. However, it is not clear that an event that is itself the first moment of a temporal

---

[10] Ibid., p. 13.
[11] Ibid., p. 14.
[12] Morriston exploits a parallel ambiguity in his critique of the *Kalām* Cosmological Argument. Wes Morriston, 'Must the Beginning of the Universe Have a Personal Cause?' *Faith and Philosophy* 16 (2000), p. 154.

series is a temporal event in the relevant sense. It could be, then, that both the cause and the effect should be regarded as timeless. And if this is the case, no good account of atemporal causation of a temporal effect would be needed, so Davis's argument would not succeed.

I believe there is a way to patch up this problem and make the argument sound, but I shall defer my solution for the final chapter. Note, however, that the criticism just outlined only applies to God's action of creation *ex nihilo*. I think that Davis's argument does succeed when applied to any other act of God within the temporal order, such as speaking, punishing, warning, or forgiving, to use Davis's examples. This second form of the argument is quite similar to Wolterstorff's discussed above, so I will not say any more about it here.

Atemporal Persons?

If arguments like those of Wolterstorff and Davis succeed, they serve to demonstrate that, given dynamic time, God must be temporal. But the existence of the universe and physical, dynamic time, are contingent facts. Almost all theologians and philosophers have agreed that creation is an entirely free divine act; God need not have created anything. So it remains a possibility that God is only contingently temporal, or that, in the view of William Lane Craig, God is atemporal *sans* creation, and temporal since creation.[13] But still, we might ask, is there anything in the concept of personhood that renders the notion of a timeless person incoherent?

We must be careful to disentangle two related ideas here. In one sense, on the B-theory of time, persons exist tenselessly as four-dimensional 'space-time worms.' And it is possible to speak of such extended entities in space-time as existing timelessly. For the sake of argument, I shall grant that the notion of a four-dimensional person is coherent.[14] Then, within the restrictions of a static theory of time, one might propose that God exists (tenselessly) just as a person (as a space-time worm) exists (tenselessly), and both, from the tenseless perspective, are timeless.

I see no incoherence with this claim. I do, however, believe it to be wrong. But that is not the idea at issue in this section. The relevant idea here is that of God existing in a Boethian 'all-at-once' non-durational eternity, in which time does not exist at all, so there is not a temporal dimension. An argument against the possibility of a timeless person will take the following form (where Px = 'x is a person,' Ax = 'x is atemporal,' and Fx = 'x is F,' where F constitutes the property or set of properties necessary and sufficient for personhood):

    A. $(x)(Px \leftrightarrow Fx)$

---

[13] See the discussion below devoted to Craig's views.
[14] While I do not agree that the view is correct, four-dimensionalism has able defenders. See, for example, Hud Hudson, *A Materialist Metaphysics of the Human Person* (Ithaca, NY: Cornell University Press, 2001).

B. (x)(Ax ⊃ ~Fx)
C. Therefore, (x)(Ax ⊃ ~Px)

If atemporal personhood is to be shown to be incoherent, then, it must be clear that F constitutes a necessary condition(s) for personhood, and also that an atemporal being could not exemplify F.

This looks to be a daunting task. The criteria of personhood are very much in debate in contemporary philosophy, from applied ethics, dealing with beginning- and end-of-life issues, to cognitive science, dealing with artificial intelligence. Perhaps any account of personhood would include, as necessary (but insufficient) conditions, such mental properties as consciousness, self-consciousness, or intentionality. But it is not clear that such properties are incompatible with timeless existence. Craig has offered an argument that timeless personhood, as defined by any criteria similar to these, is not incoherent.[15]

There are other possible strategies that might be deployed. Trenton Merricks claims that only perduring entities can exist in static time, and only enduring entities can exist in dynamic time. He then suggests that if persons are not composed of person-stages, then persons must exist in dynamic time.[16] But this does not address the question of possible atemporal persons.

Delmas Lewis mounts a stronger argument to the effect that static time is incompatible with the moral point of view. This is because (i) moral responsibility for an act can only be assigned to the same person who committed the act, not to person-stages or bundles of temporal parts, which is what persons are under tenseless time; and (ii) no person-stage or bundle of temporal parts is capable of performing morally significant actions.[17] Now, presumably, at least the capacity to have a moral point of view is a necessary criterion of personhood, but, again, Lewis's argument does not address the possibility of a possible atemporal person.

While suggestive, I do not think that either of these arguments will ultimately be successful in showing the incoherence of the concept of an atemporal person. But at the very least, I must say that it is not at all clear in what sense an atemporal being, which could not enter into temporal relations with other persons and experience temporal change, would be at all analogous to what we commonly understand by 'person.' The explanatory burden on the atemporalist is indeed heavy.

A final line of argument attempts to show that being a person entails being alive, and that being alive is necessarily a temporal property. But life, like personhood, is notoriously hard to define. Indeed, the most rigorous accounts are attempts to define biological life, so that extensions of the concept to artificial life

---

[15] William Lane Craig, *God, Time and Eternity: The Coherence of Theism II: Eternity* (Dordrecht: Kluwer Academic Publishers, 2001), pp. 43-44, hereafter *GT&E*.

[16] Trenton Merricks, 'On the Incompatibility of Enduring and Perduring Entities,' *Mind* 104 (1995), pp. 523-31.

[17] Delmas Lewis, 'Persons, Morality, and Tenselessness,' *Philosophy and Phenomenological Research* 48 (1986), pp. 305-9.

or atemporal divine life will probably be by way of analogical predication. So it is not at all straightforward to show that the concept of life entails temporality.

Summary

Contemporary arguments from personality to God's temporality have a certain intuitive appeal. However, I am unaware of any defender of such an argument who subscribes to the B-theory of time, perhaps reflecting a certain pre-theoretical ordering of commitments. Of course, if dynamic time can be established independently, as I have tried to do in Part I, then a rather rigorous account of what it means to be a person, together with a demonstration that such an account is incompatible with dynamic time, would be a strong argument.

But the prospects for a rigorous account of personhood, which is clearly incompatible with divine timelessness, do not seem to be promising. Consequently, I do not think the claim can be sustained that a timeless person is an incoherent concept.

The Argument from Dynamic Time

A number of philosophers have begun from the premise of dynamic time and inferred that God must be temporal. There are, of course, many variations in the structure of the arguments; in what follows only the most interesting features of several different philosophers' arguments are examined.

There is a strange asymmetry between the atemporalist and the temporalist theories. For the most part, the atemporalists, both medieval and modern, not only argue for God's atemporality based on premises other than the nature of time, but also seek to give an account of God's relations to the created order. On the other hand, many temporalists simply argue for God's temporality, and seem to take it for granted that a temporal God's relations to the created order are in need of no further explication. William Hasker, Richard Swinburne, Alan Padgett and William Lane Craig are exceptions in this regard, and merit serious attention.

William Hasker

The problem of divine foreknowledge and human freedom has been lurking at nearly every turn in this study. It is this problem which draws the focus of William Hasker in *God, Time and Knowledge*.[18]

In summary form, Hasker's argument is this.

4. If humans enjoy genuine freedom, then the future must be genuinely contingent.

---

[18] William Hasker, *God, Time, and Knowledge* (Ithaca, NY: Cornell University Press, 1989), hereafter *GT&K*.

5. If the future is genuinely contingent, then God cannot know it.
6. If God cannot know the future, then he must be temporal.

Since Hasker does in fact believe that humans are genuinely free, the conclusion of God's temporality follows.

> **Human Freedom, Divine Temporality (William Hasker)**
>
> Hasker argues that there is no satisfactory resolution to the dilemma of freedom and foreknowledge, so God must lack foreknowledge if humans are free in a libertarian sense. But only if time is dynamic and God is temporal could God lack foreknowledge (knowledge of future contingents is strictly impossible, since future contingent propositions have an indeterminate truth-value).

Hasker's argument takes as its first premise that humans have genuine freedom. He argues at length that a proper concept of genuine freedom entails a commitment to incompatibilism and a strong libertarian view of free will. It is well beyond the scope of this work to broach these issues. Since both incompatibilism and libertarianism are coherent concepts, and indeed have enjoyed very able and spirited defenses in recent years, I shall simply grant their truth. Now, Hasker thinks that genuine freedom entails that the future is open—that is, genuinely contingent. Although he does not discuss the issue of static versus dynamic time here, he has raised the issue elsewhere and committed himself to a theory of dynamic time.[19] So I take it that when he speaks of genuine contingency, Hasker has in mind A-theoretic time. So I think there are adequate reasons to accept (4).

Next, Hasker claims (5) that if the future is open, then God cannot know it. This claim reinforces the assumption that he is assuming dynamic time. But before reaching this conclusion, he considers and rejects Augustinian and Boethian theories of God's foreknowledge. He also goes to some length to argue against Molina's theory of middle knowledge, and the Ockhamist theory of counterfactual power over the past with its hard facts/soft facts distinction. Believing he has considered all traditional theories as to how God can know future contingents (and I would agree that he has), he concludes that even God cannot know them. I am not persuaded by his arguments against either Ockhamism or Molinism, but I shall not engage those arguments here. (See Chapter 10 where I briefly defend Molinism.) But for the sake of argument, I'll grant (5).

What follows in Hasker's discussion of (6) is the move from the premise that God cannot know the future to the conclusion that God is temporal. This move is less of an argument than it is an autobiographical confession. He begins with a demonstration that 'Whether or not there are creatures endowed with libertarian free will, it is impossible that God should use a foreknowledge derived from the actual occurrence of future events to determine his own prior actions in

---

[19] William Hasker, *Metaphysics: Constructing a World View* (Downers Grove, IL: InterVarsity Press, 1983), p. 53.

the providential governance of the world.'[20] So if simple foreknowledge does exist, it will be good for nothing, says Hasker. (By simple foreknowledge he means foreknowledge of actual free choices that will be made, not Middle Knowledge of counterfactuals of free choices.) Next, Hasker shows that a rephrasal of the above statement in tenseless terms will make no difference to its truth; hence, 'The doctrine of divine timelessness affords no help whatever in understanding God's providential governance of the world.'[21] Since Hasker thinks that the primary motive behind most contemporary defenses of divine timelessness is a desire to avoid the fatalist dilemma, he asserts that his argument has removed any such motive. The only other significant motive he can discern to maintain divine timelessness is a value judgment, traceable to the influences of Neoplatonism, that it is better to be beyond the influence of time. But this motive he finds much less than compelling.

At this point Hasker begins a somewhat autobiographical segment. He confesses to being a 'lapsed Boethian' for whom divine timelessness was incorporated early into his theological thinking. He found his philosophical home, however, in the analytical tradition, which is, he claims, not particularly congenial to the metaphysical apparatus necessary for timelessness. He finds the concept of timelessness extremely difficult to give any sense to, although not ultimately incoherent. Hasker finds unsuccessful the analogies between space and time deployed to elucidate the concept of timelessness, because the analogies only hold in those respects in which both space and time are continua, but fail otherwise. And finally he finds the most difficult aspect of theories of timelessness the idea that time is somehow present in eternity:

> ... it does not seem to me that the theory of divine timelessness can reasonably be accepted merely as a solution to the problem of foreknowledge and free will. As we have seen, the conceptual structures required for eternalism are complex and difficult, and amount to a conception of the nature of ultimate reality which is fundamentally different than one that might be held by a theist apart from this doctrine. To adopt such an elaborate metaphysic merely as a solution to a problem, even a fairly significant problem, is disproportionate.[22]

In the end, then, it is a methodological principle of parsimony combined with the coherence of intuitions and a commitment to a certain philosophical approach that drives Hasker to a temporalist thesis. Just how compelling this part of his argument is will naturally be relative to the philosophical intuitions and commitments of each individual. Freddoso, for example, counters that Hasker's amounts to a 'try it and see if it works' approach to theistic metaphysics, and fires two shots from his rhetorical cannon in defense of the medieval consensus:

---

[20] Ibid., p. 63.
[21] Ibid., p. 177.
[22] Ibid., p. 181.

> First, without the sort of systematic study that we are not generally trained for either linguistically or philosophically, we contemporary Christian philosophers are not in a position even to understand, much less to criticize, most of the work of those classical metaphysicians whom Hasker takes to task.... Second, given the realities of our peculiar historical situation [in the embryonic stage of analytical metaphysics], we have no good reason to believe that, with respect to the conception of God, the metaphysical predispositions engendered by contemporary culture and philosophical training are more reliable than those of classical Christian thinkers.[23]

Certainly Freddoso has a point; it is too easy to seek notoriety in novelty, and tradition should only be questioned with due respect. Chapters 5 and 7 of this work seek to take seriously Freddoso's concerns. However, Hasker, too, has a point. If progress is made in philosophy, then there may well be good reasons to prefer the analytic approach to the Scholastic Aristotelian tradition. Tradition cannot be allowed to play tyrant, and sufficient justification for overthrowing tradition may arise from philosophical intuitions sharpened within the context of analytical metaphysics.

So where does that leave us in considering Hasker's views? Clearly he has a valid argument, but different premises will commend themselves to different readers with greater or lesser force. Especially in drawing the conclusion that God is temporal, one could wish for more rigor.

Richard Swinburne

Perhaps the rigor will be found in that of a most analytical philosopher of religion, Richard Swinburne. In his book, *The Coherence of Theism*,[24] Swinburne makes a case for God's temporality that deploys both the argument from personhood and the argument from dynamic time. Swinburne's argument from personhood is very similar to the two already examined. My interest at this point is with his argument from dynamic time. This argument is stated more succinctly in his more recent work, *The Christian God*.[25]

Swinburne begins by arguing for dynamic time.[26] I shall not detail his arguments; they are rather different than the ones I used in Chapter 2. However it is important to note four principles concerning time which Swinburne develops, since these lie at the heart of his arguments for a temporal God. The four principles are as follows. (i) Events happen at periods. Swinburne argues that instants are bounding points of periods; periods are not composed of instants. Any event or state of affairs persists through a period. (ii) Time has a linear topology regardless of the laws of nature, but it has a metric only if there are laws of nature

---

[23] Alfred J. Freddoso, 'The "Openness" of God: A Reply to William Hasker,' *Christian Scholars Review* 28 (1998), pp. 124-33.
[24] Richard Swinburne, *The Coherence of Theism*, rev. ed. (Oxford: Clarendon Press, 1993, originally published 1977).
[25] Richard Swinburne, *The Christian God* (Oxford: Clarendon Press, 1994).
[26] Ibid., chapter 4, pp. 72-95.

that take a unique simplest form on the assumption that certain periodic processes in nature measure intervals of equal time.[27] (iii) A causal theory of time, developed by Swinburne elsewhere,[28] states that a period of time is future if it is possible now to causally affect what happens then, and past if it was possible then causally to affect what happens now. The present is an instant that is the boundary between past and future. Finally, (iv) there are tensed, indexed facts which are different from tenseless, non-indexical facts.

From these four principles Swinburne draws conclusions about the nature of time. Since his is a causal theory of time, he warns against hypostatizing time. He concludes:

> Time is something involved in the very existence of substances, the necessary framework for them, an amorphous realm of causal possibilities which surrounds actual substances. Since everything that happens, including the existence of substances, occurs (by my first principle) over a period of time, then if—impossibly—there were no time, there would be no substances.[29]

(Since Swinburne believes that God is a metaphysically necessary being, and is logically necessarily a substance, it follows that he believes that time is metaphysically necessary.)

Divine Causation; Infinite Time (Richard Swinburne)

Swinburne argues that causes necessarily preceded their effects in time; hence, God's causal acts must be located in time prior to their effects. But Swinburne's support for this, based on STR, is easily defeated, since if God is omnipresent (in the Thomist sense), then his actions would not be limited by constraints imposed by STR on the propagation of causal signals.

However, in his account of how a temporal God could have existed for an infinitely long duration, Swinburne makes a significant contribution.

When Swinburne moves to his discussion of the temporal mode of God's being, he claims that a temporalist view is both explicit and implicit in Scripture and in the writings of all the Church Fathers of the first three centuries. As I have done above, he traces the atemporalist view to Augustine and the influence of Neoplatonism.

Swinburne minces no words when he declares, 'The results of Chapter 4 [the four principles noted above] do, however, have the consequence that the "timeless" view is incoherent.'[30] How does he show the incoherence? First, he suggests that if God's timelessness is understood to have no temporal duration,

---

[27] Swinburne makes explicit later that this refers to physical time: ibid., p. 140.
[28] Richard Swinburne, *Space and Time*, 2nd ed. (London: Macmillan, 1981, originally published 1968).
[29] Swinburne, *The Christian God*, p. 95.
[30] Ibid., p. 139.

then it is natural to understand timelessness as a single instant, and this is in conflict with the first principle that requires states of affairs to persist through a period of time. But, acknowledging that some atemporalists maintain that the divine 'moment' has duration, Swinburne allows that the objection from the first principle is not decisive.

He does, however, claim that the third principle provides a conclusive objection to divine timelessness. Relying on a causal theory of time, he claims that 'if God causes the beginning or continuing existence of the world, and perhaps interferes in its operations from time to time, his acting must be prior to the effects which his action causes. Similarly, his perception of events in the world must be later than those events.'[31] Swinburne offers no argument for this latter claim, perhaps assuming that a causal theory of perception needs no argument. Of course, it is a moot question whether or not God's knowledge of events in the created world is perceptual in any ordinary sense of the word. Thomas Aquinas argued that one corollary of saying that God is omnipresent is that he is immediately present in his knowledge to every point in space. His knowledge of what is occupying those points is not mediated by anything.[32] If this is correct, then Swinburne's last point seems wrong.

The former claim is that a causal action must precede its effects. In support of this claim, two arguments against the possibility of simultaneous causation are offered. The first is a simple consequence of the Special Theory of Relativity that all causal action is propagated at a finite velocity. The second argument is similar to the one I offered in Chapter 2. Assume that A simultaneously causes B, and B simultaneously causes C. Given the Humean dictum that what causes what is logically contingent, it is logically possible that rather than C, B has as its effect not-A. But this situation is, of course, an impossible causal sequence, so the assumption of the possibility of simultaneous causation is faulty.[33]

I do not think that this is a sound argument. To see why, we need to consider again what God's omnipresence means for his action. Aquinas wrote,

> God is present in all things; not indeed as part of their essence, nor as an accident; but as an agent is present to that upon which it works. . . . Therefore as long as a thing has being, God must be present to it. . . . No action of an agent, however powerful it may be, acts at a distance, except through a medium. But it belongs to the great power of God that He acts immediately in all things.[34]

If this is correct, and it certainly seems plausible to me, then (at least some of) God's actions are not mediated through physical causal chains. So objections from STR are not relevant. Further, Hume's dictum cannot apply to God's actions.

---

[31] Ibid., pp. 139-40.
[32] Thomas Aquinas, *Summa Theologica*, tr. The Fathers of the English Dominican Province (New York: Benzinger Brothers, 1948), Ia.8.3, hereafter *ST*
[33] Swinburne, *The Christian God*, p. 82.
[34] Aquinas, *ST*, 1a.8.1.

Assuming God's omnipotence, God infallibly can bring it about that his causal activity brings about his desired effect. If God does A immediately to bring about B, and either knows or intends that B brings about C, then *ex hypothesi* it is not possible that B should have the effect not-C. So the argument against simultaneous causation fails just in the case of God's actions. Since, then, in the case of God's actions, neither empirical nor conceptual objections rule out the possibility of simultaneous causation, I conclude that Swinburne's argument is unsound.

The correct premises for the argument, I suggest, might be something along this line: since there is no model of atemporal causation where a timeless cause has an effect in time, there is no reason to believe that such a species of causation is possible. Since time is dynamic, any causal action—even immediate action by an immaterial, omnipresent God—must be located in the temporal series.

With the conclusion in hand that God is everlasting—that is, temporal, Swinburne next addresses the question of whether God is 'a prisoner of time.' While I do not find the question itself very interesting (since, as I indicated in Chapter 1, being 'in time' is not the correct way to think about being temporal), nevertheless his discussion has merit, in my view, and is germane to the question of God's relation to physical time.

Swinburne's discussion establishes several claims. First, given his second principle (concerning time's metric and the laws of nature), is the claim that if God had not (or before he had) created a physical universe, there would have been no temporal metric. There would simply be a single event or a sequence of events in God's consciousness. Assume that God is the subject of just one single conscious event with no qualitative distinguishable temporal parts. While (by principle one) this event would last some period of time, in the absence of a temporal metric there would be no truth as to how long it lasted. 'There would be no difference between a divine act of self-awareness which lasted one millisecond and one which lasted a million years.'[35]

The next claim seems preposterous on the face of it. There would be no difference either between a divine conscious act of finite length and one of infinite length. But Swinburne's arguments are compelling, in my view. He asks us to imagine two time lines, as it were, representing respectively God's 'supposedly finite' act, SF, and his act of 'supposedly infinite duration,' SID (see Figure 8.1).

Using any arbitrary metric, the SF line, bounded by instants A and B, can be subdivided into two equal segments by marking the midpoint Z. Take any point on the SID line, Q. Now divide AZ at the midpoint Y. Mark on the SID line an interval PQ equal in length to YZ. Then divide AY at the midpoint W, and mark on the SID line another interval KP equal to PQ. Divide AW at the midpoint S, and so on, ad infinitum. Then perform the same process for the other half of the lines. The result will be that all intervals of the SF line have been put into a one-to-one correspondence with the intervals on the SID line. Since the intervals themselves, by previous argument, are qualitatively indistinguishable, the

---

[35]Swinburne, *The Christian God*, p. 140.

surprising conclusion is that there is no difference between a divine act of finite duration and one of infinite duration.[36] According to Swinburne, this shows that unless God decides of his own free will to create a physical world,

> ... the aspects of time that seem to threaten his sovereignty would not hold. ... The past would not be getting lengthier by any period of time which could be picked out, let alone measured. ... God cannot affect the past—but all there would ever be to the past is his having his one divine act. ... The future, however, remains under God's total control; he need not make free creatures—in which case nothing will surprise him; or anything at all—in which case nothing will loom upon him.[37]

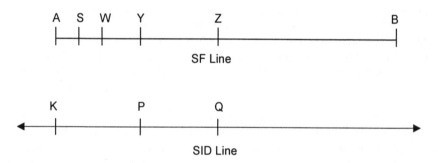

Figure 8.1 The 'metric' of God's consciousness

Source: Adapted from Swinburne, The Christian God, p. 141.

Further, God might instead have chosen to have a succession of qualitatively distinct mental acts, and thereby establish temporal order, but still not a temporal metric. If God brings about some effect A in his own mental life, then he has divided time into the realm of the past (earlier than A) in which not even God can bring it about that A not occur, and the future, rife with logical possibilities which, should God choose to actualize, would create a rich temporal series.

If God decides to create a universe distinct from himself, it could be one in which the laws of nature make regular clocks, and so a temporal metric, possible. Then God's acts could be dated by the events in the universe with which they are simultaneous.

These claims and supporting arguments about God's relation to physical time seem right to me. Perhaps within this conceptual framework, certain puzzles related to the Kalām argument might find a solution. But I shall not pursue that

---

[36] The intervals in question are both countably infinite, of cardinality $\aleph_0$ and ordinality ω.
[37] Richard Swinburne, The Christian God, p. 142.

matter here. All in all, I judge Swinburne's theory to be quite helpful, once the initial argument is modified appropriately.

Alan Padgett

In his book, *God, Eternity and the Nature of Time*,[38] Alan Padgett offers both a strong argument for divine temporality, grounded in a dynamic theory of time, and a substantial model of how God relates to the created world. I shall consider first the argument and then the model.

Metaphysical Time: Relatively Timeless (Alan Padgett)

Padgett's model involves three theses:
1. God's time is metaphysical time, as distinct from physical time.
2. Metaphysical time has no metric, hence is 'relatively timeless.'
3. Any divine action is 'Zero Time Related' to its effects in physical time.

Padgett's argument runs as follows:

6. If God sustains the universe, then he is timeless only if time is static.
7. Time cannot be static.
8. Therefore, if God sustains the universe, then God cannot be timeless.
9. God's sustaining the universe is essential to our concept of God.
10. Therefore, God cannot be timeless.

The argument is valid, provided we grant a traditional concept of God as sustaining the universe. Padgett's view of God's sustaining of the universe is broadly Thomist, with a change from Aristotelian to modern science as a working assumption. Incorporated in this view are at least the following ideas. God is the cause of all matter/energy in the universe, and any natural laws that obtain. God is the cause of the essential properties, including causal properties, which any object has at any episode of its existence. Further, God is the cause of the continuing in existence of any persisting object. Padgett takes as axiomatic the theological conclusion that the universe is contingent and depends at all times upon God for its continued existence. So far, this is a standard orthodox view of God's providential sustenance of the world.

Padgett says that God's act of sustaining the universe is to be conceived as a direct act—that is, an act that produces its effect with no intermediary causes. As an example of such direct action he proposes the direct causation of brain states by humans.

Padgett then introduces the term 'Zero Time Related': 'Two events are Zero Time Related if and only if no duration occurs between them.'[39] All direct

---
[38] Alan G. Padgett, *God, Eternity and the Nature of Time* (New York: St Martin's Press, 1992).
[39] Ibid., p. 21.

action is Zero Time Related to its effect, not merely divine action.[40] This is so because once all the sufficient causes are present, no causal gap can intervene. If all causes are present at t, then the effect should take place at (or Zero Time Related to) t. I believe that Padgett needs to do a bit more here in terms of discussing the possibility of simultaneous causation and the limitations of STR, which require that the propagation of causal action be at a finite velocity. But perhaps Padgett would reply that only those actions that involve a non-physical cause are direct, and hence Zero Time Related to their effects. Since his only example is of mental causation, this seems to be the most probable reading. With this understanding, we can ask how Padgett supports (6).

Padgett asks whether the following principle, often deployed in defense of divine temporality, is true of logical necessity:

> 11. The occurrence of any temporal effect brought about by an agent implies that the agent in question is temporal.

It can be argued in support of (11) that in order for an agent to bring about change, then the agent itself must change—that is, act differently at a time. But this is not true, for Padgett shows that the logical possibility of a timeless God's timelessly sustaining an angel is a counter-example. So he proposes a revision of (11).

> 11′. The occurrence of an effect (which is itself a change) implies a change in some cause of the effect.

Surely (11′) is true, he claims. But suppose the world is a four-dimensional space-time block. Then God could timelessly sustain the universe, as he did the timeless angel. But this would not be a counter-example to (11′); it would simply fall under a different principle. Therefore, concludes Padgett, a doctrine of timeless causation is not incoherent, provided time is static. So his conclusion is this: if God is timeless and he sustains the universe, then time is static.[41]

The next stage in Padgett's argument is a critique of the theory of static time.[42] He devotes considerable space to a rebuttal of arguments that the relativity of simultaneity under STR entails static time (arguments advanced by Grünbaum, Putnam and Weingard), and to a rebuttal of philosophical arguments for static time (arguments offered by Smart, McTaggart and Mellor). While his arguments against static time differ from those I offered in Chapters 2 and 3, the conclusion reached is the same. Therefore I am in agreement with Padgett when he concludes that having rejected the static theory of time, any theory of divine timelessness must also be rejected.

---

[40] Compare Padgett on this point to Swinburne's claim that there can be no simultaneous causation.

[41] Ibid., pp. 74-6. His conclusion, although not his argument, is the same as I reached at the end of Chapter 6.

[42] Ibid., chapter 5, 'The Stasis Theory of Time: A Critique,' pp. 82-121.

Next, Padgett offers his proposal for understanding God's relation to the temporal world.[43] Having shown that, in sustaining the world, God experiences change (by [11']), he claims that necessarily, if a change occurs, then a duration occurs. Hence God is in some sense temporal. But this conflicts, he suggests, with our deep theological intuitions that God is absolutely transcendent over all things, including time itself. So he attempts to unpack the idea of God's transcending time in terms of three theses: first, God's life is the ground of time; second, God is the Lord of time; and, third, God is relatively timeless. It is the third thesis that is of greatest interest, but a word or two must be said about the first two.

In expounding the first thesis, Padgett says that God eternally (that is, timelessly) chose 'to live the kind of life he does, and has chosen eternally to have a temporal universe in which to live.'[44] Perhaps this is just infelicitous language, but surely Padgett cannot mean that God chose to 'live in' the universe. Be that as it may, he claims that since the world could have been different, God's free choice must be seen as the ground of time. Clearly the time he has in view here is physical time. So God's eternal choice to experience succession in his own being is the metaphysical ground of the physical time of the actual universe.

In saying that God is Lord of time, Padgett acknowledges speaking metaphorically. The implications of the metaphor are that everything that happens in time is according to God's will; that God himself is not fundamentally changed by the passage of time; that he is a (metaphysically) necessary being who lives forever.

As I said above, it is the third thesis that is most interesting. In developing the idea that God is 'relatively timeless,' Padgett begins by claiming that it is a logically necessary truth that 'when both cause and effect are temporal, a cause must be temporally related to its effect.'[45] In the case of God and effects in this world, the temporal relation is the Zero Time Relation. That means that there must be some relation as well between eternity (God's time) and the physical time of our world. Padgett proposes that it is proper to say that physical time is in God's time, not vice versa. Another way of saying this is that God is 'timeless' in the sense that his time is measureless, or that God's time has no metric. For Padgett this means that 'Measured Time Words' (that is, any term in a natural language which designates some measure of duration[46]) would not truly apply to eternity. This is because (i) God is not subject to the laws of nature, as anything in measured time must be, and so the metric of time based on those laws does not apply to God; and (ii) any measured time is relative to a particular frame of reference, which need not apply to God's time.

---

[43] Ibid., chapter 6, 'A New Doctrine of Eternity,' pp. 122-46. See also Padgett's 'God and Time: Toward a New Doctrine of Divine Timeless Eternity,' *Religious Studies* 25 (1989), pp. 209-15.
[44] Padgett, *God, Eternity and the Nature of Time*, pp. 122-3.
[45] Ibid., p. 125.
[46] Ibid., p. 12.

Now, (i) is clear enough. But what about (ii)? This point is intended to answer the question arising from STR as to which relativistic reference frame is God's. By disconnecting God's time from physical or measured time, Padgett has shown the question to be meaningless.[47]

Nevertheless, I have some difficulty in knowing just how to take Padgett's claim that all this means that God is 'relatively timeless.' 'Relatively' here does not refer to STR; it must be used in the ordinary sense of 'in relation to.' So Padgett's claim is that, in relation to physical time, God is timeless. But what does that mean? Since he asserts repeatedly that God changes and so is temporal, this cannot mean what is normally meant by 'timeless.' In the end, it seems that what Padgett means by 'relatively timeless' is that the metric of physical time cannot be applied to God's time. But this can be understood two ways. On a rather harmless reading, this means no more than that the metric of time which depends upon the contingent laws of nature in the actual universe is not the intrinsic metric of God's time. This is uncontroversial.

But, on another reading, problems arise.[48] By this construal, Padgett is denying that there is any congruence between God's time and physical time. Recall Padgett's point (i), which claims that God is not subject to the laws of nature, as anything in measured time must be, and so the metric of time based on those laws does not apply to God. But it is not the case that everything 'in' measured time (physical time) must be subject to the laws of nature; it is rather that the clock that is used to measure time must be subject to the laws of nature. So it does not follow that a metric derived from the physical world cannot be applied to God. In fact, it seems incoherent to deny that there is congruence between God's time and physical time. A method of establishing such congruence is indicated by Swinburne's geometrical method outlined in the previous section. And the topological conclusions reached in Chapter 3 seem to require both that time is linear and that if two temporal series are related, then their respective intervals can be placed in a one-to-one correspondence. Further, if time is dynamic, then it follows that God's 'now' (or the present of metaphysical time) must correspond to the 'now' of physical time. While Padgett affirms this last conclusion, he denies both former claims.

A final difficulty may be surfaced by comparing Padgett's Zero Time Relation with Leftow's Zero Thesis. The Zero Thesis claimed that the distance between God and any spatial object was zero. In Chapter 6 I raised several arguments against Leftow's Zero Thesis (a distance relation), one of which alleged that it violated a type restriction on the relata, since it ascribes a spatial property to

---

[47] Craig comments, 'This distinction between ontological time and measured (or empirical) time seems to me to be an extremely important insight, which is a salutary counterbalance to the universally repeated and extravagant assertions that STR has forced us to abandon the classical views of time and space. This erroneous inference is rooted precisely in the failure to draw the sorts of distinction which Padgett has emphasized.' William Lane Craig, 'God and Real Time,' *Religious Studies* 26 (1990), p. 339.

[48] Padgett himself defends this reading: Alan Padgett, 'Can History Measure Eternity? A Reply to William Craig,' *Religious Studies* 27 (1991), pp. 333-5.

God, who is by definition a non-spatial entity. Now unless Padgett's Zero Time Relation is also guilty of a similar violation of a type restriction, then it cannot be the case that God is timeless. Padgett, as indicated, does believe that God is temporal, so it seems to introduce unnecessary confusion to use language like 'relatively timeless' and to deny that God's time is in any way congruent with our time.

Perhaps not much of significance hangs on this point, but it is at this juncture that my exposition of omnitemporality will differ from Padgett's claim that God is 'relatively timeless.' Still, Padgett's work seems to me to be significant both in its argument for God's temporality and in the model offered of God's relation to the temporal order.

William Lane Craig

Since 1978 William Lane Craig has published over 50 journal articles and five books dealing with the cluster of issues that involve God's relation to time.[49] It is almost certain that no other philosophical theologian has ever dealt more thoroughly with the issues. Certainly no one has attempted to integrate the biblical, theological, historical, metaphysical and scientific implications on the scale of Craig's work.[50]

Craig is committed to the dynamic theory of time. His arguments for dynamic time are detailed and sophisticated. He argues that the A-theory is both philosophically and theologically superior to the B-theory. While not developing the arguments, he claims that:

> In favour of the A-theory, one might argue that it gives the most adequate analysis of personal identity and that the tensed-ness of our language and experience is uneliminable.... Theologically, there seems to be a decisive reason for Christian theists' rejecting the B-theory, namely, that it cannot give an adequate analysis of the biblical doctrine of *creatio ex nihilo*.[51]

In an interesting passage Craig ponders the contingency of time. The fact that the temporal series might be different across possible worlds shows only that the temporal series is contingent, not that time itself is. While the shape of any object is contingent, it does not follow that it is contingent that every object have a shape. But in the end he concludes that, if God did not choose freely to experience temporal succession in his mental life and did not choose freely to create a

---

[49] Only a fraction of Craig's output on this subject is represented in the bibliography. I have learned much from his work.
[50] William Lane Craig, *The Tenseless Theory of Time: A Critical Examination* (Dordrecht: Kluwer Academic Publishers, 2000); its companion volume, *The Tensed Theory of Time: A Critical Examination* (Dordrecht: Kluwer Academic Publishers, 2000); *Time and Eternity: Exploring God's Relationship to Time* (Wheaton, IL: Crossway Books, 2001), hereafter *T&E*; *Time and the Metaphysics of Relativity* (Dordrecht: Kluwer Academic Publishers, 2001); *GT&E*.
[51] Craig, 'God and Real Time,' p. 337.

temporal universe, then there would have been no time, so time is in fact contingent.[52]

> **Temporal Since Creation (William Lane Craig)**
>
> Craig argues that time must be dynamic; hence, divine temporality is required by (i) God's relations with the temporal world, and (ii) God's knowledge of tensed facts.
>
> Craig makes significant contributions to the philosophy of time and philosophy of science, showing that a neo-Lorentzian interpretation of STR is preferable to an Einsteinian interpretation, thus retaining the concepts of absolute simultaneity and dynamic time (see the discussion in Chapter 3).
>
> In his most innovative move, Craig argues that 'apart from' or '*sans*' creation, God is timeless. (Consideration of this claim will be deferred until Chapter 10.)

Craig's philosophical (and theological) commitment to dynamic time underwrites his commitment to divine temporality. He states:

> I am convinced that the decision between an A- or a B-theory of time constitutes a fundamental watershed for our conception of divine eternity. For if we adopt a B-theory of time, most of the typical arguments against divine timelessness. . . are doomed to failure. . . . By contrast, on the A-theory of time, the concept of a timeless God who is really related to the world does seem incoherent.[53]

So we may assume that the overall structure of Craig's argument would be something like this: the concept of a timeless God together with dynamic time is incoherent. But time is dynamic; therefore, God must be temporal.[54]

Craig calls the time experienced by God 'metaphysical,' or 'ontological,' or 'real' time, not to be identified with physical time.[55] Indeed, the failure to distinguish metaphysical time from physical time is the root of many misapplications of STR, from Stump and Kretzmann to Leftow, and a host of other writers who use STR to argue for static time.[56] Metaphysical time is A-theoretic,

---

[52] William Lane Craig, 'Divine Timelessness and Necessary Existence,' *International Philosophical Quarterly* 37 (1997). p. 219.

[53] 'God and Real Time,' p. 336.

[54] Almost incomprehensibly, in one place Craig says, 'A persuasive case can be made, I think, that physical time is a tenseless, B-theoretical time which has been abstracted from the richer, A-theoretical metaphysical time in order to rid scientific theories of indexical elements and thus render them universalizable.' William Lane Craig, 'The Special Theory of Relativity and Theories of Divine Eternity,' *Faith and Philosophy* 11 (1994), p. 25. The only way I can interpret this, in light of Craig's insistence that physical time is A-theoretic, is that in this one passage he is using the term 'physical time' here in a restricted sense to refer to the use of time as a parameter or a coordinate in theories of physics, and not to measured or clock time (what I have called 'physical time').

[55] See above, footnote 47.

[56] Craig, 'God and Real Time,' p. 339; 'Divine Timelessness,' pp. 218-19; 'The Special Theory of Relativity,' pp. 19-37.

but has no intrinsic temporal metric since a metric would either come from physical time where the laws of nature permit a regular clock, or from God's fiat.

Craig follows in the classical Newtonian tradition of positing absolute time.[57] I have already indicated my reasons for preferring a relational to an absolute theory of time (see Chapter 2), and will not rehearse them here.

Concerning the relation of a temporal God to the created order, Craig has more to say, primarily about creation itself. Much of Craig's concern stems from his defense of the Kalām Cosmological Argument. By arguing that it is logically impossible that an infinite series of events has transpired, Craig concludes that the universe had a beginning. Various opponents of the Kalām argument have raised a number of questions about the meaning of the concept of a first moment of time. Craig has offered several alternatives in response in his early articles, one alternative being an Ockhamist-inspired one whereby God's having had an infinite past at the time of creation is a soft fact about God. If God had refrained from creating a temporal world, then there would not have been time but, by creating, God non-causally brought it about that time existed before creation.[58]

Alternatively, he suggests that perhaps there is, after all, an infinite series of events in God's consciousness.

> Given the existence of this cosmic time, it is my contention that the moments of God's real, A-series time, while not perhaps *identical* with the moments of measured, cosmic time, are nonetheless *coincident* with them. God's ontological time clearly exceeds cosmic time in that the former may have preceded the latter (imagine God leading up to creation by counting '1, 2, 3, . . . , *fiat lux*!'), but once cosmic time comes to exist, its moments would coincide with the moments of real time.[59]

Yet another alternative offered by Craig is that of 'God's being timeless without creation and in time subsequent to creation.'[60] Elsewhere he elaborates, claiming that God is causally but not temporally prior to the origin of the universe, 'For the Creator sans the universe, there simply is not time because there are no events; time begins with the first event, not only for the universe, but also for God, in virtue of his real relation to the universe.'[61] More recently, Craig has made it clear that he now adheres to this third view.[62]

---

[57] Craig, *GT&E,* chapter 5, 'The Classical Concept of Time,' pp. 143-62, and chapter 6, 'God's Time and Relativistic Time,' pp. 163-96.
[58] This alternative is suggested in William Lane Craig, 'Timelessness and Creation,' *Australasian Journal of Philosophy* 74 (1996), p. 652.
[59] Craig, 'God and Real Time,' p. 343.
[60] William Lane Craig, 'Graham Oppy on the Kalām Cosmological Argument.' *Sophia* 32 (1993), p. 8. This position is spelled out in more detail in *T&E*, pp. 217-37, and in *GT&E*, pp. 256-80.
[61] William Lane Craig, 'Prof. Grünbaum on Creation,' *Erkenntnis* 40 (1994), p. 327.
[62] Craig, *T&E*, pp. 217-37; and *GT&E*, pp. 256-80.

I find Craig's final theory of God's relation to the first moment of created time to be highly problematic. But because this is such a crucial issue, I shall reserve discussion for chapter 10 where I consider God's relation to creation.

Craig's work is thorough and compelling, but ultimately I find his substantial view of time inferior to a causal theory in which time is a relation grounded in causation. And I find his suggestion that God is atemporal *sans* creation, but temporal since creation, rather problematic.

Summary

The arguments from dynamic time to God's temporality seem to me to be compelling. In addition, some proponents of God's temporality have offered models of God's relation to physical time. The models vary, but all affirm two common themes: (i) God genuinely experiences succession in his mental states as well as in his relations to creation; and (ii) God is temporally present at every moment of physical time. The omnitemporal model that I shall present in the next chapter also affirms these two themes.

The Argument from the Incarnation

One additional argument for divine temporality must be noted briefly. While it has not figured large in the literature, it will, if sound, add additional support to the conclusion that God is temporal. This is the argument from the Incarnation

Orthodox Christians believe that God became incarnate. The Second Person of the Trinity took on human nature and became the man, Jesus of Nazareth, but in doing so did not cease to be fully God. Since the Council of Chalcedon (AD 451), this has been expressed by saying that Jesus Christ is 'one person' with 'two natures.' It is beyond the scope of this book to explore or defend the Chalcedonian formula.[63] So although I believe the antecedent is true, the argument may be construed as a conditional argument: if Jesus Christ is one person with both a divine and a human nature, then the Incarnation poses certain objections to the doctrine of God's timelessness.

How is the Incarnation supposed to count against God's timelessness? Thomas Senor puts the argument succinctly:

12. Jesus Christ was the bearer of temporal properties.
13. No bearer of temporal properties is atemporal.
14. Jesus Christ = God the Son (a divine person).
15. God the Son is not atemporal.

Senor continues:

---

[63] I will note only one recent reference, notable for its thoroughness: Stephen T. Davis, Daniel Kendall and Gerald O'Collins, ed., *The Incarnation: An Interdisciplinary Symposium on the Incarnation of the Son of God* (Oxford: Oxford University Press, 2002).

Now, Christian theology does not want to identify one member of the Trinity with any other, much less with them all. So the atemporality [*sic*; read: temporality] of the Godhead does not obviously follow from [(15)]. It does follow, however, that there exists a temporal divine being and, a fortiori, atemporality is not essential for divinity.[64]

As I shall argue in the next chapter, if an entity is atemporal, it is permanently or immutably so.[65] But it stretches the theological imagination to attribute permanent timelessness to two members of the Trinity and temporality to another. Although this does not prove that God is temporal, it does make atemporality look suspect.

How might an atemporalist respond to the problem raised for her view by the Incarnation? The first response would be to retreat to 'mystery' or 'paradox.' So Nelson Pike, after noting that 'it could hardly escape notice that the doctrine of God's timelessness does not square well with the standard Christian belief that God once assumed finite, human form,' and allowing that 'as a man, of course, God had both temporal extension and temporal location,' simply sidesteps the issue: 'The claim that god assumed finite and temporal form is not supposed to fit well with other things that Christians believe about the nature of God.'[66] Now, the category of 'mystery' is appropriate when finite and fallible human minds address awesome questions about God, but I believe it is a category to be invoked later rather than sooner. And Pike's retreat to 'mystery' is too soon.

A second response is the so-called 'qua move.'[67] Consider the following two propositions:

A. Jesus Christ was omnipotent.
B. Jesus Christ was not omnipotent.

To avoid the blatant inconsistency, some Christian philosophers have claimed that (A) and (B) are ambiguous. Christians, they say, are committed only to the following:

---

[64] Thomas D. Senor, 'Incarnation, Timelessness, and Leibniz's Law Problems,' in *God and Time*, ed. Gregory E. Ganssle and David M. Woodruff (New York: Oxford University Press, 2002), p. 220.

[65] Roughly, if an entity were to change its temporal mode of being from atemporal to temporal, then in some difficult-to-express sense, the entity has undergone a *temporal* change. I say 'difficult to express' because temporal indicators such as 'before' and 'earlier' don't apply in a straightforward sense. Assuming the case in the text is possible, it would not be correct to say that the entity was atemporal *before* it was temporal, since that would be to ascribe an A-predicate to an entity that by definition possesses no A-properties. Nevertheless, there would have been a change involving time. For more, see the discussion of atemporality in the following chapter.

[66] Nelson Pike, *God and Timelessness* (London: Routledge & Kegan Paul, 1970), pp. 172-3.

[67] The 'qua move' is argued for by Eleonore Stump and Norman Kretzmann, 'Eternity,' *Journal of Philosophy* 78 (1981): 429-58, and by Douglas K. Blount, 'On the Incarnation of a Timeless God,' in *God and Time*, ed. Ganssle and Woodruff, pp. 236-248.

A*. Qua God, Jesus Christ was omnipotent.
B*. Qua man, Jesus Christ was not omnipotent.

Conjunctive propositions of the type 'x is F qua M and x is not F qua N' are known as reduplicative propositions. Rephrasing (A) and (B) as reduplicative sentences is not sufficient in itself, however. Senor identifies three ways of understanding the semantics of reduplicative sentences, and claims that the only understanding that avoids the problem involves both a commitment to nominalism and a denial of the Law of the Excluded Middle for properties.[68] Whether the atemporalist should be willing to pay such a high price to retain atemporality while avoiding the problem posed by the Incarnation is not for me to say.

Brian Leftow offers a twist on the 'qua move.'[69] He begins by defending an account of the Incarnation in which the Son of God, the Second Person of the Trinity, assumed a human body (B) and a human soul (S). In a normal case, the mereological sum S+B constitutes a human being, but in the Incarnation it is the Son+S+B that does.[70] So, claims Leftow, we may legitimately predicate something of a part of the whole that is not true of the other part or of the whole. Thus, it is true that S+B began to exist, but false that the Son began to exist or that the mereological whole Son+S+B began to exist.

The obvious objection is that substances are not simply mereological sums; the Son+S+B is a single substance (on pain of heresy). Leftow anticipates this, arguing that it is often causal relations that unite parts into substances and not mereological sums. Now, I would agree with Leftow on this; the unifying relation of compound or composite substances is often causal.[71] However, this response puts Leftow in the position of arguing that the timeless Son stands in a unifying causal relation with the temporal parts S+B such that a substance (the Son+S+B) with both timeless and temporal parts results. Of course, this relies on the success of an account of timeless causation, which is, as we have seen, highly problematic. If I am right about that, then Leftow's attempt to save the 'qua move' does not succeed.

At the very least, the argument from the Incarnation raises a serious problem for the atemporalist, and the available responses to it are beset with significant difficulties.[72]

---

[68] Senor, 'Incarnation, Timelessness, and Leibniz's Law Problems,' p. 231.

[69] Brian Leftow, 'A Timeless God Incarnate,' in Davis *et al.*, *The Incarnation*, pp. 273-99.

[70] This amounts to a tripartite claim about the Incarnation, a claim that must do some fancy footwork to avoid falling into the heresy of Nestorianism. See Alvin Plantinga, 'On Heresy, Mind and Truth,' *Faith and Philosophy* 16 (1999), pp. 182-93, for a careful discussion of tripartite versus bipartite views.

[71] See Joshua Hoffman and Gary S. Rosenkrantz, *Substance: Its Nature and Existence* (New York: Routledge, 1997), pp. 73-90; E.J. Lowe, *The Possibility of Metaphysics: Substance, Identity, and Time* (Oxford: Clarendon Press, 1998), pp. 164-9.

[72] See also Wolterstorff, 'Unqualified Divine Temporality,' pp. 209-10.

Conclusion

Chapters 5 through 8 have examined the philosophical theological tradition. We have seen that as the medieval tradition grew, culminating in the work of Thomas Aquinas, it became more and more difficult to maintain Neoplatonism along with either a psychological or a dynamic view of time that could incorporate God's atemporality in a solution to the foreknowledge/freedom dilemma. That dilemma was the driving issue for the medieval philosophers, and so it is not surprising that later philosophers such as Duns Scotus and Ockham began to consider both the reality of temporal becoming and some sort of temporality of God's being.

Indeed, this seems to me to be a clear case where contemporary metaphysical analysis shows that a traditional theological concept stands in need of revision. With the contemporary formulations of serious theological and metaphysical objections to divine timelessness and significant proposals of models of divine temporality, the current balance seems to be tilting towards a temporalist view. The metaphysical sophistication of the models, coupled with a greater explanatory power relative to issues in philosophical theology (see Chapter 10), makes the temporalist view worthy of serious consideration.

PART III
OMNITEMPORAL GOD

Chapter 9

Omnitemporality

In this chapter I shall give an account of the concept of omnitemporality as the temporal mode of God's being. I shall begin by trying to motivate the concept in a rather basic, non-analytical way. Then I shall derive and defend definitions of four crucial concepts: temporality, atemporality, sempiternity and omnitemporality. Finally I shall apply the concept of omnitemporality to God's temporal mode of being.

Omnitemporality: An Informal Account

Theists are unanimous in affirming that God is in some sense eternal. But what does it mean to say that God is eternal? At the very least, it is to say that God has no beginning or end, that God's existence is temporally unbounded. Now clearly if God is atemporal, his existence has no temporal bounds. But the argument of the book to this point is that an atemporal conception of God's eternity is open to rather serious objections, and some additional objections will emerge below. Omnitemporality, which I explicate in this chapter, is a way of being eternal in the sense of being temporally everlasting.

To motivate the concept of omnitemporality, I shall begin with the spatial analogue, omnipresence. The Bible clearly claims that God is a spirit,[1] not a physical being, and philosophical theology uniformly agrees that God, not being a physical entity, does not have spatial location or extension.[2] God transcends space; he is not in space.

Now, the claim that an entity is not in space may be construed in two ways. First, it may mean that the entity has no spatial properties and stands in no spatial relations. Clearly, such entities would have no spatial extension and would occupy no spatial region. Abstracta such as numbers would seem to be paradigmatic examples of entities that are not in space in this first way.

However, the Bible claims that God is immanent in creation,[3] and philosophical theology agrees that the being of God is 'present to' all points in the universe. There is no point in physical space at which God is absent. But if God is not in space, how can he be 'present to' spatial locations? And if God is indeed

---

[1] For example, 2 Chronicles 6:18; John 4:24.
[2] Cf. Richard Swinburne, *The Coherence of Theism*, rev. ed. (Oxford: Clarendon Press, 1993), pp. 104-7.
[3] For example, Psalm 139:7-12; Isaiah 57:15.

'present to' spatial locations, how can he not be located at those points? The claim clearly is not that God is so big that he cannot be confined in the physical universe. Were that the case, we could speak of 'spatial parts' of God being at spatial locations. But God is non-physical, so this is not what is meant.

Omnispatial, Omnitemporal

A rough analogy may motivate the concept of omnitemporality:
- An omnispatial being is one that is present to every actual point in space, without thereby being located in physical space.
- An omnitemporal being is one that is present to every actual moment of time, without thereby being located in physical time.

In God's case, then, the claim that he is not in space must be construed in a second way. It is conceivable that an entity which does not occupy space, or equivalently which is not limited by a continuous set of spatial points which define its surface, nevertheless can be present to space. (This is in fact the way in which most cosmologists conceive of space-time singularities in the framework of General Relativity.[4]) Now any entity that is in space is present in the region of space which is enclosed by its surface, and present to those points in space adjacent to the surface. An entity that is not in space in this second way is not present in any spatial region but may be present to a spatial point or region.

If God were this kind of entity, just what kinds of spatial property and relation would be proper to him? A certain 'rough and ready' answer can be given to the question.[5] As noted, since God does not occupy space, he would not have properties of spatial dimension. But it would be proper to attribute spatial presence to him in a certain way. We could truly say 'God is here,' or 'God is there,' where 'here' and 'there' are indexicals relative to the speaker. Used of a physical entity, these expressions would entail spatial location or extension, but when used of God, no such entailment holds; God is wholly present to any location. Further, God himself could say, with reference to some place, 'I am here,' and with reference to some other place, 'I am there.' In this case, however, 'here' and 'there' would be indexicals relative to the designated place, and not to God as the speaker, since

---

[4] John Earman, *Bangs, Crunches, Whimpers, Shrieks: Singularities and Acausalities in Relativistic Spacetimes* (New York: Oxford University Press, 1995), chapters 2-3.

[5] This is not meant to be a thorough analysis of how an omnipresent spiritual being is related to space, but merely the drawing od certain parallels between omnipresence, conceived of as omnispatiality, and eternality, conceived of as omnitemporality. For a fuller account, see Paul Helm, 'God and Spacelessness,' in *Contemporary Philosophy of Religion*, ed. Steven M. Cahn and David Shatz (Oxford: Oxford University Press, 1982), pp. 99-110, or J. R. Lucas, *A Treatise on Space and Time* (London: Methuen, 1973), Section 3, 'The Theology of Space,' pp. 141-88. Helm holds to a timeless God, while Lucas defends a temporal God. Their accounts of omnipresence are clearly influenced by their conceptions of God's temporal mode of being, and the issues of God's relation to space and to time are much more closely linked than most studies acknowledge. But I shall not pursue the concept of omnipresence further here.

God is wholly present to every spatial location. Additionally, God would know the referent of any reference to a spatial location given in terms of any appropriate metric. He would know the spatial coordinates of any spatial point relative to any reference frame, or the trajectories of objects moving in space. Omnispatiality does not entail epistemic befuddlement concerning spatial location.

St Thomas Aquinas helps clarify these notions. He writes:

> God is present in all things; not indeed as part of their essence, nor as an accident; but as an agent is present to that upon which it works. . . . Therefore as long as a thing has being, God must be present to it. . . . No action of an agent, however powerful it may be, acts at a distance, except through a medium. But it belongs to the great power of God that He acts immediately in all things.[6]

Here Aquinas draws on the intuition that no agent can act in a place where it is not.[7] Since I am localized, I can act only on those points in proximal space. I can directly act on this keyboard (by means of my fingers, ignoring for present purposes associated and unconscious neurological events). This kind of action is direct. But to close the door over there I either have to change physical locations (so as to be able to act directly) or else use some tool—throw a book at it, perhaps. Such action is non-direct. Since God can act at any spatial location, he must be at, or present to, that location, without thereby being located in space. But God, being omnipresent, does not need to use tools or change location. He can act directly (and simultaneously) at every point in space.

Further, Aquinas maintains that since God is an incorporeal spirit, he is 'in place not by contact of dimensive quantity, as bodies are.' That is, one cannot locate the point where a physical body contacts God by specifying a point 'on God' using physical dimensions. But since God's indivisible incorporeal substance is 'outside the whole genus of the continuous' (as are angels and souls, and also universals such as whiteness), he concludes that 'as the soul is whole in every part of the body, so is God whole in all things and in each one.'[8]

Aquinas suggests three ways in which God's omnipresence is to be understood. He writes, 'God is in all things by His power, inasmuch as all things are subject to His power; He is by His presence in all things, as all things are bare and open to His eyes; He is in all things by His essence, inasmuch as He is present to all as the cause of their being.'[9]

First, to say that 'all things are subject to His power' as part of an explication of omnipresence is to say that God can act directly at any point in the universe.

---

[6] St Thomas Aquinas, *Summa Theologica*, tr. The Fathers of the English Dominican Province (New York: Benzinger Brothers, 1948), 1a.8.1, hereafter *ST*.

[7] Aquinas would doubtless have regarded this as a necessary truth. I leave to the reader the question of what Aquinas would have to say about the notion of non-locality in quantum physics.

[8] Aquinas, *ST*, 1a.8.2.

[9] Ibid., 1a.8.3.

Second, if God is omnipresent, then he can directly know every point in space. This is what Aquinas means when he says that 'all things are bare and open to His eyes.'[10] I can directly observe only what is in my visual field, and with a decreasing acuity as the distance increases. To know who is at the door, I must either change locations or request the information from my daughter who answered the doorbell. But God, being omnipresent, does not need to change locations or have information from another location transmitted to him in order to know who or what is at that location. He knows directly (and simultaneously) every point in space.

Finally, Aquinas maintains that God is omnipresent 'by His essence, inasmuch as He is present to all as the cause of their being.' To say that God is omnipresent is to affirm that God is the ultimate cause of every contingent thing. But if space is substantial, then God is the cause of space itself, and so he cannot be confined to a location in space, nor even to all of space itself. As the cause of space (and its contents), he can continue to exist perfectly well should the physical universe cease to exist.[11]

The picture that emerges is of a being who, while being the creator of space, and transcending space, still is present in his whole being to every point in space. And indeed, even while retaining God's spatially-transcendent nature, we could call attention to his immanence by speaking of him as omnispatial. Although the concept is not altogether straightforward, neither is it completely opaque. To be omnispatial would be to exist in a metaphysical analogue of space, the 'spiritual' mode of being.

Now while far from perfect, this suggests to us an analogy with God's relation to time. An entity that is eternal may be said to be not subject to temporal limits (which I shall leave deliberately vague for the present), but as with the spatial analogue, this may be construed in two ways. If an entity is not in time in the first way, then it has no temporal properties and stands in no temporal relations. This way of being eternal would be atemporal or timeless existence. Again, abstracta would qualify as paradigmatic examples.

But an eternal entity may exist but not be subject to temporal limits in a second way if it is not contained in, but is present to, time. As such it would be a temporal analogue of an omnispatial entity—call it omnitemporal. To be omnitemporal would be to exist in the metaphysical analogue of time, which I have called metaphysical time (more on this shortly).

But immediately a disanalogy with the spatial case surfaces. An omnispatial being does not have spatial limits. But if time is dynamic, as I have argued, it is not possible that an omnitemporal being will not have the present instant as a temporal limit. Even if it would be correct to understand its past

---

[10] Whether God 'sees,' 'hears,' 'smells,' or not is of no concern here. The point is that God has direct, noninferential knowledge. Aquinas uses the term 'knowledge of sight' in this sense.

[11] Aquinas means more than this in saying that 'God exists in everything . . . by His essence,' but he means at least this much.

existence as without a temporal limit, surely it cannot extend temporally into the (unreal) future. Physical time is bounded by the Big Bang as its earliest point, and the present instant as its latest. Metaphysical time need not be bound but an earliest moment (more on this in Chapter 10). But metaphysical time is bounded by the present instant. It thus would be a metaphysical necessity for anything that is present to time to have the present instant as a temporal limit.

But this is as it should be. For on dynamic time, the past does not exist now, although it did once exist (all past moments once were actual). But an omnitemporal being cannot now be present to nonexisting moments, even though that being was present to those moments when they were actual. This is analogous to saying, of an omnispatial entity, that it is only present to existing spatial locations, not to nonexistent ones.

As with the concept of omnispatiality, we can offer a rough and ready account of the kinds of temporal property and relation that are proper to God if he is indeed an omnitemporal entity. It would be proper for God to use temporal indexicals to describe his existence and actions. For any action R, God could truly say, using A-determinations, 'I am doing R now,' 'I did R in the past,' or 'I will do R tomorrow.' He could also truly say, using B-determinations, 'I did R before (simultaneous with, after) such-and-such a date.' Further, if God is omnitemporal, he could know the referent of any temporal location given in terms of any appropriate temporal metric. So he would know what time it is now, and he would know the length of any temporal interval relative to the appropriate reference frame in which it is measured, including the effects of relativistic time dilation.

But why call this 'omnitemporal?' a critic might ask. Perhaps omnispatial makes sense, since there are many spatial locations that are actual at any moment. But according to the theory of dynamic time, there is only one temporal location that is actual: the present. The answer is this. It is certainly possible that there are other worlds each with its own time, and an omnitemporal being would be temporally present in those worlds also. In particular, it is possible that the sum total of all created beings includes spiritual beings, angels, whose time would not be the physical time of this universe. So by claiming that God is omnitemporal, I am saying that he is present to the 'now' of the *aevum*, as the medievals called the time of the angels.

Analysis of Key Concepts

The informal account just offered lays out the general idea of what it would be for God to be omnitemporal. In this section I shall offer analyses of the concepts of temporality, atemporality, sempiternity and omnitemporality. I shall show that omnitemporality is a coherent concept, distinct from other proposed temporal modes of divine being, and suggest that this concept is thus the preferred temporal mode of God's being.[12]

Temporality

We naturally understand by 'temporal' the mode of existence of an object that is in time. But what is meant by 'in time'? Generally, I think we understand by this expression that the entity in question exists in the past or the present or the future, or exists earlier than or simultaneous with or later than some other entity that is also in time. So a temporal entity either possesses a monadic A-property (of pastness, presentness, or futurity), or it stands in a dyadic B-relation (of earlier than, simultaneous with, or later than).

Temporal Entities

Temporal entities share the following features:
- location in a temporal series
- concreteness (i.e. can stand in a causal relation)
- possibly contingent (if the relevant time is contingent; for example, physical time).

Thus we have:

> 1. T is a temporal entity iff$_{def}$ T possesses an A-property or stands in a B-relation to some other entity.

Several important consequences follow from this definition. First, a temporal entity is located in a (metaphysical or physical) temporal series in virtue of its A-properties or B-relations. That is, it has temporal location. To exist 'in time' then is to exist 'at a time.'[13]

A second important consequence is that (metaphysically or physically) temporal entities are concrete. Why? The reason has to do with the relation between concrete entities and causation. It is well known that the abstract/concrete distinction is notoriously difficult to draw, yet it will suffice for present purposes to give what is generally accepted to be a sufficient condition for concreteness: a concrete entity is one which is possibly the direct terminus of a causal relation. On

---

[12] What follows in this chapter is adapted from Garrett DeWeese, 'Atemporal, Sempiternal, or Omnitemporal: God's Temporal Mode of Being,' in *God and Time: Essays on the Divine Nature*, ed. Gregory E. Ganssle and David M. Woodruff (New York: Oxford University Press, 2002); used by permission.

[13] Strictly speaking there is a difference between a time and an instant: an instant is a metaphysical 'slice' of time, without duration, but with temporal location. A time is the complete collection of all events and entities that exist at an instant in a given world. But for my purposes here the distinction is not relevant; I shall use the two terms interchangeably.

the plausible interpretation of the conservation laws as causal laws,[14] every point of space-time is directly caused by an earlier point of space-time. Thus every space-time point is concrete even if unoccupied by a material object.[15]

Now, every space-time point has A-properties or stands in B-relations. Hence every space-time point is a temporal entity. Similar reasoning shows that space-time regions, as well as events and event-boundaries, are likewise possible termini of a causal relation. (And indeed, if we consider an example of a putative uncaused event, such as the emission of an alpha particle by a radioactive isotope, there is no reason to hold a priori that such an event could not be the possible terminus of a causal relation.) So it follows from these considerations that every temporal entity is also concrete.

A third consequence of (1) is that every physically temporal entity is contingent. An entity can have temporal location iff time exists, and there is no reason to believe the existence of physical time is logically necessary. To the contrary, it can be argued that physical time is contingent. States of affairs consist in an entity's having a property or standing in a relation. Since temporal entities have A-properties or stand in B-relations, they constitute temporal states of affairs. But we can conceive of a possible world with no temporal states of affairs, a (physically) timeless world.[16] If a possible world had, say, only three spatial dimensions and no temporal dimension, the world would be timeless, changeless, and contain no temporal entities. Now if conceivability is prima facie evidence of logical possibility, then the existence of physical time is not logically necessary, and temporal entities are contingent.[17]

Is this claim compatible with the claim that God is temporal? Yes, since if God is temporal, God's time would be metaphysical rather than physical.

Fourth, (1) has as a corollary that since a temporal entity is one that has temporal location, and a temporal location may be a point in time, a durationless instant, it follows that a temporal entity need not be an enduring entity.

---

[14] For the argument, see Michael Tooley, *Time, Tense and Causation* (Oxford: Oxford University Press, 1997), pp. 341-4, hereafter *TT&C*.

[15] This, of course, assumes that space is substantial, arguably a corollary of a causal theory of dynamic time. See ibid., pp. 258-64. If the reader rejects the thesis of substantival space, then the weaker claim is that actual space-time entities occupying space-time regions are possibly the termini of causal relations, which still yields the desired conclusion regarding the concreteness of temporal entities.

[16] Kant, of course, maintained in the 'Transcendental Aesthetic' of his first *Critique* (B46/A31) that time is the 'form of inner sense,' a 'necessary representation that underlies all intuitions.' He claims that while we can imagine eventless time, we can never conceive of the absence of time. I think Kant was just wrong about this. The possibility of timeless worlds is discussed below. See Immanuel Kant, *Critique of Pure Reason*, unabridged ed., tr. Norman Kemp Smith (New York: St Martin's Press, 1929).

[17] A timeless world W thus conceived would be *physically* timeless. W could be conceived as occupying an instant of metaphysical time, and although W *contained* no causal relations, the entire world would be the terminus of God's causal activity in that metaphysical instant. W thus would be intrinsically timeless but extrinsically temporal. As temporal, God would be 'relatively timeless' in W (since everything in W is timeless).

It may seem strange to hold that an entity can be said to exist in the case that its existence has no duration. But it seems to me that this is correct. Consider the event of my starting this paper. Given that I have started, the event of my starting must have existed. Clearly, that was an event located in time (it is past—an A-property; it is earlier than midnight—a B-relation), but, equally clearly, my starting did not persist through a span of time. The event of starting was an instantaneous event. Similarly, beginning a class, winning a race, ending a lecture, or losing a game, are all instantaneous events. Even though the same terms may often refer to a process, the process begins or ends at—is bounded by—a point in time. And these events are located in time; they have A-properties or stand in B-relations. Thus there would seem to be no reason not to regard them as temporal entities.

Brian Leftow, however, disagrees: 'A being is intrinsically timeless iff it does not 'contain' time, i.e., does not endure through time. If there are events without duration, they are intrinsically timeless even if they are located in time.'[18] But by my definition (1), being located in time is what constitutes an entity as temporal. However, Leftow regards an argument similar to the one I gave above as insufficient to show that instantaneous events are temporal and not timeless entities. So I shall offer another based on a spatial analogy. By definition, a point in geometry has no extension, but it has location. The point exists in space. It

---

[18] Brian Leftow, *Time and Eternity* (Ithaca, NY: Cornell University Press, 1991), p. 31, hereafter *T&E*. Leftow's reasons for taking this position are, I think, related to his own definitions of timelessness and eternity. Leftow desires to maintain that God is timeless, that is, he exists in an eternal (or atemporal) reference frame. But he also wants to allow that temporal entities may be present in eternity. Thus he defines an entity K as temporal 'iff K can have a location in a B-series,' and as timeless (atemporal) 'iff K can exist now but cannot be located in a B-series.' He continues:
> If these definitions hold no nasty surprises, they let us say that a temporal thing can occur within an atemporal reference frame without compromising the absolute distinction between temporal and timeless things or reference frames. . . . They also let us say that events A-occur in both eternity and time, but B-occur only in time. Yet temporal events that A-occur in eternity also B-occur in time, and so occur in eternity as ordered in timeless analogues of their B-relations. (p. 241).

Admittedly, this passage is not entirely clear, especially out of the context of the intricate argument of the book (see my discussion of Leftow's argument in Chapter 6 above). But I will make two observations about his argument. First, while Leftow desires to be neutral throughout as regards tensed or tenseless theories of time, there seem to be several places where his argument must assume a tenseless view or else it will not work. And that is consistent with my second observation. Leftow accepts—too easily, in my opinion—the Minkowski space-time manifold of STR as a true metaphysical description of the universe. Thus he cannot allow any absolute reference frame; he sees all reference frames—temporal and timeless alike—as relative. If, however, we reject the tacit acceptance of a tenseless view of time, and reject also the Einsteinian interpretation of STR, then there is no reason to adopt Leftow's more restrictive definitions. See Tooley, *TT&C*, chapter 11; Quentin Smith, *Language and Time* (Oxford: Oxford University Press, 1993), ch. 7; and William Lane Craig, *Divine Foreknowledge and Human Freedom* (Leiden: E.J. Brill, 1991), Appendix I.

would seem at least very odd to argue that the point was 'spaceless' even though located in space; rather, we would want to say it was a spatial point. And it would be most surprising if what was plausible in the spatial case was not analogous with what was plausible in the temporal case. So it is plausible to accept that a point in time is a temporal point, not 'timeless.' But this leads to a clarification of what it is to be timeless or atemporal.

Atemporality

Many objects have historically been thought of as being atemporal—that is, as not being part of the temporal world.

Atemporal Entities

Atemporal entities share the following features:
- no temporal location (no A- or B-properties)
- abstractness (cannot stand in a causal relation)
- necessity
- immutability.

It seems straightforward to define atemporal as the negation of (1):

    2.   A is an atemporal entity iff$_{def}$ A has no A-properties and stands in no B-relations.

Atemporal entities thus do not exist at any time.[19] How, then, do they exist, if indeed any do?

First, it can be argued that atemporal entities must be abstract. We saw above that temporal entities are concrete; can it be shown that atemporal entities are not concrete? On a causal theory of time, temporal relations are causal relations.[20] So if an entity is possibly the terminus of a causal relation, it possibly stands in a B-relation.[21] But by (2), atemporal entities stand in no B-relations, and

---

[19] Compare Nelson Pike's definition of 'timeless' as an entity that lacks temporal location and temporal extension: *God and Timelessness* (London: Routledge and Kegan Paul, 1970), p. 7.

[20] Any theory of time, whether dynamic or static, must give an account of three features of time: flow, direction and distance (temporal relations). If time is dynamic, a causal theory offers the most satisfactory account of all three features, and especially of direction—'time's arrow.' Explanations of time's arrow grounded in time-asymmetric physical processes (e.g. entropic processes) or in 'indelible traces' left in the world can both be reduced to causal relations. See my discussion in Chapter 2.

[21] This assumes that causation is a temporal relation. At this point the atemporalist will object that this begs the question, that God's causal relation to the world is atemporal. But no satisfactory account of atemporal causation is available. Paul Helm attempts such an account in 'Eternal Creation,' *Tyndale Bulletin* 45 (1994), pp. 321-38, but admits that in his account '"causation" is used in an analogical or stretched sense.' The burden is on the

therefore atemporal entities cannot be concrete. If something must be either concrete or abstract (a reasonable supposition), atemporal entities must be abstract.

Second, there are good reasons to think that any atemporal entities that exist must exist necessarily.[22] The contingency of physically temporal entities is linked to the contingency of physical time in which they are located. So it seems reasonable to suppose that anything not located in physical time would not be infected by time's contingency. The shortest way to show the reasonableness of this is to produce candidates for the category 'necessary abstract atemporal entity.'

Platonic forms or universals, and mathematical and logical entities, are standardly offered as examples of atemporal entities. I cannot devote the time or space here to defending realism about universals. So we'll consider mathematical entities and assume that whatever we conclude of them we could also say of logical entities. William Kneale, for example, argues that numbers are atemporal:

> An assertion such as 'There is a prime number between five and ten' can never be countered sensibly by the remark, 'You are out of date: things have altered recently.' And this is the reason why the entities discussed in mathematics can properly be said to have a timeless existence. To say only that they have a sempiternal or omnitemporal existence (i.e. an existence at all times) would be unsatisfactory because this way of talking might suggest that it is at least conceivable that they should at some time cease to exist, and that is an absurdity we want to exclude.[23]

However, many philosophers do indeed want to say that numbers exist eternally but not atemporally. Steven Davis, for one, says that the number seven is eternal but not atemporal:

> But if the number seven is not just eternal but timeless, then . . . the following statements cannot be meaningfully made:
> The number seven existed on 27 July 1883.
> The number seven was greater than the number six during the whole of the Punic Wars.
> The number seven existed yesterday and will exist tomorrow.
> But the number seven is not a timeless being; all three of these sentences, in my opinion, are not only meaningful but true.[24]

---

atemporalist to provide a plausible account of atemporal causation that does not beg the question against a dynamic theory of time. Indeed, I have argued elsewhere that any theory of a timeless God who is related to the temporal series is committed to the B theory of time. See Garrett DeWeese, 'Timeless God, Tenseless Time,' *Philosophia Christi*, 2:2:1 (2000), pp. 53-9.

[22] The modality is metaphysical, not logical, necessity.

[23] William Kneale, 'Time and Eternity in Theology,' *Proceedings of the Aristotelian Society* 61 (1960-61), p. 98.

[24] Stephen T. Davis, *Logic and the Nature of God* (Grand Rapids, MI: Eerdmans, 1983), p. 17.

I believe that Kneale is right and Davis is wrong about this. If numbers were temporal entities, then, as has been shown, they would be concrete entities and would be possible termini of causal relations. But it is very odd to speak of causal relations with numbers; what could possibly be meant by saying that something caused (or was caused by) seven? Further, temporal entities are contingent. Although individual exemplifications of numbers may indeed be contingent, the numbers themselves, together with their fundamental logical and mathematical relations, would seem to remain necessarily unchanged.[25] So, since we have good reason to believe that numbers are not contingent but are necessary entities, they must be atemporal.

How, then, should we account for Davis's examples? I believe the simplest answer is to maintain that all three examples are ill-formed statements that seem to express truth not because numbers have temporal location but precisely because they don't. Statements made about atemporal entities are made at a time, and a truth value may be assigned at that time, but this does not entail that the subject of the statement itself be located in time. A proposition about timeless entities that is tenselessly true is true now, but this does not entail that the subject of the statement itself be located at the present.

These considerations regarding numbers lead to the following: Mathematical statements are necessarily true if true, and necessarily false if false. The only way a mathematical statement could fail to have a truth value would be if the numbers involved in mathematical statements failed to exist. But since it has been shown that numbers necessarily exist, mathematical truths are necessarily true, hence abstract and atemporal. I believe similar reasoning could be applied to logical truths, since the foregoing argument tacitly rests on a logicist analysis of mathematics.

A third statement we can make about atemporal entities is that, in addition to being abstract and logically necessary, it is logically impossible that they change. This follows directly from (2). For x to change is for x to have a property P at t_1 that x does not have at t_2. But for this to be true, x must occupy a location in a B-series (that is, stand in a B-relation) such that the state of affairs x's-having-P-at-t_1 is earlier than the state of affairs x's-not-having-P-at-t_2. But by (2) atemporal entities do not stand in B-relations. Therefore atemporal entities must be changeless.

Given that atemporal entities are necessary, changeless things, it is clear why many philosophical theologians have wanted to say that God is atemporal. An atemporal God would be immutable, immaterial, and necessary. All seem to be attributes of God that a traditional theist would want to maintain. The tradition, as we saw in Chapters 5 and 6, has a fine historical pedigree. But surely no theistic philosopher or theologian would want to say that God was an abstract entity! So if my argument is correct, there is good reason to suspect that God is not atemporal, or if he is, then the defender of atemporality must make plausible the claim that God is the unique atemporal concrete entity.

---

[25] See George Bealer, *Quality and Concept* (Oxford: Oxford University Press, 1982), p. 123.

The atemporal view of God's temporal mode of being is not without further difficulties, which have led the majority of contemporary philosophers of religion and philosophical theologians to hold to divine temporality, as we saw in Chapters 7 and 8. I shall not summarize the arguments here.

Proponents of divine temporality have had to propose some temporal mode of God's being which nevertheless maintains his essential attributes. They have typically used words such as 'everlasting,' 'sempiternal,' 'relatively timeless,' or 'omnitemporal' to describe God's being. The final two sections of this chapter will be devoted to defining and drawing the necessary distinctions between two of these terms. Although my focus is on philosophical theology, the definitions and distinctions are as general as those given above of temporality and atemporality.

Sempiternity

The concept of a sempiternal entity may be derived from the etymology of the term: an entity that exists at all times. But this, of course, is ambiguous. What is generally meant is an entity that, once it has begun to exist, cannot, as a matter of nomological necessity, cease to exist at any subsequent moment.

Sempiternal Entities

Sempiternal entities share the following features:
- temporal location
- possibly stands in a causal relation
- nomologically necessary.

As a definition,

    3.   S is a sempiternal entity iff$_{def}$ if S exists, then (i) S possesses an A-property or stands in a B-relation, and (ii) S is nomologically necessary.

It might appear that (3) contradicts what was said above regarding the contingency of temporal entities. But this is not so. S is still contingent; only if S comes into existence is S nomologically necessary. If time were to cease, so would S, for there would be no time in which S could be located. In other words, what (3) attempts to capture is the concept of an entity that, once it has begun to exist in time, will necessarily exist as long as the time in which it is located exists. To distinguish the concept of sempiternity from temporality, we may make use of the notion of possible futures. It seems reasonable to believe (*pace* determinists) that the future can go any one of a number of possible ways. The future is 'a garden of forked paths.' If we picture time as something like a tree structure, then what we have is a universe in which the past is singular and real, the future is a branching tree of possibilities, and the present is the edge of growing reality as future possibilities collapse into the one reality.²⁶ Each different branch represents a possible future (or possible world) that gets 'cut off' as time advances.

Now, any particular entity that is physically temporal will exist in some of those possible futures, and not exist in others. However, a sempiternal entity, once it begins to exist, will continue to exist at all times subsequent to its coming into being, but only in those possible worlds in which the laws that render the sempiternal entity nomologically necessary continue to obtain. For example, suppose one family of branches of possible worlds represents those futures in which the universe, being sub-critical, collapses on itself in a Big Crunch. Some of these possible futures are such that no rebound universe results, so the space-time points of those crunched worlds would no longer exist. But other possible futures are such that some crunched worlds would result in rebound universes, and it is possible that in some of these rebound universes, different laws will obtain. In these possible worlds, therefore, some sempiternal entities, which existed in the past history of these worlds, will no longer exist. Thus sempiternal entities, being only nomologically necessary, do not exist in all possible worlds.²⁷

What sorts of entity might be sempiternal? The most likely candidates are contingent entities such as Planck's constant, the charge of an electron, the net sum of mass/energy in the universe (given that conservation laws are true natural laws),²⁸ and perhaps also fundamental particles such as protons (if they do not decay) and possibly quarks or strings. There are possible worlds in which these things never exist, or have different expressions. Further, they have not all existed from the beginning of time, assuming some version of Big Bang cosmology is correct. But it seems to be nomologically necessary that once they began to exist in the actual world, they necessarily exist as long as time—the physical time of their respective worlds—exists. Indeed the laws themselves would seem to be sempiternal.

Sempiternal entities share the property of contingency with temporal entities, and the property of changelessness with atemporal entities. Some sempiternal entities clearly are concrete things (fundamental particles); and while

---

²⁶ This notion of time has been advocated by Storrs McCall, *A Model of the Universe: Space-Time, Probability, and Decision* (Oxford: Clarendon Press, 1994). McCall's realism about possible futures seems to have grown over the years since he first adumbrated his concept of branching futures in 'Objective Time Flow,' *Philosophy of Science* 43 (1976), pp. 337-62.
²⁷ While this might satisfy some process theologians, most in the Judeo-Christian tradition would want to say something stronger of God.
²⁸ That these entities are represented by numbers does not count against their contingency, since the numbers express a relation or a determinate value in certain units. I do not mean to imply that the numerical value of physical constants is immutable. Certainly different units of measurement yield different values. Further, Field has argued that it is a mistake to regard scientific theories as entailing the existence of numbers. See Hartry Field, *Science Without Numbers: A Defense of Nominalism* (Princeton, NJ: Princeton University Press, 1980). Field may or may not be correct in this. The point at issue, though, stated more precisely, is that the mathematical expression of the constant is fixed.

some would seem to be abstract things (Planck's constant, statements of physical laws), it is the numerical value or the mathematical expression that is abstract, not the entity itself.

Omnitemporality

The final temporal mode of existence to be discussed is omnitemporality. Often one finds 'omnitemporal' in the literature used as equivalent to sempiternity, but one can also find it used as equivalent to atemporal. I propose to use 'omnitemporal' for an entity that is metaphysically temporal and exists necessarily.

If we understand this entity as metaphysically (as opposed to physically) temporal, and if 'existing at all times' introduces a metaphysical modality, then the entity exists of metaphysical necessity.[29] But it follows from the topology of dynamic time that the 'now' of metaphysical time coincides with the 'now' of any possible physical time,[30] so an omnitemporal entity will be temporally present at every present moment of any possible physical time.

Omnitemporal Entities

Omnitemporal entities share the following features:
- temporal properties with respect to metaphysical time
- metaphysically necessary
- present to all actual moments of any temporal world.

As a definition, we have:

---

[29] Philosophical theologians debate whether the proper modality to ascribe to God's existence is logical or metaphysical. I believe that the latter is correct. See Richard Swinburne, *The Christian God* (Oxford: Clarendon Press, 1994), pp. 144-9, for a defense.

[30] Greg Ganssle has suggested to me that an objection to this claim is the relativity of simultaneity entailed by the standard Einsteinian interpretation of STR, which would have the consequence of 'a strange fracturing of God's consciousness' which follows from the loss of the transitivity of simultaneity on the Einsteinian interpretation. This seems to be a problem for any view of God's temporality if simultaneity is relative.

The heart of the problem is Einstein's operationalist definition of time, as shown by Lawrence Sklar, 'Time, Reality, and Relativity,' in *Reduction, Time and Relativity*, ed. Richard Healey (Cambridge: Cambridge University Press, 1981). This leads to a failure to keep separate the concepts of metaphysical time, which grounds all causal successions, and the measurement of physical time, which will be affected by relevant physical laws.

However, alternative mathematical formulations of STR are available which are empirically equivalent to the standard interpretation, and which maintain absolute simultaneity. Certain of these formulations are arguably superior to the standard interpretation for reasons independent of the question of the relativity of simultaneity. See John A. Winnie, 'Special Relativity without One-Way Velocity Assumptions,' *Philosophy of Science* 37 (1970), pp. 223-38.

4. O is an omnitemporal entity iff$_{def}$ (i) O is necessarily metaphysically temporal; and (ii) O necessarily exists.

What this definition of omnitemporality tries to capture is what is intuitively contained in the notion of 'everlasting' as the temporal mode of divine existence. For a traditional theist, sempiternality will not do, since sempiternal things only exist contingently, while the theist wants to insist that God exists necessarily. And further, many objections to the notion that God is temporal are grounded in understanding God's time as physical time, thereby making God, in some sense, dependent upon the physical time of this universe. So if the objections to atemporality as the mode of divine existence are found to be compelling, then another mode must be found which includes necessary existence and metaphysical time. Omnitemporality as defined by (4) does this.

Stating that an entity is metaphysically temporal is to say that it is a temporal entity, but the temporal properties and relations that belong to it are defined with reference to metaphysical and not physical time. What constitutes metaphysical temporality is the same relation that constitutes any other temporality: causation. My suggestion is that the causal succession of mental states in God's conscious life grounds the flow and direction of metaphysical time.[31] And, given that God is creator and sustainer of the contingent order, his causal sustenance of every world will ground the time of that world.

As it is possible that there might not be an intrinsic metric to metaphysical time, it is possible that no quantitative temporal relations hold for O. What this means is that, although moments of a temporal world can be placed in a one-to-one correspondence with moments of metaphysical time, one could give no sense to the statement that a certain duration of metaphysical time lasted a certain number of seconds (days, years, and so on).[32]

The modality in both clauses of the definiens captures the intuition that if an entity is truly omnitemporal—that is, if it is present to all actual times—then it must exist in all possible worlds. It is easy to understand an omnitemporal entity existing in all temporal worlds. But what about the possibility of timeless worlds? It follows from (4) that metaphysical time is necessary, so can there be any timeless worlds? I suggest that there are two possible ways to think of timeless worlds. First, one could argue that atemporal worlds were simply abstract objects that exist timelessly, just as numbers do. This would have the consequence, as

---

[31] It is an interesting question of philosophical theology whether God's mental life consists essentially in a causal succession of mental states. If this were so, it would not be a limitation on God, since he is the cause of his own being, including his own mentality. Further, it would aid our understanding of the dynamic relations among the persons of the Trinity prior to creation (cf. the doctrine of *perichoresis*, the 'interpenetration' of the persons of the Trinity).

[32] Swinburne concludes, 'There would be no difference between a divine act of self-awareness which lasted one millisecond and one which lasted a million years': *The Christian God*, p. 140. Compare Psalm 90:4 and 2 Peter 3:8: 'With the Lord a day is like a thousand years, and a thousand years are like a day.'

seen above, that timeless worlds were necessary. But this is a consequence that not all would be willing to accept.

Second, one could argue that a timeless world is one that is intrinsically timeless but extrinsically temporal. That is, if W_A is an atemporal world, and M is the metaphysical temporal series of O, then W_A will have temporal location in M but will not have temporal duration in M. So a timeless world would be intrinsically timeless but extrinsically temporal, in that it would possess an A-property or stand in a B-relation to an entity in M. By showing that a timeless world is extrinsically temporal, we avoid the claim that timeless worlds are necessary.

It follows that if W_A were a non-actual possible timeless world, it would be an abstract entity and exist necessarily. But if W_A were an actual timeless world, it would be extrinsically temporal by virtue of having location in the metaphysical temporal order, and would be contingent (as indeed all concrete objects other than God are).

The claim that O is temporally present at every present moment of any possible time may now be stated more carefully. For any temporal world W and any time t that lies in the present at metaphysical time m, t is an instantaneous state of affairs, t is actual as of time t, and no state of affairs in W that is earlier than or later than t is actual as of time m.[33] Now, since simultaneity is a transitive and reflexive relation, any time or event which is simultaneous with time m will also lie in the present at time t. Consequently, to be present at time t is to be present at metaphysical time m and vice versa. So it is correct to say of O that 'O is (temporally) present,' while it would not be correct to say that of an atemporal entity such as, say, the number seven.

If omnitemporality is to meet the objections that have been raised against atemporality, then God as an omnitemporal being must be able to sustain temporal relations. As an omnitemporal being, God would be 'above' physical time but still temporally present and thus able to enter into relations with temporal entities. As I see it, this means not only that the temporal entity must in some sense be present to God, but also, contra such atemporalists as Leftow, Stump and Kretzmann, that God must have the A-property of presentness, in the sense just given. This is because I do not see that any account offered so far of atemporal causation is satisfactory, and I doubt that a satisfactory account can be formulated. And if my arguments above are correct, an atemporal entity is abstract and so could not enter into causal relations.

---

[33] This analysis is stated in tenseless terms so as to avoid possible ambiguities between metaphysical and physical temporal indexicals. The rationale for tenseless analysis of tensed statements is found in Tooley, *TT&C*, pp. 190-95.

Conclusion

Omnitemporality, as God's temporal mode of being, offers conceptual resources for understanding many theological claims. As omnitemporal, God would experience succession in his mental states, thus being much more in line with what we understand by 'person.' As omnitemporal, God could experience change in his relational properties, so that an individual would at one time be an object of God's wrath and at a later time, after redemption, an adopted child of God. God's providence is also more easily understood if God is omnitemporal, as is the efficacy of petitionary prayer. And if one desires to maintain a form of libertarian free will, an omnitemporal God coheres well with Ockhamist or Molinist solutions to the dilemma of foreknowledge and free will. Finally, in Trinitarian theology, the individual persons of the Trinity would exist in dynamic relationship before the creation of the temporal world, thus grounding the dynamic emphasis of the *perichoresis* or mutual indwelling, interpenetration, of the Trinity. And an omnitemporal God certainly makes understanding of the temporality of the Incarnation much easier.

Chapter 10

Implications of Omnitemporality

If the arguments of the previous nine chapters are cogent, then the notions of dynamic time and divine temporality will commend themselves to us in a strong way. I shall assume that the notion of omnitemporality, analyzed in the previous chapter, offers sufficient conceptual clarity to serve as the model of divine temporality, and I shall proceed on that assumption.

Clearly, the concept of divine omnitemporality is consistent with dynamic time. But an objector might well dissent. 'Look at the costs of your view,' she might say. 'Not only are we required to reinterpret the best-confirmed theories of modern physics along neo-Lorentzian lines, but we are required as well to set aside the general consensus of nearly 2000 years of philosophical theology that God is simple, strongly immutable, and atemporal. Those costs are too high to accept.' The objector, then, is urging something like a cost/benefit analysis of omnitemporality.

Fair enough. It is just such an analysis that I shall undertake, at least in a programmatic way, in this final chapter. There are indeed trade-offs—costs versus benefits—associated with both views. What I hope to show in this chapter is that the costs of adopting omnitemporality are more than acceptable in light of the benefits. I shall not revisit the discussions of the metaphysics of time and time in physics in Chapters 2 and 3, but rather shall concentrate on the theological costs and benefits of omnitemporality. The costs are related to certain traditional doctrines that cannot consistently be held along with omnitemporality. The benefits may be subsumed under the rubric of greater explanatory power.

The two attributes of simplicity and strong immutability, which have been ascribed to (and thought to entail the timelessness of) God, are clearly inconsistent with divine temporality, so the loss of these traditional doctrines is a cost of accepting omnitemporality. Further, the affirmation of divine temporality raises some difficulties for the traditional doctrine of God's absolute and complete (fore)knowledge of future contingent propositions. If these difficulties are irresolvable, they too will be a cost that omnitemporality will have to bear. I shall argue, however, that these costs are neither unacceptable nor irresolvable.

On the benefit side of the analysis, omnitemporality suggests approaches to certain traditional problems in philosophical theology. In particular, I believe that the concept is applicable to the doctrines of *creatio ex nihilo*, petitionary prayer, and divine providence. I shall demonstrate that omnitemporality offers satisfying solutions to traditional conundrums in these areas. I believe that this will show that the concept has significant explanatory power and so, at the very least, is a tenable theory meriting serious attention. More optimistically, I hope that this will show that the concept has greater explanatory power and so commends itself to our acceptance.

What Cannot Be Affirmed

If God is temporal and enters into real temporal relations with his creation, then the attributes of divine simplicity and strong immutability cannot be ascribed to God. I argued above in Chapter 5 that these attributes flowed more form a Neoplatonic metaphysical system than from biblical exegesis. It is clear that Neoplatonism heavily informed the philosophical context within which many of the central Christian doctrines were formulated. This in itself does not invalidate the doctrinal formulations, of course; all theology is done within some metaphysical context or other.[1] But it should be clear that if there are good reasons to reject the metaphysical background of certain doctrinal formulas, then it is probable that the formulas themselves will need revision, or may even prove untenable.

It is clear to me that the views of dynamic time and divine temporality that I have defended are in conflict with certain elements of Neoplatonic Christianity, in particular, the attributes of divine simplicity and absolute immutability under consideration. While some might argue that the loss of these two doctrines mortally wounds Christian theology, I do not believe that is the case, and in this section I will discuss both issues.

The Doctrine of Divine Simplicity

As I have argued, the motivation behind much of the medieval philosophers' attempts to explicate divine atemporality seems to be the doctrine of divine simplicity. Stump and Kretzmann write:

> The doctrine of God's absolute simplicity denies the possibility of real distinctions in God. It is, e.g., impossible that God have any kind of parts or any intrinsic accidental properties, or that there be any real distinctions among God's essential properties or between any of them and God himself.[2]

---

[1] See Nicholas Wolterstorff, Unqualified Divine Temporality,' in *God and Time: Four Views*, ed.Gregory E. Ganssle (Downers Grove, IL: InterVarsity Press, 2001), pp. 210-11.

[2] Eleonore Stump and Norman Kretzmann, 'Absolute Simplicity,' *Faith and Philosophy* 2 (1985), p. 353. Stump and Kretzmann make a valiant (but, in my opinion, ultimately unsuccessful) effort to explicate and defend the doctrine of simplicity. See also James Ross, 'Comments on "Absolute Simplicity",' *Faith and Philosophy* 2 (1985), pp. 383-91, and the response by Stump and Kretzmann, 'Simplicity Made Plainer: A Reply to Ross,' *Faith and Philosophy* 4 (1987), pp.198-201. Brian Leftow explicitly ties his theory of divine timelessness to the doctrine of simplicity; see his *Time and Eternity* (Ithaca, NY: Cornell University Press, 1991), pp. 66-71, where he offers a promissory note of a thorough defense of the concept in a forthcoming book on divine simplicity. I remain unconvinced that the classical doctrine of simplicity is defensible.

But I suggested that this doctrine is virtually unintelligible outside the context of Neoplatonic (or some similar) metaphysics. It seems highly paradoxical to claim that God, a substance, is identical with his essential properties, or that each of his essential properties are identical with one another.[3] Granted that when we predicate a quality of God, the predication is analogical. Still, it would seem to be the height of equivocation to claim that omnipotence is identical with omniscience.

Divine Simplicity
The notion of a simple being is inconsistent with temporality, and so must be abandoned if omnitemporality is the correct model of God's temporal mode of being. However, at least some of what has intuitive appeal in simplicity can still be maintained. And what is lost may be only a vestige of Neoplatonic metaphysics.

In addition to problems of conceptual clarity, the classical doctrine also lacks motivation without the background of Neoplatonic metaphysics. The lack of motivation is perhaps behind the apparent indifference to the doctrine, especially among Protestants. Theologian Donald Bloesch devotes less than one page to the concept,[4] and philosopher Edward Wieringa, in a book devoted to a philosophical analysis of the divine attributes, says:

> Several authors have argued that the doctrines of divine eternity and immutability have their source in Greek and neo-Platonic philosophy.... In particular, eternity is often derived from another philosophical doctrine, that of divine simplicity.... We shall not investigate the doctrine of divine simplicity; I have nothing constructive to say about it.[5]

For reasons made explicit by Stump and Kretzmann, and Leftow, who are reflecting the tradition dating at least to Anselm, divine simplicity is inconsistent with divine temporality, since it is clearly at odds with any possibility of God's experiencing real change. Omnitemporality, therefore, entails the rejection of the classical doctrine of simplicity.

However, the basic intuition embodied in the classical formulation might not be incompatible with omnitemporality. That intuition, I suggest, amounts to this: there is a natural unity of God's essential attributes, so that any single attribute

---

[3] In fairness, Aquinas tries to finesse this by denying that God's attributes are identical to each other, and claiming rather that *instances* of God's properties are identical to each other and to God. See *Summa Theologica*, tr. The Fathers of the English Dominican Province (New York: Benzinger Brothers, 1948), 1a.3.7. Whether or not this makes the concept any more acceptable I do not know; it remains inconsistent with divine temporality.

[4] And that in the volume entirely devoted to God's nature, out of a projected seven volume systematic theology. Donald G. Bloesch, *God the Almighty*, volume 3 of *Christian Foundations* (Downers Grove, IL: InterVarsity Press, 1995), p. 90.

[5] Edward R. Wieringa, *The Nature of God: An Inquiry into Divine Attributes* (Ithaca, NY: Cornell University Press, 1989), p. 173.

cannot be hypostatized and abstracted from God's essential nature.[6] While conceptually we may discuss God's wisdom apart from his omnipotence, divine wisdom cannot exist apart from divine power. Now this intuition, sometimes expressed by saying that God has no proper parts, stands in need of further analysis, but that is not the subject of this study. The point is that it is not at all clear that such analysis is inconsistent with divine temporality in the way that the classical formulation of simplicity is. Consequently the loss of the classical concept of divine simplicity, heavily dependent upon Neoplatonism, is not a loss I regard as serious.

The Doctrine of Strong Immutability

The Platonic motivation for the notion of God's immutability is clear. In Chapter 5 I distinguished two senses of immutability. First is the strong sense of immutability:

> SI: An entity is strongly immutable iff necessarily, none of its properties can change.

This would be opposed to a weak sense of immutability:

> WI: An entity is weakly immutable iff necessarily, none of its non-relational properties can change.[7]

Strong Immutability or Weak Immutability?

A timeless God is strongly immutable—he necessarily cannot change in any respect. However, an omnitemporal God possibly can change in certain respects, and will change in his relations with the temporal world. But this is consistent with the understanding that God does not change in his essential moral and ontological nature.

To say God is weakly immutable is not to impute weakness to him; it is merely in contrast to the strong immutability view.

---

[6] See Richard Swinburne, *The Christian God* (Oxford: Clarendon Press, 1994), pp. 160-63.

[7] For my purposes here, a relational property should be understood as one that entails the actual existence of something other than the entity that exemplifies the property. Clearly, standing in a relation would be a relational property. But so would a property such as 'regarding x as F.' So if I am in the mental state of regarding that book on my desk as being blue, then I exemplify a certain relational property. But suppose the book were not there at all; suppose it were an illusion or a hallucination. Depending on one's theory of perception, I might be in a real relation with a sense-datum (which would actually exist). But in God's case, it is not possible that he is under an illusion or hallucinating. Other theories of perception generally deny that illusions or hallucinations are genuine instances of perceptions. So I take it that God's regarding x as F entails the actual existence of x (as well as x's being F).

Clearly, (SI) is incompatible with divine temporality. Consider the following example, which is consistent with orthodox Christian theology as well as biblical language.[8] During the time interval T1, God regards Paul as a rebellious sinner, and so rightly an object of his wrath. At moment t, which bounds T1 as its *terminus ad quem*, Paul accepts God's forgiveness through Jesus Christ as his savior. During interval T2, which is bounded by t as its *terminus a quo*, God regards Paul as his adopted son and rightly an object of his covenantal love. Now, if God is timeless and time is static, this story can be easily retold in tenseless terms. For God is (tenselessly) eternally related to Paul's space-time segment S1 (throughout T1) by the 'angry with' relation. And God is (tenselessly) eternally related to Paul's space-time segment S2 (throughout T2) by the 'loves' relation. Further, God regards (tenselessly) S1 as rebellious, and S2 as forgiven. This is compatible with (SI).

But if time is dynamic, the same story cannot be told. As I argued at the end of Chapter 7, if God is timeless, then time is tenseless. So, if time is dynamic and God is temporal, there has been a real change in God's internal mental state (his 'regarding' of Paul as rebellious and later as forgiven) as well as in his relation to Paul (from 'angry with' to 'loving'). And these changes in properties and relations are incompatible with (SI).

But it is unlikely that (SI) is the proper construal of divine immutability. As seen in Chapter 4, the biblical writers did not balk at writing of God in ways that imply change. And more recent authors have also rejected the classical notion of (SI) for something along the lines of (WI).[9]

Is (WI) compatible with omnitemporality? I believe that it is. The changes God experiences in the story are relational in the sense that they require the actual existence of someone other than God if they are to be truly meaningful (that is, God is not 'regarding' merely the concept of Paul as rebellious, nor is he 'angry with' that concept). Consequently, none of God's essential, non-relational properties changes.

I believe that it is a reasonable to claim that the properties that are essential to deity are not relational. One reason why this is so is this. It is possible that God should never have created anything, in which case he would not possess real relational properties   There is nothing contradictory in the idea that, necessarily, the only properties of God that change are relational properties.

---

[8] Ephesians 2:1-5; Romans 8:15-17. Paul's story, of course, is told in Acts 9:1-10.
[9] For example, theologian Bruce Ware, in 'An Evangelical Reformulation of the Doctrine of the Immutability of God,' *Journal of the Evangelical Theological Society* 29 (1986), pp. 431-46, argues that God is ontologically and ethically immutable, but is relationally and emotionally mutable. See also Nicholas Wolterstorff, 'God Everlasting,' in *Contemporary Philosophy of Religion*, ed. Steven M. Cahn and David Shatz (Oxford: Oxford University Press, 1982). For a defense of (SI) see Jonathan L. Kvanvig and Hugh J. McCann, 'Divine Conservation and the Persistence of the World,' in *Divine and Human Action: Essays in the Metaphysics of Theism*, ed. Thomas V. Morris (Ithaca, NY: Cornell University Press, 1988), pp. 13-49.

This study is not, however, an analysis of the divine attributes. Philosophical theology can carry the project forward. Suffice it to say here that changes in God's relational properties are not the sorts of change that affect his perfect, divine nature, and so there should be no difficulty in reconceptualizing immutability as (WI) suggests, in line with a temporal God and dynamic time.

What is Problematic: God's Knowledge of Future Contingents

Traditionally orthodoxy has maintained that God, being omniscient, has absolute knowledge of the future, just as he has of the past and the present. Indeed, absolute foreknowledge (knowledge of what will happen in the future) is often thought to be an essential attribute of God.[10] Of course, if the B-theory of time is correct and if God is atemporal, then the future is real and God can know it. However, if time is dynamic and the future is not real, and if determinism or fatalism is not true, then how can even God know it? But I have rejected the B-theory in favor of a dynamic theory of time in which the future does not (yet) exist. Further, I reject the idea of divine atemporality. So must I give up God's absolute knowledge of the future?

God's Knowledge of the Future

If time is dynamic and God is temporal, some account must be given of how God can know now future contingent propositions (or states of affairs) that do not now exist.

Part of the discussion turns on whether or not future contingent propositions conform to the Principle of Bivalence. If they do, then either Middle Knowledge or Simple Foreknowledge may offer a satisfactory account.

If such an account fails, or if future contingents are logically indeterminate, then we must abandon the idea that omniscience entails knowledge of future contingents.

First, it should be clear that God's knowledge of the future is not rendered indeterminate simply by the contingency of the future. While any proposition about the future will always be contingent on the continuation of time, it is God who, in upholding the causal structure of the universe, determines whether or not physical time continues. So since the continuation of time is something God himself causes, and since he knows his own intentions infallibly, then he knows whether or not he will continue to sustain the universe. Thus his knowledge of future contingents is not undermined by the contingency of time.

But problems remain. Assume for a moment that it is true that the future cannot be known because it does not exist. Assume, that is, that bivalence must be rejected for tensed statements, and that the truth value of future-tense propositions

---

[10] The biblical record surely seems clear on this. For example, in Isaiah 40-49, God challenges the idolatry of Israel. His challenge to the idols in 45:21 is typical of the section: 'Who foretold this long ago, who declared it from the distant past? Was it not I, the Lord? And there is no God apart from me.' The ability to foretell what will come (and thus to foreknow it) is the distinguishing mark of the true deity.

is indeterminate. Would that count against God's omniscience? No. This is because omniscience means that God knows whatever it is possible to know, not that he knows everything. And if it is not, in fact, possible to know the future, then it cannot count against God's omniscience that he doesn't know the future. If future contingents are unknowable in principle, then God cannot be expected to know them any more than he could know the name of the first married bachelor or the color of a square circle.

But there is more to the story than that. There would be much about the future that God could know, even on a dynamic theory of time. First, having infallible knowledge of his own will, if God decided that at some future time t he would bring about some state of affairs S, regardless of the outcome of any other future contingents, then he would know that S would obtain at t. Or, if he decided that he would bring about S only if certain other events had already transpired, and if he knew that those events were certain to transpire at some time or another, then he would know that S would obtain in the future, although he would not know precisely when. Further, if he knew that a deterministic causal chain had been initiated, he would know when the final effect would certainly come about, should he not intervene. So there would be a class of future contingent propositions, perhaps even a significantly large class, which God would know. And as for most other future contingents, God would have probabilistic knowledge far more precise than any available to human forecasters. So if omnitemporality forces upon us a revised understanding of God's omniscience, the result is not a God who is totally in the dark concerning the future.

The view outlined in the previous paragraphs has come to be known as 'Freewill Theism,' or the 'Openness of God.'[11] The 'Openness' position has been the subject of extensive recent discussion,[12] and it is not my intent to enter into that debate here. I simply note it to show that the cost of giving up the classical view, according to which God's omniscience entails exhaustive knowledge of future contingents, may be acceptable to some Christians.

But that is not the end of the story. Ockhamist or Molinist solutions to the foreknowledge problem were outlined in Chapter 7, and it is clear that both are compatible with omnitemporality. If, as Ockham suggests, God knows all timelessly true propositions just in virtue of his divine essence, then he also knows

---

[11] This is the position defended by William Hasker, *God, Time, and Knowledge* (Ithaca, NY: Cornell University Press, 1989). On a more popular level, see Clark Pinnock, Richard Rice, John Sanders, William Hasker, and David Basinger, *The Openness of God: A Biblical Challenge to the Traditional Understanding of God* (Downers Grove, IL: InterVarsity Press, 1994); David Basinger, *The Case for Freewill Theism: A Philosophical Assessment* (Downers Grove, IL: InterVarsity Press, 1996); Clark H, Pinnock, *Most Moved Mover* (Grand Rapids, MI: Baker, 2001).

[12] For critical discussion see Bruce Ware, *God's Lesser Glory: The Diminished God of Open Theism* (Wheaton, IL: Crossway, 2000); or William Lane Craig *Time and Eternity: Exploring God's Relationship to Time* (Wheaton, IL: Crossway, 2001), hereafter *T&E*; Norman Geisler and H. Wayne House, *The Battle for God: Responding to the Challenge of Neotheism* (Grand Rapids, MI: Kregel, 2001).

the truth of future contingents in virtue of their timeless truth.[13] While not endorsing such a theory, William Lane Craig seems to believe such an approach is possible. Although Craig holds to a dynamic theory of time (see Chapter 8), he denies that future contingent propositions are exempt from the Principle of Bivalence. Thus the future is determinate, although not determined. Being determinate, there is (tenselessly) a truth value for future contingent propositions. As the being who infallibly knows all true propositions, then, God has infallible knowledge even of the future. Still, with respect to Ockham's view, Craig wonders, 'The question is, how does God know true future-tense propositions about causally undetermined events?'[14]

If an answer is to be given to Craig's question, it might well be located in Molina's theory of Middle Knowledge. In fact, Craig himself opts for this solution.[15] The obvious problem with the Middle Knowledge solution is whether or not there are true counterfactuals of freedom. I am inclined to think that there are,[16] and so I am quite congenial towards the Middle Knowledge solution. If this is correct, then it turns out that even though time is dynamic and God is temporal, he can still have absolute knowledge of the future.

No matter which solution is adopted, it is clear that omnitemporality is not as problematic as it might at first have seemed for the orthodox affirmation of God's omniscience.

What Can Be Asserted

Having shown that the costs associated with omnitemporality are minimal and should be acceptable, I shall now show its benefits. I claim that a conception of dynamic time, along with a temporal God, provides an explanatory context in

---

[13] For a discussion of the 'Thomistic Ockhamist' view, see Linda Trinkaus Zagzebski, *The Dilemma of Freedom and Foreknowledge* (New York: Oxford University Press, 1991), chapter 3; in opposition, see Michael Tooley, 'Freedom and Foreknowledge,' *Faith and Philosophy* 17 (2002), pp. 212-24.

[14] William Lane Craig, *Divine Foreknowledge and Human Freedom* (Leiden: E. J. Brill, 1991), p. 227; 'William of Ockham on Divine Foreknowledge and Future Contingency,' *Pacific Philosophical Quarterly* 69 (1988), pp. 117-35. Craig has indicated (personal conversation) that he sees nothing incoherent in saying that since truth logically supervenes upon all states of affairs that (tenselessly) obtain, and since God knows all truth, then God knows the truth about future contingent states of affairs. One could then say, as did Ockham, 'I just don't know how God knows this.' But Craig does not seem comfortable with this fideistic solution. Nor am I.

[15] Craig, *Divine Foreknowledge*, pp. 236-78; see also 'Middle Knowledge,' in *Four Views on Divine Knowledge*, ed. James Beilby and Paul Eddy (Downers Grove, IL: InterVarsity Press, 2001).

[16] See Alvin Plantinga, *The Nature of Necessity* (Oxford: Clarendon Press, 1974), pp. 174-80; Alfred J. Freddoso, 'Introduction,' in Luis de Molina, *On Divine Foreknowledge (Part IV of the Concordia)*, tr. Alfred J. Freddoso (Ithaca, NY: Cornell University Press, 1988), pp. 68-81; Craig, *Divine Foreknowledge*, pp. 246-69.

which three areas of philosophical theology can find simpler and more powerful explanations than is possible either on the assumption of static time or divine timelessness. The three areas that I have in mind are the doctrines of creation, providence, and petitionary prayer.

Creation

Augustine considered the question of what God was doing before he created the world, and asked why God did not create the world sooner.[17] The concept of omnitemporality offers certain answers to both these questions.

God and Creation

Absent an account of atemporal causation, the creation of the universe *ex nihilo* is better explained on omnitemporality.
However, temporalists must answer the question of God's temporal relation to the time of creation. Each of the three available answers has associated costs.
1. Metrically amorphous time. God's time prior to creation can be considered as one single moment of infinite duration; creation thus is the 'second moment' of God's time.
2. Timeless *sans* creation, temporal since creation. God's life has two phases. Apart from creation he is timeless and changeless, but at creation he freely becomes temporal due to his relations with the created temporal order.
3. Infinite metaphysical time. The persons of the Trinity have lived through an infinite series of past moments of metaphysical time, where such moments are defined by the causal succession of God's mental states.

Omnitemporality is consistent with all three options; the question is, which brings the greatest conceptual clarity to the issue?

Only a theory of divine temporality can make much sense of Augustine's first question. For if time began with creation at t_0, then there was no time prior to t_0 in which God could have been doing anything. The only sense of 'before' that would be relevant to such a question would be a logical or conceptual priority, but then the dynamic sense of 'doing' is lost. If God is temporal, however, there is a sense in which God exists before t_0, since t_0 picks out an instant in physical time, and God's metaphysical time exists prior to that instant. Given the conclusions reached in Chapter 3 about the topology of time, some moment in metaphysical time—call

---

[17] Actually, the first question is not really Augustine's. He writes,
> My answer to those who ask, 'What was God doing before he made heaven and earth?' is not 'He was preparing Hell for people who pry into mysteries.' This frivolous retort has been made before now, so we are told, in order to evade the point of the question. But it is one thing to make fun of the questioner, and another to find the answer. So I shall refrain from giving this reply (St Augustine, *Confessions*, tr. R. S. Pine Coffin [London: Penguin Books, 1961], XI.12).

Nevertheless, the question and the flippant retort are often attributed to Augustine; see, for example, Douglas J. Soccio, *Archetypes of Wisdom: An Introduction to Philosophy* (Belmont, CA: Wadsworth, 1992), p. 243. Still, Augustine does attempt an answer, as indicated above.

it T_c (for the time of creation)—would coincide with t_0, so there would be at least one moment of metaphysical time prior to T_c but no moment of physical time prior to t_0.[18]

But, it could be responded, giving sense to Augustine's first question only invites his second, and that turns out to be a distinct problem for any theory of divine temporality. The argument runs like this. Suppose that a series of moments (whether finite or infinite makes no difference at this point) had elapsed prior to God's creation of the world at T_c. Suppose, further, that the following principle is true:

> P. For any action A, God performs A only if God has sufficient reason to perform A.

Then:

1. God creates the world at T_c only if God has sufficient reason to create the world at T_c and not at some other time T^*.
2. Since prior to T_c nothing but God exists, the sufficient reason for God's creating at T_c cannot be external to God.
3. Therefore, the sufficient reason for God's creating at T_c must be internal to God.
4. But since God is immutable, there can be no intrinsic difference between T_c and T^*.
5. Therefore, if God had sufficient reason to create at T_c, he also had sufficient reason to create at T^*.
6. Therefore, there is no reason why God created at T_c rather than at T^*.

But (6) contradicts (1). So either (P) is false, or the assumption that a series of moments had elapsed prior to T_c is false. But no traditional theist would deny (P). Hence it is not the case that a series of moments had elapsed prior to T_c.

A proponent of temporality has one clear response at this stage of the argument, and that is to deny that (P) is being used properly. If (1) is supposed to be an inference from (P), then what is needed is (P*):

> P*. For any action A and any time t, God performs A at t only if God has sufficient reason to perform A at t.

---

[18] Here the instant picked out by t_0 would be the instant that constitutes the limit of physical time. There are different ways to conceive of this instant depending upon different cosmological models. See Robert C. Russell, 'T=0: Is it Theologically Significant?' in *Religion and Science: History, Method, Dialogue*, ed. W. Mark Richardson and Wesley J. Wildman (New York: Routledge, 1996), pp. 201-43.

But why should a theist believe this version? (P*) seems to rely on another principle that every decision to act is determined by a reason that exists (lies in the past) at the time of the decision. And while that principle might be applicable to Buridan's ass, it is not at all clear that it applies to free agents, let alone to God. In particular, why assume that God's decision to create at T_c is not free? In other words, why should God's decision at T_c require a sufficient reason that lies in the past at T_c? Nicholas Wolterstorff asks:

> Why *must* there be such a reason? When I reflect on my getting out of bed in the morning, I find that after lying awake for a while I just, at a certain moment, decide to get up. I don't have a reason for deciding to get up precisely when I do; I just do at a certain moment decide to get up.... Why does it have to be different for God?[19]

Brian Leftow uses an argument similar to this to reach a different conclusion.[20] He argues that the denial of a series of moments prior to God's act of creation means that God's act must be timeless. But this conclusion does not follow. If Richard Swinburne is correct (see the discussion in Chapter 8), then God's existence prior to the creation of the universe is temporally undifferentiated and has no intrinsic metric. Consequently while God's existence has temporal duration, there is no truth of the matter as to how long the duration lasts, and nothing more can be said than that it is one moment in God's (metaphysical) time. Thus God's decision to create the universe occurs in the first moment of metaphysical time, and T_c is the second moment. So the answer to Augustine's second question is that, strictly speaking, God created the world as soon as he could—that is, at the second moment of his existence.

Be that as it may, the discussion raises a difficult problem for all temporalists—namely, how to conceive of God's temporality prior to creation, or, alternatively, God's relation to the time of creation. I claimed in the previous chapter that God's time is metaphysical time, not physical time, so the fact that physical time began at creation forces the question of how long—how many moments of metaphysical time—passed before T_c. It seems that there are three possible responses available to the temporalist, each of which brings with it some thorny consequences. But if the consequences are not incoherent, and if divine temporality is to be preferred on other grounds, then the temporalist will have to choose which bullet to bite.

The first option would be to follow Swinburne and claim that exactly one amorphous moment of metaphysical time had passed before T_c. But Craig is not happy with Swinburne's response.[21] Craig's objection to Swinburne's view is that it does not block the inference that God has had an infinite past: 'Thus the

---

[19] Nicholas Wolterstorff, 'Unqualified Divine Temporality,' p. 238.
[20] Brian Leftow, 'Why Didn't God Create the World Sooner?' *Religious Studies* 27 (1991), pp. 157-72.
[21] Indeed, Craig critiques what he refers to as 'the Oxford school' on this issue, including Swinburne, J. R. Lucas and Alan Padgett: Craig, *T&E*, pp. 233-5.

amorphous time prior to creation would be infinite, even though we cannot compare the lengths of separate intervals within it. Thus, all the difficulties of an infinite past return to haunt us.'[22]

Why does Craig desire to avoid an infinite past? Because he believes that an actually infinite past is impossible. As the foremost contemporary defender of the Kalām Cosmological Argument, Craig argues forcefully for the impossibility of an infinite past—a crucial premise of the argument.[23]

According to Craig, Swinburne's 'metric conventionalism' is the fatal flaw in the approach:

> According to metric conventionalism, there is no objective fact of the matter concerning the comparative lengths of separate temporal intervals. But metric conventionalism does not hold that there really are no intervals of time or that no intervals can be objectively compared with respect to length. . . . [But] in the case of intervals which are proper parts of other intervals, the proper parts are factually shorter than their encompassing parts. But this implies that prior to creation God has endured through a beginningless series of longer and longer intervals. In fact we can even say that such a time must be infinite. For the past is finite if and only if there is a first interval of time. (An interval is first if there exists no interval earlier than it, or if there exists no interval greater than it but having the same end point.)[24]

There are two objections here. First is the argument that even metrically amorphous time allows for objective differences in the lengths of temporal intervals, and second is the argument that such amorphous time would lack a first interval and so would be infinite.

Are Craig's objections decisive? I don't think so. Consider first his argument that amorphous time still allows for objective differences in the lengths of temporal intervals. Craig supports the conclusion in this way. Let T_c be the moment of creation, and let t_1, t_2, and t_3 be points (instants?) earlier than T_c (see Figure 10.1). Craig's argument is that there is an objective difference in the length of the interval t_3-t_1 and t_2-t_1. This difference is objectively real because interval t_2-t_1 is a proper part of interval t_3-t_1.

Figure 10.1 Intervals in metrically amorphous time

Source: Adapted from Craig, *T&E*, p. 234.

---

[22] Ibid., pp. 234.
[23] William Lane Craig, *The Kalām Cosmological Argument* (New York: Macmillan, 1979; reprint Eugene, OR: Wipf and Stock, 2000), pp. 79ff., hereafter *Kalām*.
[24] Craig, *T&E*, pp. 234-5.

Now, Swinburne would reply that such judgments could not be made since they rely on specifying points in the temporal series as boundaries of the intervals, which *ex hypothesi* one cannot do in amorphous time. Craig's rejoinder would be that the epistemic inability to make the judgments in question does not alter the mereological principle that any whole must be greater than any proper part of itself. But this rejoinder is unsuccessful, in my opinion. Why? The points t_1, t_2, and t_3 must be abstract points, and any interval would contain an infinite number of such points. As in the analogy of a line, which 'contains' an infinite number of points, the temporal points are imposed upon the continuum of time logically subsequent to the existence of the continuum. So the points would not themselves entail the existence of an actual infinite.[25] The intervals themselves would also be abstract (as are the spatial intervals that generate many of Zeno's paradoxes). It is a fundamental axiom of transfinite mathematics that an infinite set can be put into a one-to-one correspondence with a proper subset of itself (both sets having the same cardinality). So Craig's claim that any one of the intervals is larger than one of its proper subsets fails.

Absent the argument that there are objective truths about temporal intervals in metrically amorphous time, it becomes difficult to see how the second argument goes through. For there would then be only one interval—the amorphous moment—and that would meet Craig's definition: 'An interval is first if there exists no interval earlier than it, or if there exists no interval greater than it but having the same end point.'

I conclude, then, that Craig has not succeeded in showing Swinburne's view to be incoherent, entailing the same problems as those accompanying the existence of an actual infinite in Craig's defense of the Kalām argument. So the temporalist might well agree to 'bite the bullet' of a single amorphous moment of metaphysical time prior to creation.[26]

Craig has offered his own proposal, which, if successful, gives the temporalist a second way to conceive of God's relationship to the moment of creation. He proposes that God exists 'timelessly sans creation and temporally from the moment of creation':

> But why could there not be two phases of God's life, one atemporal and one temporal, which are not related to each other as earlier and later?. . . In reality,

---

[25] In a defense of the Kalām argument, Moreland argues that 'an actual infinite cannot exist in the spatio-temporal cosmos with finite, moveable, relevantly extended members, not that an actual infinite cannot exist in the realm of abstract objects.' J. P. Moreland, 'A Response to a Platonistic and a Set-Theoretic Objection to the Kalām Cosmological Argument,' *Religious Studies* 39 (2003), pp. 373-90.

[26] Craig allows that there is a sense in which this amorphous moment counts as temporal because 'it exists literally before God's creation of the world and the inception of metric time. That fact may be advantage enough for some thinkers to embrace such a conception of divine eternity sans the world; it is not to be downplayed.' William Lane Craig, *God, Time and Eternity: The Coherence of Theism II: Eternity* (Dordrecht: Kluwer Academic Publishers, 2001), p. 270, hereafter *GT&E*.

> God existing sans creation is entirely alone, utterly changeless, and perfect, and not a single event disturbs his immobility. There is no before, no after, no temporal passage, no future phase of His life. There is just God, changeless and solitary.... [T]he state of affairs of God existing changelessly sans creation is timeless.[27]

I confess that I'm baffled by Craig's view. It seems to entail the following conjunction: 'God is (tenselessly) timeless and God is (present tense) temporal.' I argued in Chapter 9 that statements made about timeless entities are made at a time, and a truth value may be assigned at that time; a proposition about timeless entities that is tenselessly true is true now. If this is correct, then if God is (tenselessly) timeless, then it is true that God is timeless now, and that clearly contradicts the second conjunct.

Of course, we could avoid the contradiction if we could restate the first conjunct as past tense. But we could not say that 'God is timeless before creation,' or that 'God's timeless phase is earlier than his temporal phase,' since a timeless entity neither possesses A-properties nor stands in B-relations. So how are the two phases of God's existence to be related? Craig offers an analogy from cosmology. In the standard Big Bang model, the initial singularity is not 'in' space-time, but instead is at the boundary of space-time (see the discussion in Chapter 3). Similarly, Craig suggests, 'the envisioned state is a boundary of time which is causally, but not temporally, prior to the origin of the universe.'[28] But I can't see how this helps avoid the contradiction—certainly not on a causal theory of dynamic time such as the one I have defended.

Given the opaque nature of the relation between the two phases of God's existence, we may well ask if Craig has given a coherent view. It seems that God both is and is not in different mental states, is and is not temporally relating to is creatures, is and is not engaged in the eternal act of creation.... The confusion only grows deeper.

There are further difficulties with Craig's view. I argued in Chapter 9 that a timeless entity is necessarily changeless. I claimed that for x to change is for x to have a property P at t_1 that x does not have at t_2. But for this to be true, x must occupy a location in a B-series (that is, stand in a B-relation) such that the state of affairs x's-having-P-at-t_1 is earlier than the state of affairs x's-not-having-P-at-t_2. However, by my definition of atemporality, as well as by Craig's characterization of God's timeless phase, God would stand in no B-relations. Consequently he necessarily could not change.

I could, of course, be wrong about this; in fact, Craig argues that 'God existing timelessly sans creation must be de facto changeless, but there is no reason to think Him immutable.'[29] But intuitively, existing timelessly, and hence changelessly, seems to be at least coextensive with existing immutably. And intuitively, any change, whether intrinsic of extrinsic, seems to entail temporal

---

[27] Ibid., pp. 270-71.
[28] Ibid., p. 272.
[29] Ibid., p. 278.

location. So the burden is on Craig to articulate a sense of change according to which a timeless being could undergo change that is not temporal (that does not entail location in a temporal series).

But I'll grant, for the sake of argument, that de facto changelessness does not entail immutability. Still, how could such a changeless God bring about anything? The problem of atemporal causation, which I argued remains unsolved by atemporalists, will plague Craig as well.

It is tempting to charge that a concept as opaque as 'timeless sans creation, temporal since creation' is incoherent. But I can't demonstrate that Craig's view clearly entails a contradiction. So I'll simply note that if the temporalist takes this route to solving the problem of God's relation to the time of creation, she is biting a very large bullet indeed. Or so I say.

In addition to Swinburne's and Craig's solutions to the problem of God's relation to creation, I shall now suggest a third option. Why not simply deny that it is impossible for God to have endured through an infinite series of past moments? Suppose, as I suggested in the previous chapter, that metaphysical time, prior to creation, is constituted by the causal succession of God's mental states? This picture might be reinforced in Christian Trinitarianism. The doctrine of *perichoresis* (περιχώησις ὑποστάτων), the interpenetration, mutual indwelling, or co-inherence, of the persons of the Trinity, is often interpreted dynamically. For the Father, the Son, and the Holy Spirit, to enjoy dynamic relationality in any sense analogous to the relationality of human persons would be to experience change.

The barrier to this proposal, of course, is the Kalām-style argument that the past cannot be infinite. How does Craig support this thesis? His two central arguments are (i) the impossibility of an actual infinite, and (ii) the impossibility of forming an actual infinite by successive addition.[30] I have already discussed (i) and found it to be lacking as an objection to the existence of an actual infinity of abstract objects. But my preferred theory of dynamic time is causal, and so the states in God's mind the succession of which constitute metaphysical time cannot be abstract objects; they stand in causal relations and so are concrete. But God's mental states are not spatially extended, nor are they extended in physical time. Consequently the argument against an actual infinite series of past time must apply to moments of metaphysical time.

A *reductio* is commonly given to prove that an actual infinite cannot exist in reality. The assumption that an actual infinite can exist in reality is shown to generate absurdities.[31] Wes Morriston rejects the Kalām argument, and has argued that the kinds of thing which generate the absurdities involve 'a collection of coexistent objects. . . whose physical relationship to one another can be changed. It is only when these features are combined with the property of having infinitely

---

[30] Craig, *Kalām*; *GT&E*, p. 260.
[31] Examples are David Hilbert's mythical Hotel, or Craig's mythical Library. Both Hilbert's Hotel and Craig's Library appear frequently in the literature; for a classic statement, see Craig, *Kalām*, pp. 82-7.

many members that we get [absurdities].'[32] But in a recent defense of the Kalām argument, J.P. Moreland disagrees that the objects need to be coexistent, but further qualifies the relevant kind of objects: 'It is the fact that the elements of the infinite set are finite, contingent, moveable entities that can be added or subtracted and that exemplify the relevant sort of extension that generates the absurdities.'[33] And it is not at all clear that a causal succession of God's mental states would satisfy the conjunction of qualifications that either Morriston or Moreland find necessary. Clearly, further argument is needed

What about (ii)? Clearly, as stated, (ii) seems unassailable. But if (ii) is to be employed here, it needs an additional premise. The argument would be:

7. If the past is an actual infinite, it has been formed by successive addition
8. An actual infinite cannot be formed by successive addition.
9. Therefore, the past is not an actual infinite.

Premise (7) does not seem clearly to be true. That is, in forming something by successive addition, we begin somewhere. But the assertion that the past is infinite is the denial of that. So perhaps we should reformulate the argument in terms of successive subtraction. We might be tempted to think that there is no real change in the situation; if we begin at infinity and proceed by successive subtraction, we'll never arrive at 0.[34]

But that isn't what the claim of a past infinity amounts to. Both attempts so far have included the idea of beginning somewhere. But to say the past is infinite is to say there was no beginning of the series; we might imagine that God has always been counting down to t = 0 (the first moment of physical time). So even if an infinite series cannot be formed by successive addition *or* successive subtraction if we begin somewhere, it is not so clear that someone who has always been subtracting—that is, who never began to subtract—won't be able to complete the countdown.

Craig offers two additional considerations. The first, I will acknowledge, tells in favor of a finite past for God. Briefly, the first is Leibniz's question to Clarke: 'Why didn't God create the world sooner?'[35] I have shown above that this does not lead to a contradiction. Yet there is force to the objection, although not as strong as Craig claims when he concludes, 'Accordingly, the Leibnizian challenge seems to me to furnish a cogent and persuasive argument for thinking that the past is finite, God's idling away eternity, continually delaying His creation of the world

---

[32] Wes Morriston, 'Craig on the Actual Infinite,' *Religious Studies* 38 (2002), p. 148.
[33] See Moreland, 'A Response to a Platonistic and a Set-Theoretic Objection to the Kalām Cosmological Argument.'
[34] Graham Oppy considers this alternative: 'Time, Successive Addition, and *Kalām* Cosmological Arguments,' *Philosophia Christi* 2:3:1 (2001), pp. 185, 188.
[35] G. W. Leibniz, 'Mr. Leibniz's Third Paper,' §6, in *The Leibniz-Clarke Correspondence*, ed. H. G. Alexander (New York: Manchester University Press, 1956), pp. 26-7.

throughout infinite past time, seems to be an unintelligible conception.'[36] Whether the dynamic relationship within the Trinity described by *perichoresis* is sufficient to blunt the last charge will, of course, depend on the individual.

The last additional consideration Craig offers seems to me to be beside the point. He says:

> In some difficult to articulate way, it does seem unintelligible why, say, 2000 is present if the past is infinite. If time had a beginning, then a firm foothold is gained which makes 2000's being present more intelligible. For a certain number of years earlier than 2000, a certain first moment was present, and it makes no sense to ask of it at that time why it, as the first, was present.[37]

But surely this confuses physical time with metaphysical time! My claim is not that past physical time is infinite, but that past metaphysical time is. So this consideration has no force at all.

The temporalist thus has three clearly distinguished options regarding God's temporal status prior to (*sans*) creation: God's existence was a single, amorphous, temporal moment, or God's existence has two phases—one timeless *sans* creation, the other temporal with creation—or God's existence is from the infinite past in metaphysical time, although physical time did indeed have a first moment. Each view has its own difficulties. But I submit that the positive reasons for accepting divine temporality outweigh the costs of these views. The temporalist then is left having to decide which seems to be the smaller bullet to bite.

Providence

The conjunction of dynamic time and divine temporality offers a plausible conceptual scheme for understanding divine providence. The etymology of the term suggests foresight but, in John Calvin's memorable phrase, God's providence 'belongs no less to his hands than to his eyes.'[38] The Westminster Confession of 1647 describes the idea of providence as it has traditionally been conceived:

> God, the great Creator of all things, doth uphold, direct, dispose, and govern all creatures, actions and things, from the greatest even to the least, by his most wise and holy providence, according to his infallible foreknowledge, and the free and immutable counsel of his own will, to the praise of the glory of his wisdom, power, justice, goodness, and mercy.[39]

---

[36] Craig, *GT&E*, p. 265.
[37] Ibid., pp. 265-6.
[38] John Calvin, *Institutes of the Christian Religion*, tr. Henry Beveridge (Grand Rapids, MI: Eerdmans, 1970), I, xvi, 4.
[39] Cited in Paul Helm, *The Providence of God* (Downers Grove, IL: InterVarsity Press, 1994), p. 42.

Traditionally, providence has been seen as consisting in two, and possibly three, conceptually distinct aspects: sustenance, governance, and possibly concurrence.[40] While all three aspects can be explained within the framework of divine timelessness and static time, I suggest that all three take on a much more plausible, even more robust, character if divine omnitemporality is correct.

Divine Providence

God's interactions with the created order, or providence, has been considered under three aspects:
1. *Sustenance*: God's action by which he preserves all contingent things.
2. *Governance*: God's sovereign direction of the course of history for his own ends.
3. *Concurrence*: more controversial than the first two aspects, concurrence seeks to explain how God can be the first cause of all that occurs and still be free of causing sin.

Sustenance is the providential act of God by which he maintains or preserves all contingent things in existence. It is this aspect of God's action that is envisioned in the 'contingency' version of the Cosmological Argument in the tradition of Aristotle or Aquinas. There is nothing contradictory in the notion of an atemporal God timelessly causing the four-dimensional universe of static time. However, if causation is the explanation for the flow and direction of dynamic time, it is irresistible to see God's sustenance as the cause of the continued existence of the substantial space and the matter/energy that it contains, together with the properties and dispositions that enter into causal relations. Physical time itself is sustained by God, and as God's 'now' in metaphysical time is simultaneous with the 'now' of physical time, God sustains (present tense) the world at the present moment as it yields seamlessly to the next moment of time. Seen this way, the uniformity of the laws of nature, and the continuation of the flow of time, are due to the ongoing providential activity of a rational and good God. The dynamic implications of descriptions such as 'sustaining,' 'maintaining,' or 'preserving,' is not merely figurative.

Governance is the aspect of providence by which God achieves his purpose in the world.[41] Traditionally, this has been interpreted to mean that God providentially controls temporal circumstances and events to ensure that his ultimate will is accomplished. In some cases this might involve his exercising an overriding determination; in others, a non-determining influence; and in still others, non-intervention. While, again, there is nothing contradictory in the notion of an atemporal God timelessly governing a tenseless world, governance seems much more cogent if time is dynamic and God is temporal. On omnitemporality, God's governance is based on his foreknowledge or Middle Knowledge of all

---

[40] G. C. Berkouwer, *The Providence of God*, tr. Louis B. Smedes (Grand Rapids, MI: Eerdmans, 1952).
[41] Ibid., pp. 83-124.

possible futures, not his foreseeing of the actual future that already exists (as on the B-theory).[42]

The more controversial third aspect of providence, concurrence, has to do with the relation between divine and human activity, and developed specifically as an attempt to understand the relationship between God's action and human sin.[43] Often elucidated in terms of primary and secondary causes, concurrence strives to give an account of how God can be the first cause of the existence of all things (sustenance), including the power to act, and yet not himself be culpably involved in secondary causes and thus be 'guilty' of sin. In the Calvinist tradition, it is sometimes claimed that only through the notion of concurrence can a proper distinction between God and his creation be maintained. Pantheism identifies secondary causes with God, whereas deism divorces secondary causes from the first cause, God. Whether or not concurrence is a coherent notion is not my concern here. I would only observe that if it is, then like sustenance and governance, concurrence seems much more coherent on the assumption of dynamic time and divine omnitemporality than on static time and divine timelessness.

Certainly, other accounts of providence have been offered which rest on the notion of divine atemporality, perhaps most notably the Thomist account. However, all founder on the intractable problem of a timeless God's dynamic interactions with his temporal creation.[44] In short, if God's providence is to have the kind of meaning that its defenders claim for it, then it must be seen as a dynamic activity of God involving relations to tensed time. But we have seen in several places above that we can only make sense of God's relations with dynamic, temporal events if he himself is temporal. Thus omnitemporality can give a much more robust understanding of divine providence than can a timeless conception of God's being.

Petitionary Prayer

Anyone who desires to retain a strong sense of God's sovereignty, let alone of strong immutability or impassibility, must confess mystification as to how petitionary prayer can be efficacious. If we take both the biblical and the traditional teaching on prayer seriously, then it seems quite likely that in at least some circumstances, if we pray, A will come about, and if we do not, A will not. That is to say, it is within our power indirectly to bring about A's obtaining. On the static view of time, however, the future is real, so it is now the case that either

---

[42] Middle Knowledge has been discussed above and in Chapter 7; see also Thomas P. Flint, *Divine Providence: The Molinist Account* (Ithaca, NY: Cornell University Press, 1998). On simple foreknowledge, see David P. Hunt, 'Divine Providence and Simple Foreknowledge,' *Faith and Philosophy* 10 (1993), pp. 394-414, and the ensuing discussion in the same journal issue.
[43] Ibid., pp. 125-60.
[44] See discussion in Flint, *Divine Providence*, pp. 82-94.

A or not-A obtains. But if this is the case, then either prayer is only a formal feature of the timeless history of the world, or else its value is therapeutic and not efficacious. Omnitemporality gives a view in which God, being temporal, 'hears' prayers and so acts as to bring about the answers in the as-yet contingent future—effects which are genuinely contingent upon the prayer. If the biblical and traditional portrayal of the importance of prayer in the believer's life is to be taken at face value, then it seems that divine temporality offers the best explanation.

Conclusion

I argued in this book that if time is dynamic rather than static, then the traditional concept of a timeless God must be abandoned in favor of a temporal conception of God. Further, I argued that the antecedent of this conditional is true. To make sense of the consequent, then, I offered an account of omnitemporality as the preferred mode of God's temporal nature. But I have attempted to do more, in showing not only that omnitemporality is a consistent concept, but, beyond that, the theory of omnitemporality can provide better explanations of certain claims in philosophical theology than any of its atemporalist competitors. As a theory of God's temporal mode of being which takes seriously both the theory of dynamic time and the traditional concept of God, omnitemporality can be seen to have much to commend it.

But as I conclude this study, I must end with a disclaimer. It is one thing to have confidence in philosophical arguments, but quite another to claim confidence that what one ascribes to God is wholly accurate. The Judeo-Christian tradition has uniformly maintained that there is much more to God than is accessible to our reason, or even than is revealed to us. So let me insist that the assertions and conclusions I have reached concerning God are offered in a spirit of humility and reverence. In short, the previous pages have been an exercise in 'faith seeking understanding,' not one in reason dictating to faith. My sincere desire is that, as time passes, God will allow me to see and correct errors as I come to a better understanding of our world and of him.

Bibliography

Achtner, Wolfgang, Stefan Kunz and Thomas Walter. *Dimensions of Time: The Structures of the Time of Humans, of the World, and of God.* Tr. Arthur H. Williams, Jr. Grand Rapids, MI: Eerdmans, 2002.

Adams, Marilyn McCord. 'The Problem of God's Foreknowledge and Free Will in Boethius and William Ockham.' Unpublished Ph.D. dissertation, Cornell University, 1967.

_____. 'Introduction.' In *William Ockham, Predestination, God's Foreknowledge, and Future Contingents*, 2nd edition. Tr. Marilyn Adams and Norman Kretzmann. Indianapolis: Hackett Publishing, 1983.

_____. *William Ockham*. 2 volumes. Notre Dame, IN: University of Notre Dame Press, 1987.

Adams, Robert M. 'Middle Knowledge and the Problem of Evil.' *American Philosophical Quarterly* 14 (1977): 109-17.

Akhundov, Murad D. *Conceptions of Space and Time: Sources, Evolution, Directions.* Tr. Charles Rougle. Cambridge, MA: The MIT Press, 1986.

Alexander, H. G., ed. *The Leibniz-Clarke Correspondence.* New York: Manchester University Press, 1956.

Alston, William P. 'Does God Have Beliefs?' In *Divine Nature and Human Language: Essays in Philosophical Theology.* Ithaca, NY: Cornell University Press, 1989.

Anselm, St. *Saint Anselm: Basic Writings*, 2nd edition. Tr. S. W. Deane. La Salle, IL: Open Court, 1962.

Aquinas, St Thomas. *Summa Theologica.* Tr. The Fathers of the English Dominican Province. New York: Benzinger Brothers, 1948.

_____. *Summa Contra Gentiles*, University of Notre Dame edition. Tr. Anton C. Pegis. Notre Dame, IN: University of Notre Dame Press, 1975.

Aristotle. *The Physics.* Loeb Classical Library, tr. Philip H. Wicksteed and Francis M. Comfort. New York: G. P. Putnam's Sons, 1929.

Armstrong, D. M. *What is a Law of Nature?* Cambridge: Cambridge University Press, 1983.

_____. *A World of States of Affairs.* Cambridge: Cambridge University Press, 1997.

Augustine, St. *The City of God.* Tr. Marcus Dods. New York: Random House, 1950.

_____. *De Libero arbitrio.* In *Augustine: Earlier Writings.* Tr. John H.S. Burleigh. Philadelphia: Westminster Press, 1953.

_____. *De vera religione.* In *Augustine: Earlier Writings.* Tr. John H.S. Burleigh. Philadelphia: Westminster Press, 1953.

_____. *Confessions.* Tr. R. S. Pine-Coffin. London: Penguin Books, 1961.

_____. *De Trinitate.* Tr. A.W. Hadden. In *A Select Library of Nicene and Post-Nicene Fathers*, ed. Phillip Schaff. Grand Rapids, MI: Eerdmans, 1980.

_____. *De Genesis ad litteram.* Tr. John Hammond Taylor, S.J. In *Ancient Christian Writers.* Vol. 41. New York: Newman Press, 1982.

Banner, Michael C. *The Justification of Science and the Rationality of Religious Belief.* Oxford: Oxford University Press, 1990.

Barr, James. *The Semantics of Biblical Language.* Oxford: Oxford University Press, 1961.

_____. *Biblical Words for Time*, revised edition. Naperville, IL: Alec R. Allenson, 1969.
Barrow, John D. *The World within the World*. New York: Clarendon Press, 1988.
Barrow, John D. *Theories of Everything*. London: Vintage, 1991.
Basinger, David. *The Case for Freewill Theism: A Philosophical Assessment*. Downers Grove, IL: InterVarsity Press, 1996.
Bealer, George. 'Predication and Matter.' *Synthese* 31 (1975): 493-508.
_____. *Quality and Concept*. New York: Oxford University Press, 1982.
_____. 'Mental Properties.' *Journal of Philosophy* 91 (1984): 185-208.
Bergson, Henri. *Duration and Simultaneity*. Tr. Leon Jacobson. Indianapolis: Bobbs-Merrill, 1965.
Berkouwer, G. C. *The Providence of God*. Tr. Louis B. Smedes. Grand Rapids, MI: Eerdmans, 1952.
Bloesch, Donald G. *God the Almighty*. Volume 3 of *Christian Foundations*. Downers Grove, IL: InterVarsity Press, 1995.
Blount, Douglas K. 'On the Incarnation of a Timeless God.' In *God and Time*, ed. Gregory E. Ganssle and David M. Woodruff. New York: Oxford University Press, 2002.
Boethius. *De Trinitate*. Tr. H. F. Stewart and E. K. Rand. In *Boethius: The Theological Tractates*, ed. H. F. Stewart. Cambridge, MA: Harvard University Press, 1946.
_____. *The Consolation of Philosophy*. Tr. Richard Green. Indianapolis: Bobbs-Merrill, 1962.
Bondi, Hermann. *Relativity and Common Sense: A New Approach to Einstein*. New York: Dover, 1964.
Boorstin, Daniel J. *The Discoverers: A History of Man's Search to Know His World and Himself.* New York: Random House, 1983.
Brown, Francis, S. R. Driver and Charles A. Briggs, eds. *A Herbrew and English Lexicon of the Old Testament*. Oxford: Clarendon Press, 1907.
Bruce, F. F. *Tradition: Old and New*. Grand Rapids, MI: Zondervan, 1970.
Callahan, John F. *Four Views of Time in Ancient Philosophy*, revised edition. Cambridge, MA: Harvard University Press, 1979.
Calvin, John. *Institutes of the Christian Religion*. Tr. Henry Beveridge. Grand Rapids, MI: Eerdmans, 1970.
Carter, William S., and H. Scott Hestevold. 'On Passage and Persistence.' *American Philosophical Quarterly* 31 (1994): 269-84.
Chalmers, David J. *The Conscious Mind: In Search of a Fundamental Theory*. New York: Oxford University Press, 1996.
Chisholm, Robert B. Jr. 'Does God "Change His Mind"?' *Bibliotheca Sacra* 152 (1995): 387-99.
Chisholm, Roderick. *On Metaphysics*. Minneapolis: University of Minnesota Press, 1989.
Cleland, Carol. 'On the Individuation of Events.' *Synthese* 86 (1991): 229-54.
Connor, Steven. *Postmodern Culture: An Introduction to Theories of the Contemporary*. Cambridge, MA: Blackwell, 1989.
Cook, Robert R. 'God, Time and Freedom.' *Religious Studies* 23 (1987): 81-94.
Craig, William Lane. 'Was Thomas Aquinas a B-Theorist of Time?' *New Scholasticism* 59 (1985): 475-83.
_____. *The Only Wise God*. Grand Rapids, MI: Baker, 1987.
_____. 'God, Time, and Eternity.' *Religious Studies* 14 (1987): 497-503.
_____. 'John Duns Scotus on God's Foreknowledge and Future Contingents.' *Franciscan Studies* 25 (1987): 98-122.

_____. *The Problem of Divine Foreknowledge and Future Continents from Aristotle to Suarez*. Leiden: E. J. Brill, 1988.

_____. 'William Ockham on Divine Foreknowledge and Future Contingency.' *Pacific Philosophical Quarterly* 69 (1988): 117-35.

_____. '"What Place, Then, for a Creator?": Hawking on God and Creation.' *British Journal for the Philosophy of Science* 41 (1990): 229-34.

_____. 'God and Real Time.' *Religious Studies* 26 (1990): 335-47.

_____. *Divine Foreknowledge and Human Freedom*. Leiden: E. J. Brill, 1991.

_____. 'Time and Infinity.' *International Philosophical Quarterly* 31 (1991): 387-401.

_____. 'Graham Oppy on the Kalām Cosmological Argument.' *Sophia* 32 (1993): 1-11.

_____. 'Prof. Grünbaum on Creation.' *Erkenntnis* 40 (1994): 325-41.

_____. 'The Special Theory of Relativity and Theories of Divine Eternity.' *Faith and Philosophy* 11 (1994): 19-37.

_____. 'Tense and the New B-Theory of Language.' *Philosophy* 71 (1996): 5-26.

_____. 'Timelessness and Creation.' *Australasian Journal of Philosophy* 74 (1996): 646-56.

_____. 'Divine Timelessness and Necessary Existence.' *International Philosophical Quarterly* 37 (1997): 217-24.

_____. 'Hartle-Hawking Cosmology and Atheism.' *Analysis* 57 (1997): 291-5.

_____. *The Kalām Cosmological Argument*. New York: Macmillan, 1979. Reprint Eugene, OR: Wipf and Stock, 2000.

_____. *The Tensed Theory of Time: A Critical Examination*. Dordrecht: Kluwer Academic Publishers, 2000.

_____. *The Tenseless Theory of Time: A Critical Examination*. Dordrecht: Kluwer Academic Publishers, 2000.

_____. *God, Time, and Eternity: The Coherence of Theism II: Eternity*. Dordrecht: Kluwer Academic Publishers, 2001.

_____. 'Middle Knowledge, Truth-Makers, and the Grounding Objection.' *Faith and Philosophy* 18 (2001): 337-52.

_____. 'Middle Knowledge.' In *Four Views on Divine Knowledge*, ed. James Beilby and Paul Eddy. Downers Grove, IL: InterVarsity Press, 2001.

_____. *Time and Eternity: Exploring God's Relationship to Time*. Wheaton, IL: Crossway Books, 2001.

_____. *Time and the Metaphysics of Relativity*. Philosophical Studies Series 84. Dordrecht: Kluwer Academic Publishers, 2001.

Craig, William Lane and Quentin Smith. *Theism, Atheism, and Big Bang Cosmology*. Oxford: Clarendon Press, 1993.

Cross, Richard. 'Duns Scotus on Eternity and Timelessness.' *Faith and Philosophy* 14 (1997): 3-25.

Cullmann, Oscar. *Christ and Time: The Primitive Christian Conception of Time and History*. Tr. Floyd V. Filson. Philadelphia: Westminster Press, 1950.

Danto, Arthur C. 'Basic Actions.' *American Philosophical Quarterly* 2 (1965): 141-48.

_____. *Connections to the World: The Basic Concepts of Philosophy*. Berkeley: University of California Press, 1989.

Davidson, Donald. 'The Individuation of Events.' In *Actions and Events*, ed. Ernest LePore and Brian P. McLaughlin. New York: Blackwell, 1985.

Davies, Paul. *About Time: Einstein's Unfinished Revolution*. New York: Simon and Schuster, 1995.

Davies, P. C. W. 'Time and Reality,' in *Reduction, Time and Reality: Studies in the Philosophy of the Natural Sciences*, ed. Richard Healey. Cambridge: Cambridge University Press, 1981.

Davis, Stephen T. *Logic and the Nature of God*. Grand Rapids, MI: Eerdmans, 1983.

Davis, Stephen T., Daniel Kendall and Gerald O'Collins, eds., *The Incarnation: An Interdisciplinary Symposium on the Incarnation of the Son of God*. Oxford: Oxford University Press, 2002.

Deltete, Robert J. and A. Guy Reed. 'Hartle-Hawking Cosmology and Unconditional Probabilities.' *Analysis* 57 (1997): 304-15.

De Vries, Simon J. *Yesterday, Today and Tomorrow: Time and History in the Old Testament*. Grand Rapids, MI: Eerdmans, 1975.

DeWeese, Garrett. 'Timeless God, Tenseless Time.' *Philosophia Christi* 2:2:1 (2000): 53-9.

_____. 'Atemporal, Sempiternal, or Omnitemporal: God's Temporal Mode of Being.' In *God and Time*, ed. Gregory E. Ganssle and David M. Woodruff. New York: Oxford University Press, 2002.

Duns Scotus, John. *A Treatise on God as First Principle*, 2nd edition. Tr. and ed. Allan B. Wolter. Chicago: Franciscan Herald Press, 1966.

_____. *Contingency and Freedom: Lectura I 39*. Tr. A. Vos Jaczn., H. Veldhuis, A.H. Looman-Graaskamp, E. Dekker, and N.W. den Bok. Dordrecht: Kluwer, 1994.

Earman, John. 'An Attempt to Add a Little Direction to "The Problem of the Direction of Time".' *Philosophy of Science* 41 (1974): 15-47.

_____. *World Enough and Space-Time: Absolute versus Relational Theories of Space and Time*. Cambridge, MA: MIT Press, 1989.

_____. *Bangs, Crunches, Whimpers, Shreiks: Singularities and Acausalities in Relativistic Spacetimes*. New York: Oxford University Press, 1995.

Eddington, Sir Arthur S. *The Nature of the Physical World*. Cambridge: Cambridge University Press, 1928.

Einstein, Albert. *Relativity: The Special and the General Theory*. Tr. Robert W. Lawson. New York: Bonanza Books, 1961.

Eliade, Mircea. *The Myth of the Eternal Return, or, Cosmos and History*. Tr. Willard R. Trask. Princeton, NJ: Princeton University Press, 1971.

Ellis, George F. R., and Ruth M. Williams. *Flat and Curved Space-Times*. Oxford: Clarendon Press, 1988.

Ferguson, Kitty. *The Fire in the Equations*. Grand Rapids, MI: Eerdmans, 1994.

Field, Hartry H. *Science without Numbers: A Defence of Nominalism*. Princeton, NJ: Princeton University Press, 1980.

Fitzgerald, Paul. 'The Truth About Tomorrow's Sea Fight.' *Journal of Philosophy* 66 (1969): 307-29

_____. 'Stump and Kretzmann on Time and Eternity.' *Journal of Philosophy* 82 (1985): 260-69.

Flint, Thomas P. *Divine Providence: The Molinist Account*. Ithaca, NY: Cornell University Press, 1998.

Freddoso, Alfred J. 'Accidental Necessity and Logical Determinism.' *Journal of Philosophy* 80 (1983): 257-78.

_____. 'The Necessity of Nature.' *Midwest Studies in Philosophy* 11 (1986): 215-42.

_____. 'Introduction.' In Luis de Molina, *On Divine Foreknowledge (Part IV of the Concordia)*. Ithaca, NY: Cornell University Press, 1988.

_____. Review of William Hasker's *God, Time and Knowledge*. *Faith and Philosophy* 10 (1993): 99-107.

_____. 'The "Openness" of God: A Reply to William Hasker.' *Christian Scholars Review* 28 (1998): 124-33.

_____. 'St. Thomas's Rejection of Ockham's Way Out.' Unpublished paper, <http://www.nd.edu/~afreddos/papers/futcon.htm>.

Friedman, Michael. *Foundations of Space-Time Theories: Relativistic Physics and Philosophy of Science*. Princeton, NJ: Princeton University Press, 1983.

Friedman, William. *About Time*. Cambridge, MA: MIT Press, 1990.

Gale, Richard M., ed. *The Philosophy of Time*. New Jersey: Humanities Press, 1978.

_____. 'Omniscience-Immutability Arguments.' *American Philosophical Quarterly* 23 (1986): 319-35.

Ganssle, Gregory. 'Atemporality and the Mode of Divine Knowledge.' *International Journal for Philosophy of Religion* 34 (1993): 171-80.

Ganssle, Gregory E., ed. *God and Time: Four Views*. Downers Grove, IL: InterVarsity Press, 2001.

Ganssle, Gregory E. and David M. Woodruff, eds. *God and Time: Essays on the Divine Nature*, New York: Oxford University Press, 2002.

Geisler, Norman and H. Wayne House. *The Battle for God: Responding to the Challenge of Neotheism*. Grand Rapids, MI: Kregel, 2001.

Gennaro, Rocco J. 'Kant versus Lewis on the Singularity of Space and Time.' *History of Philosophy Quarterly* 11 (1994): 205-18.

Gesenius, W. and E. Kautzsch, *Hebrew Grammar*. 2nd English edition. Tr. A. E. Cowley. Oxford: Clarendon Press, 1910.

Girdlestone, Robert B. *Synonyms of the Old Testament*. Grand Rapids, MI: Eerdmans, 1974.

Grünbaum, Adolf. *Philosophical Problems of Space and Time*, 2nd edition. Dordrecht: Reidel, 1973.

_____. 'The Status of Temporal Becoming.' In *The Philosophy of Time*, ed. Richard M. Gale. New Jersey: Humanities Press, 1978.

Harris, C. R. S. *Duns Scotus. Vol. 2: The Philosophical Doctrines of Duns Scotus*. New York: Humanities Press, 1959.

Harris, James F. *Against Relativism: A Philosophical Defense of Method*. La Salle, IL: Open Court, 1992.

Hartle, J. and S. W. Hawking. 'Wave Function of the Universe.' *Physical Review D* 28 (1983): 2960-75.

Hasker, William. *Metaphysics: Constructing a World View*. Downers Grove, IL: InterVarsity Press, 1983.

_____. 'Reply to Basinger on Power Entailment.' *Faith and Philosophy* 5 (1988): 87-90.

_____. 'Yes, God Has Beliefs!' *Religious Studies* 24 (1988): 385-94.

_____. *God, Time, and Knowledge*. Ithaca, NY: Cornell University Press, 1989.

Hawking, Stephen, and Roger Penrose. *The Nature of Space and Time*. Princeton: Princeton University Press, 1996.

Hawking, Stephen W. *A Brief History of Time*. New York: Bantam, 1988.

_____. 'Classical Theory.' In Stephen Hawking and Roger Penrose, *The Nature of Space and Time*. Princeton, NJ: Princeton University Press, 1996.

Healey, Richard, ed. *Reduction, Time and Reality: Studies in the Philosophy of the Natural Sciences*. Cambridge: Cambridge University Press, 1981.

Helm, Paul. 'Timelessness and Foreknowledge.' *Mind* 84 (1975): 516-27.

_____. 'God and Spacelessness.' In *Contemporary Philosophy of Religion*, ed. Steven M. Cahn and David Shatz. Oxford: Oxford University Press, 1982.

_____. *Eternal God*. New York: Oxford University Press, 1988.
_____. *The Providence of God*. Downers Grove, IL: InterVarsity Press, 1994.
_____. 'Eternal Creation.' *Tyndale Bulletin* 45 (1994): 321-38.
_____. *Faith and Understanding*. Grand Rapids, MI: Eerdmans, 1997.
_____. 'Divine Timeless Eternity.' In *God and Time: Four Views*, ed. Gregory E. Ganssle. Downers Grove, IL: InterVarsity Press, 2001.
Hirsch, Eli. *Dividing Reality*. New York: Oxford University Press, 1993.
Hoefer, Carl. 'The Metaphysics of Space-Time Substantivalism.' *Journal of Philosophy* 93 (1996): 5-27.
Hoffman, Joshua, and Gary S. Rosenkrantz. *Substance: Its Nature and Existence*. New York: Routledge, 1997.
Holt, Dennis C. 'Timelessness and the Metaphysics of Temporal Existence.' *American Philosophical Quarterly* 18 (1981): 149-56.
Horgan, Terrence E. 'Reduction, Reductionism.' In *A Companion to Metaphysics*, ed. Jaegwon Kim and Ernest Sosa. Cambridge, MA: Basil Blackwell, 1995.
Horwich, Paul. 'On the Existence of Time, Space and Space-Time.' *Noûs* 12 (1978): 397-419.
_____. *Asymmetries in Time*. Cambridge, MA: MIT Press, 1987.
Hudson, Hud. *A Materialist Metaphysics of the Human Person*. Ithaca, NY: Cornell University Press, 2001.
Hume, David. *A Treatise of Human Nature*, 2nd edition. Ed. P. W. Nidditch. Oxford: Clarendon Press, 1978.
Hunt, David P. 'Divine Providence and Simple Foreknowledge.' *Faith and Philosophy* 10 (1993): 394-414.
Kant, Immanuel. *Critique of Pure Reason*, unabridged edition. Tr. Norman Kemp Smith. New York: St Martin's Press, 1929.
Kaplan, David. 'Thoughts on Demonstratives.' In *Demonstratives*, ed. Palle Yourgrau. Oxford: Oxford University Press, 1990.
Kenny, Anthony. 'Divine Foreknowledge and Human Freedom.' In *Aquinas: A Collection of Critical Essays*, ed. Anthony Kenny. Garden City, NY: Anchor Books, 1969.
_____. *The God of the Philosophers*. Oxford: Clarendon Press, 1979.
Kim, Jaegwon. 'Causation, Nomic Subsumption, and the Concept of Event.' In *Supervenience and Mind: Selected Philosophical Essays*. New York: Cambridge University Press, 1993. Originally published in the *Journal of Philosophy* 70 (1973): 217-36.
King, Peter J. 'Other Times.' *Australasian Journal of Philosophy* 73 (1995): 532-47.
Klein, G. L. 'The "Prophetic Perfect".' *Journal of Northwest Semitic Languages* 16 (1990): 45-60.
Kneale, William. 'Time and Eternity in Theology.' *Proceedings of the Aristotelian Society* 61 (1960-61): 87-108.
Knight, George A. F. *A Christian Theology of the Old Testament*. Richmond, VA: John Knox Press, 1959.
Kosso, Peter. *Appearance and Reality: An Introduction to the Philosophy of Physics*. New York: Oxford University Press, 1998.
Kretzmann, Norman. 'Omniscience and Immutability.' *Journal of Philosophy* 63 (1966): 409-21.
Kvanvig, Jonathan L. and Hugh J. McCann. 'Divine Conservation and the Persistence of the World.' In *Divine and Human Action: Essays in the Metaphysics of Theism*, ed. Thomas V. Morris. Ithaca, NY: Cornell University Press, 1988.

Layzer, David. *Cosmogenesis: The Growth of Order in the Universe.* New York: Oxford University Press, 1990.

Leftow, Brian. 'Time, Actuality and Omniscience.' *Religious Studies* 26 (1990): 303-21.

———. *Time and Eternity.* Ithaca, NY: Cornell University Press, 1991.

———. 'Why Didn't God Create the World Sooner?' *Religious Studies* 27 (1991): 157-72.

———. 'Timelessness and Foreknowledge.' *Philosophical Studies* 63 (1991): 309-25.

———. 'Eternity and Simultaneity.' *Faith and Philosophy* 8 (1991): 148-79.

———. 'A Timeless God Incarnate.' In *The Incarnation*, ed. Stephen T. Davis, Daniel Kendall and Gerald O'Collins. Oxford: Oxford University Press, 2002.

Leftow, Ian. 'Timelessness and Divine Experience.' *Sophia* (1992): 43-53.

Leggett, Anthony. 'Time's Flow and the Quantum Measurement Problem,' in *Time's Arrow Today: Recent Physical and Philosophical Work on the Direction of Time*, ed. Steven F. Savitt. Cambridge: Cambridge University Press, 1995.

Lemmon, John. 'Comments on D. Davidson's "The Logical Form of Action Sentences".' In *The Logic of Decision and Action*, ed. Nicholas Rescher. Pittsburgh: University of Pittsburgh Press, 1966.

Le Poidevin, Robin. Review of *Time, Tense and Causation*, by Michael Tooley. *The British Journal for the Philosophy of Science* 49 (1998): 365-9.

Le Poidevin, Robin, and Murray MacBeath, eds. *The Philosophy of Time.* Oxford: Oxford University Press, 1993.

Lewis, C. S. *The Screwtape Letters.* New York: Macmillan, 1982.

Lewis, Delmas. 'Eternity Again.' *International Journal for Philosophy of Religion* 15 (1984): 73-79.

———. 'Persons, Morality, and Tenselessness.' *Philosophy and Phenomenological Research* 48 (1986): 305-9.

———. 'Eternity, Time and Tenselessness.' *Faith and Philosophy* 5 (1988): 72-86.

Lieb, Irwin C. *Past, Present, and Future: A Philosophical Essay About Time.* Urbana, IL: University of Illinois Press, 1991.

Ljungberg, Bo-Krister. 'Tense, Aspect, and Modality in Some Theories of the Biblical Hebrew Verbal System.' *Journal of Translation and Textlinguistics* 7:3 (1995): 82-96.

Loux, Michael J. *Metaphysics: A Contemporary Introduction.* New York: Routledge, 1998.

Lowe, E. J. 'The Metaphysics of Abstract Objects.' *Journal of Philosophy* 92 (1995): 509-24.

———. *The Possibility of Metaphysics: Substance, Identity, and Time.* Oxford: Clarendon Press, 1998.

Lucas, J. R. *A Treatise on Space and Time.* London: Methuen, 1973.

Lyons, John. *Introduction to Theoretical Linguistics.* Cambridge: Cambridge University Press, 1971.

MacBeath, Murray. 'Time's Square.' In *The Philosophy of Time*, ed. Robin Le Poidevin and Murray MacBeath. Oxford: Oxford University Press, 1993.

McCall, Storrs. 'Objective Time Flow.' *Philosophy of Science* 43 (1976): 337-62.

———. 'Counterfactuals Based on Real Possible Worlds.' *Noûs* 18 (1984): 463-77.

———. *A Model of the Universe: Space-Time, Probability, and Decision.* Oxford: Clarendon Press, 1994.

McTaggart, J. M. E. 'Time.' Chapter 33 of *The Nature of Existence.* Cambridge: Cambridge University Press, 1927. Reprinted as 'The Unreality of Time.' In *The Philosophy of Time*, ed. Robin Le Poidevin and Murry MacBeath. Oxford: Oxford University Press, 1993.

Malament, David. 'Causal Theories of Time and the Conventionality of Simultaneity.' *Noûs* 11 (1977): 293-300.
Mayo, Bernard. 'Professor Smart on Temporal Asymmetry.' *Australasian Journal of Philosophy* 33 (1955): 38-44.
Mellor, D.H. 'Special Relativity and Present Truth.' *Analysis* 34 (1974): 74-77.
_____. *Real Time*. Cambridge: Cambridge University Press, 1981.
_____. *The Facts of Causation*. New York: Routledge, 1995.
Merricks, Trenton. 'On the Incompatibility of Enduring and Perduring Entities.' *Mind* 104 (1995): 523-31.
_____. *Objects and Persons*. Oxford: Clarendon Press, 2001.
Miller, Ed. L. *God and Reason*, 2nd edition. Englewood Cliffs, NJ: Prentice Hall, 1995.
Molina, Luis de. *On Divine Foreknowledge (Part IV of the Concordia)*. Tr. Alfred J. Freddoso. Ithaca, NY: Cornell University Press, 1988.
Moreland, J. P. 'A Response to a Platonistic and a Set-Theoretic Objection to the Kalām Cosmological Argument.' *Religious Studies* 39 (2003): 373-90.
Morris, Thomas V. *Our Idea of God: An Introduction to Philosophical Theology*. Downers Grove, IL: InterVarsity Press, 1991.
Morriston, Wes. 'Must the Beginning of the Universe Have a Personal Cause?' *Faith and Philosophy* 16 (2000): 149-69.
_____. 'Craig on the Actual Infinite.' *Religious Studies* 38 (2002): 147-66.
Murphy, Nancey. *Theology in the Age of Scientific Reasoning*. Ithaca, NY: Cornell University Press, 1990.
New International Dictionary of New Testament Theology, ed. Colin Brown. 3 volumes. Grand Rapids, MI: Eerdmans, 1967-73.
Newman, Andrew. *The Physical Basis of Predication*. Cambridge: Cambridge University Press, 1992.
Newton-Smith, W. H., *The Structure of Time*. London: Routledge and Kegan Paul, 1980.
Nodland, Borge. 'A Glimpse of Cosmic Anisotropy.' Unpublished paper, <http://www.cc.rochester.edu/ college/rtc/Borge/overview.html>.
Noonan, Harold. *Personal Identity*. London: Routledge, 1989.
Normore, Calvin. 'Future Contingents.' In *The Cambridge History of Later Medieval Philosophy*, ed. Norman Kretzmann, Anthony Kenny and Jan Pinborg. Cambridge: Cambridge University Press, 1982.
Novikov, Igor D. *The River of Time*. Cambridge: Cambridge University Press, 1998.
Oaklander, L. Nathan and Quentin Smith, eds. *The New Theory of Time*. New Haven, CT: Yale University Press, 1994.
Oakes, Robert. 'Temporality and Divinity: An Analytic Hurdle.' *Sophia* 31 (1992): 11-26.
Ockham, William. *Predestination, God's Foreknowledge, and Future Contingents*. Tr. Marilyn Adams and Norman Kretzmann. 2nd edition. Indianapolis: Hackett Publishing, 1983.
Oddie, Graham. 'Backwards Causation and the Permanence of the Past.' *Synthese* 85 (1990): 71-93.
Oddie, Graham and Roy W. Perrett. 'Simultaneity and God's Timelessness.' *Sophia* 31 (1992): 123-7.
Ogden, G. S. 'Time, and the Verb היה in O.T. Prose.' *Vetus Testamentum* 21 (1971): 451-69.
Oppy, Graham. 'Time, Successive Addition, and *Kalām* Cosmological Arguments.' *Philosophia Christi* 2.3.1 (2001): 181-91.

Orelli, Conrad von. *Die hebräischen Synonyma der Zeit und Ewigkeit genetisch und sprachvergleichend dargestellt.* Leipzig: A. Lorentz, 1871.
Padgett, Alan G. 'God and Time: Toward a New Doctrine of Divine Timeless Eternity.' *Religious Studies* 25 (1989): 209-15.
_____. 'Can History Measure Eternity? A Reply to William Craig.' *Religious Studies* 27 (1991): 333-35.
_____. *God, Eternity and the Nature of Time.* New York: St Martin's Press, 1992.
Parunak, H. Van Dyke. 'A Semantic Survey of *NHM*.' *Biblica* 56 (1975): 512-32.
Penrose, Roger. 'Singularities and Time-Asymmetry.' In *General Relativity: An Einstein Centenary Survey*, ed. S. W. Hawking and W. Israel. Cambridge: Cambridge University Press, 1979.
_____. 'Gravity and State Vector Reductions.' In *Quantum Concepts in Space and Time*, ed. R. Penrose and C. Isham. Oxford: Clarendon Press, 1986.
Perry, John, ed. *Personal Identity.* Berkeley: University of California Press, 1975.
Perry, John. 'Frege on Demonstratives.' In *Demonstratives*, ed. Palle Yourgrau. Oxford: Oxford University Press, 1990.
Pike, Nelson. *God and Timelessness.* London: Routledge & Kegan Paul, 1970.
Pinnock, Clark H. *Most Moved Mover.* Grand Rapids, MI: Baker, 2001.
Pinnock Clark, Richard Rice, John Sanders, William Hasker, and David Basinger. *The Openness of God.* Downers Grove, IL: InterVarsity Press, 1994.
Plass, Paul C. 'Timeless Time in Neoplatonism.' *The Modern Schoolman* 55 (1977): 1-19.
Plantinga, Alvin. *The Nature of Necessity.* Oxford: Clarendon Press, 1974.
_____. 'On Ockham's Way Out.' *Faith and Philosophy* 3 (1986): 235-69.
_____. 'On Heresy, Mind and Truth.' *Faith and Philosophy* 16 (1999): 183-93.
Plato, *Timaeus.* In *Readings in Ancient Greek Philosophy: From Thales to* Aristotle, ed. S. Marc Cohen, Patricia Curd, and C. D. C. Reeve, tr. D. J. Zeyl. Indianapolis, IN: Hackett, 1995.
_____. *Phaedo.* In *Readings in Ancient Greek Philosophy: From Thales to* Aristotle, ed. S. Marc Cohen, Patricia Curd, and C. D. C. Reeve, tr. G. M. A. Grube. Indianapolis, IN: Hackett, 1995.
Plumer, Gilbert. 'Mustn't Whatever is Referred To Exist?' *Southern Journal of Philosophy* 27 (1989): 511-28.
Plumber, Gilbert. 'A Here-Now Theory of Indexicality.' *Journal of Philosophical Research* 18 (1993): 193-211.
Popper, Karl. 'The Arrow of Time.' *Nature* 177 (1956): 538.
Popper, Karl N. *Quantum Theory and the Schism in Physics*, from *Postscript to the Logic of Scientific Discovery*, ed. W. W. Bartley, III. New York: Routledge, 1982.
Price, Huw. *Time's Arrow and Archimedes' Point: New Directions for the Physics of Time.* Oxford: Oxford University Press, 1996.
Prior, A. N. 'Thank Goodness That's Over.' *Philosophy* 34 (1959): 12-17.
_____. 'The Formalities of Omniscience.' *Philosophy* 37 (1962): 119-29. Reprinted in A. N. Prior, *Papers on Time and Tense.* Oxford: Oxford University Press, 1968.
Purtill, Richard L. 'Foreknowledge and Fatalism.' *Religious Studies* 10 (1974): 319-24.
Putnam, Hilary. 'Time and Physical Geometry.' *Journal of Philosophy* 64 (1967): 240-47.
Quinn, John M. 'Time.' In *Augustine Through the Ages: An Encyclopedia.* Ed. Allan D. Fitzgerald. Grand Rapids, MI: Eerdmans, 1999.
Quinn, Philip L. 'On the Mereology of Boethian Eternity.' *International Journal for the Philosophy of Religion* 32 (1992): 51-60.
Reichenbach, Hans. *The Direction of Time*, ed. Maria Reichenbach. Berkeley: University of California Press, 1956.

_____. *The Philosophy of Space and Time.* Tr. Maria Reichenbach and John Freund. New York: Dover, 1958.

Rietdijk, C. W. 'A Rigorous Proof of Determinism Derived from the Special Theory of Relativity.' *Philosophy of Science* 33 (1966): 341-44.

Rogers, Katherine A. 'Eternity Has No Duration.' *Religious Studies* 30 (1994): 1-16.

Ross, James. 'Comments on "Absolute Simplicity".' *Faith and Philosophy* 2 (1985): 383-91.

Rudavsky, T. M. *Time Matters: Time, Creation, and Cosmology in Medieval Jewish Philosophy.* Albany, NY: State University of New York Press, 2000.

Russell, Bertrand. *A History of Western Philosophy*, 2nd edition. New York: Simon and Schuster, 1972.

_____. *An Inquiry into Meaning and Truth.* London: George Allen and Unwin, 1940.

Russell, Robert C. 'T=0: Is it Theologically Significant?' In *Religion and Science: History, Method, Dialogue*, ed. W. Mark Richardson and Wesley J. Wildman. New York: Routledge, 1996.

Saunders, Simon. 'Time, Quantum Mechanics, and Decoherence.' *Synthese* 102 (1995): 235-66.

Savitt, Steven F., ed. *Time's Arrows Today: Recent Physical and Philosophical Work on the Direction of Time.* Cambridge: Cambridge University Press, 1995.

Schlesinger, George N. *Aspects of Time.* Indianapolis: Hackett, 1980.

_____. *Timely Topics.* New York: St Martin's Press, 1994.

Schrödinger, Erwin. *Space-Time Structure.* Cambridge: Cambridge University Press, 1950.

Seddon, Keith. *Time.* London: Methuen, 1987.

Searle, John. *The Construction of Social Reality.* New York: Free Press, 1995.

Senor, Thomas D. 'Divine Temporality and Creation Ex Nihilo.' *Faith and Philosophy* 10 (1993): 86-92.

_____. 'Incarnation, Timelessness, and Leibniz's Law Problems.' In *God and Time*, ed. Gregory E. Ganssle and David M. Woodruff. New York: Oxford University Press, 2002.

Shapiro, Herman. *Motion, Time and Place According to William Ockham.* New York: The Franciscan Institute, 1957.

Shoemaker. Sydney. 'Time without Change.' *Journal of Philosophy* 66 (1969): 363-81. Reprinted in *The Philosophy of Time*, ed. Robin Le Poidevin and Murry MacBeath. Oxford: Oxford University Press, 1993.

_____. 'Causality and Properties.' In *Time and Cause*, ed. P. van Inwagen. Dordrecht: D. Reidel, 1980. Reprinted in *Properties*, ed. D. H. Mellor and Alex Oliver. Oxford: Oxford University Press, 1997.

Sider, Theodore. *Four Dimensionalism: An Ontology of Persistence and Time.* Oxford: Clarendon Press, 2001.

Silva, M. *Biblical Words and their Meaning: An Introduction to Lexical Semantics.* Grand Rapids, MI: Zondervan, 1983.

Simons, John. 'Eternity, Omniscience and Temporal Passage: A Defence of Classical Theism.' *Review of Metaphysics* 42 (1989): 547-68.

Sklar, Lawrence. 'Time, Reality, and Relativity.' In *Reduction, Time and Reality*, ed. R. Healey. Cambridge: Cambridge University Press, 1981.

_____. *Physics and Chance: Philosophical Issues in the Foundations of Statistical Mechanics.* Cambridge: Cambridge University Press, 1993.

_____. 'Time in Experience and in Theoretical Description of the World.' In *Time's Arrows Today: Recent Physical and Philosophical Work on the Direction of Time*, ed. Steven F. Savitt. Cambridge: Cambridge University Press, 1995.

Smart, J. J. C. 'The Temporal Asymmetry of the World.' *Analysis* 14 (1954): 79-82.

_____. 'Mr. Mayo on Temporal Asymmetry.' *Australasian Journal of Philosophy* 33 (1955): 124-27.
_____. 'Time and Becoming.' In *Time and Cause*, ed. Peter van Inwagen. Boston: Reidel, 1980.
_____. 'Problems with the New Tenseless Theory of Time.' *Philosophical Studies* 52 (1987): 77-98.
_____. 'The Reality of the Future.' In *Essays Metaphysical and Moral*. New York: Basil Blackwell, 1987.
Smith, Quentin. 'The Multiple Uses of Indexicals.' *Synthesse* 78 (1989): 167-91.
_____. 'A New Typology of Temporal and Atemporal Permanence.' *Noûs* 23 (1989): 307-30.
_____. *Language and Time*. Oxford: Oxford University Press, 1993.
_____. 'Quantum Cosmology's Implication of Atheism.' *Analysis* 57 (1997):295-304.
_____. Review of *Time, Tense and Causation*, by Michael Tooley. *The Philosophical Review* 108 (1999): 123-7.
Smoot, G. F. et al. 'Preliminary Results from the COBE Differential Microwave Radiometers: Large Angular Scale Isotropy of the Cosmic Microwave Background,.' *Astrophysical Journal Letters* 371 (1991): L1-5
Soccio, Douglas J. *Archetypes of Wisdom: An Introduction to Philosophy*. Belmont, CA: Wadsworth, 1992.
Sokolowski, Robert. *Introduction to Phenomenology*. New York: Cambridge University Press, 2000.
Sorabji, Richard. *Time, Creation and the Continuum: Theories in Antiquity and the Early Middle Ages*. Ithaca, NY: Cornell University Press, 1983.
Stein, Howard. 'On Einstein-Minkowski Space-Time.' *Journal of Philosophy* 65 (1968): 5-23.
Stewart, Ian. *Does God Play Dice? The New Mathematics of Chaos*, 2nd edition. Malden, MA: Blackwell, 1997.
Strawson, Peter F. *Individuals*. London: University Paperbacks, 1979.
Stump, Eleonore, and Norman Kretzmann. 'Eternity.' *Journal of Philosophy* 78 (1981): 429-58.
_____. 'Absolute Simplicity.' *Faith and Philosophy* 2 (1985): 353-82.
_____. 'Simplicity Made Plainer: A Reply to Ross.' *Faith and Philosophy* 4 (1987): 198-201.
_____. 'Eternity, Awareness and Action.' *Faith and Philosophy* 9 (1992): 463-82.
Swinburne, Richard. 'Times.' *Analysis* 26 (1965): 185-91.
_____. *Space and Time*, 2nd edition. London: Macmillan, 1981.
_____. 'Tensed Facts.' *American Philosophical Quarterly* 27 (1990): 117-30.
_____. *The Coherence of Theism*, revised edition. Oxford: Clarendon Press, 1993.
_____. 'God and Time.' In *Reasoned Faith*, ed. Eleonore Stump. Ithaca, NY: Cornell University Press, 1993.
_____. *The Christian God*. Oxford: Clarendon Press, 1994.
Theological Dictionary of the New Testament, ed. Gerhard Kittel. Tr. Geoffrey W. Bromiley. 10 volumes. Grand Rapids, MI: Eerdmans, 1964 *seq*.
Theological Dictionary of the Old Testament, ed. G. Johannes Botterweck and Helmer Ringgren. 7 volumes. Grand Rapids, MI: Eerdmans, 1977 *seq*.
Theological Wordbook of the Old Testament, ed. R. Laird Harris, Gleason L. Archer, and Bruce K. Waltke. 2 volumes. Chicago: Moody Press, 1980.
Thorne, Kip. *Black Holes and Time Warps*. New York: W. W. Norton, 1994.

Tipler, Frank J. 'The Many-Worlds Interpretation of Quantum Mechanics in Quantum Cosmology.' In *Quantum Concepts in Space and Time*, ed. R. Penrose and C. Isham. Oxford: Clarendon Press, 1986.
Tomkinson, J. L. 'Divine Sempiternity and Atemporality.' *Religious Studies* 18 (1982): 177-89.
Tooley, Michael. *Causation: A Realist Approach*. Oxford: Clarendon Press, 1987.
_____. 'Causation: Reductionism versus Realism.' *Philosophy and Phenomenological Research*, suppl. 50 (1990): 215-36.
_____. 'The Nature of Causation: A Singularist Account.' *Canadian Journal of Philosophy*, suppl. 16 (1990): 271-322.
_____. *Time, Tense and Causation*. Oxford: Oxford University Press, 1997.
_____. 'Freedom and Foreknowledge.' *Faith and Philosophy* 17 (2002): 212-24.
Trench, Richard C. *Synonyms of the New Testament*. Grand Rapids, MI: Eerdmans, 1953.
Trusted, Jennifer. *Physics and Metaphysics: Theories of Space and Time*. New York: Routledge, 1991.
Van Frassen, Bas C. *Laws and Symmetry*. Oxford: Clarendon Press, 1989.
_____. *Quantum Mechanics: An Empiricist View*, Oxford: Clarendon Press, 1991.
van Inwagen, Peter. 'Four-Dimensional Objects.' *Noûs* 24 (1990): 245-55.
_____. *Material Objects*. Ithaca, NY: Cornell University Press, 1990.
Vos Jaczn., A, H. Veldhuis, A. F. Looman-Graaskamp, E. Dekker and N.W. den Bok. 'Introduction' and 'Commentary.' In *John Duns Scotus: Contingency and Freedom: Lectura I 39*, tr. A. Vos Jaczn. et al. Doordrecht: Kluwer, 1994.
Wang, Hao. 'Time in Philosophy and in Physics from Kant and Einstein to Gödel.' *Synthese* 102 (1995): 215-34.
Ware, Bruce. 'An Evangelical Reformulation of the Doctrine of the Immutability of God.' *Journal of the Evangelical Theological Society* 29 (1986): 431-46.
_____. *God's Lesser Glory: The Diminished God of Open Theism*. Wheaton, IL: Crossway, 2000.
Wetzel, James. 'Time after Augustine.' *Religious Studies* 31 (1995): 341-47.
Weingard, Robert. 'Relativity and the Reality of Past and Future Events.' *British Journal for the Philosophy of Science* 23 (1972): 119-21.
Whitrow, G. J. *The Natural Philosophy of Time*, 2nd edition. Oxford: Clarendon Press, 1980.
Widerker, David. 'A Problem for the Eternity Solution.' *International Journal for the Philosophy of Religion* 29 (1991): 87-95.
_____. 'Providence, Eternity, and Human Freedom.' *Faith and Philosophy* 11 (1994): 242-54.
Wieringa, Edward R. *The Nature of God: An Inquiry into Divine Attributes*. Ithaca, NY: Cornell University Press, 1989.
Williams, Donald C. 'The Myth of Passage.' *Journal of Philosophy* 48 (1951). Reprinted in *The Philosophy of Time*, ed. Richard M. Gale. New Jersey: Humanities Press, 1978.
Winnie, John A. 'Special Relativity without One-Way Velocity Assumptions.' *Philosophy of Science* 37 (1970): 81-99; 223-38.
Wolter, Alan B. *The Philosophical Theology of John Duns Scotus*, ed. Marilyn McCord Adams. Ithaca, NY: Cornell University Press, 1990.
Wolterstorff, Nicholas. 'God Everlasting.' In *Contemporary Philosophy of Religion*, ed. Steven M. Cahn and David Shatz. Oxford: Oxford University Press, 1982. Originally published in *God and the Good: Essays in Honor of Henry Stob*, ed. Clifton J. Orlebeke and Louis B. Smedes. Grand Rapids, MI: Eerdmans, 1975.

_____. *Reason within the Bounds of Religion*, 2nd edition. Grand Rapids, MI: Eerdmans, 1984.

_____. 'Unqualified Divine Temporality.' In *God and Time: Four Views*, ed. Gregory E. Ganssle. Downers Grove, IL: InterVarsity Press, 2001.

Wood, David. *The Deconstruction of Time*. Atlantic Heights, NJ: Humanities Press International, 1989.

Zagzebski, Linda. 'Individual Essence and the Creation.' In *Divine and Human Action: Essays in the Metaphysics of Theism*, ed. Thomas V. Morris. Ithaca, NY: Cornell University Press, 1988.

Zagzebski, Linda Trinkaus. *The Dilemma of Freedom and Foreknowledge*. New York: Oxford University Press, 1991.

Index

Adams, Marilyn McCord, 143, 194, 195, 196, 197
Adams, Robert M., 202
aion (αἰών), 94, 102, 108, 131
Alston, William P., 180
Ammonius, 141, 143
Anselm, St, 3, 5, 145-51, 154, 155, 156, 158, 168, 185, 259
Aristotle, 15, 101, 116, 117, 135, 141, 142, 152, 187-196 passim, 274
A-series, 38, 133, 140, 159, 211, 231
'attâ (עַתָּה), 94, 98
atemporality, 2, 3, 8, 96, 104, 110, 111-57, 159-84, 185, 205, 206, 207, 211, 215, 217, 221, 232, 233, 237, 245-8, 249, 260, 272
A-theory, 16, 17, 26, 29, 34, 45, 68, 76, 116, 172, 181, 229-31
Augustine, St, 1, 2, 8, 34, 111-32, 137, 143, 144, 150-56 passim, 187, 191, 197, 200, 221, 263, 264, 265
Aquinas, St Thomas, 3, 8, 142, 143, 144, 151-7, 161, 181, 182, 185, 192, 203, 204, 205, 222, 235, 239, 240, 257, 272

Barr, James, 94, 95, 96, 101, 104, 105, 109-10
Bealer, George, 23, 249
Big Bang, 73, 79, 82, 85, 176, 243, 251, 270
Bloesch, Donald, 210, 259
Boethius, 3, 115, 133-45, 149, 151, 153-6 passim, 158, 181, 185, 186, 205
B-series, 28, 32, 121, 140, 169-75 passim, 249, 270
B-theory, 16, 17, 32, 35, 45, 156-65, 172, 184, 185, 215, 217, 229, 230, 262, 275

Callahan, John F., 102, 118, 127
Causation, 7, 18-20, 33, 36-53, 59, 63, 76, 78, 80, 83, 84, 87, 88, 159, 162, 193, 197, 214, 222, 223, 225, 226, 232, 234, 244, 253, 254, 269, 271, 274
Chisholm, Robert B., 107
chronos (χρόνος), 103-4, 108, 117
Craig, William Lane, viii, ix, 18, 21, 28, 29, 35, 39, 55, 56, 59, 66, 71, 72, 73, 75, 81, 85, 86, 127, 129, 133, 138, 142, 143, 152, 154, 157, 166, 176, 180, 185, 187, 192, 193, 194, 197, 198, 202, 206, 216, 217, 228, 229-32, 246, 261, 263 264, 267-73
Cullmann, Oscar, 101, 103

Davies, Paul, 9, 37, 44, 46, 66, 74
Davis, Stephen, 166, 180, 209, 213-15, 232, 234, 248, 249
de Vries, Simon, 95, 98, 99, 101
Derrida, Jacques, 22
DeWeese, Garrett J., 179, 244, 248
Duns Scotus, John, 3, 8, 185, 186-94, 197, 200, 201, 202, 203, 205, 208, 235

Earman, John, 39, 48, 56, 66, 75, 76, 78, 80, 81, 82, 88, 240
Einstein, Albert, 9, 10, 15, 44, 45, 65, 66, 68-79 passim, 124, 252
Eliade, Mircea, 101
Ellis, George F. R., 66, 67, 7578
entropy, 43-47
'ēt (עֵת), 97-8, 100, 106
ET-simultaneity, 135, 137, 160-66, 168, 170

fatalism, 100, 106, 133, 142, 152-4, 185, 207, 219, 262
foreknowledge, 1, 3, 4, 8, 115, 116, 127, 133, 134, 142-56 passim, 178, 185, 187, 191, 194, 197, 200, 201, 202, 204, 206, 217, 218, 219, 235, 255, 262, 263, 273, 274
Freddoso, Alfred J., 6, 191, 194, 199, 200, 202, 205, 206, 207, 219, 220, 264

Frege, Gottlob, 23, 26, 27
Friedman, Michael, 66, 85

Gale, Richard M., 34, 37, 120, 179
Ganssle, Gregory, xi, 176, 180, 210, 233, 242, 250, 256
God
   immutability, 1, 3, 8, 106-8, 111, 113, 114, 128, 129, 130, 133, 139, 151, 155, 158, 159, 177, 185, 191, 210, 257-62, 269, 271, 275
   omnipotence, 5, 210, 222, 259, 260
   omniscience, 1, 179, 180, 259, 263, 264
   providence, 1, 8, 144, 207, 255, 265, 273-5
   simplicity, 1, 3, 8, 111, 114, 128, 129, 130, 132, 133, 138, 139, 146-7, 149, 150, 151, 154, 155, 158, 159 167, 175, 181, 185, 191, 193, 210, 255, 257-60ü
Gödel, Kurt, 77, 78, 120
Grünbaum, Adolf, 34, 112, 120-21, 226
GTR. *See* Relativity, General Theory of

Harris, James F., 22
Hasker, William, 6, 115, 133, 164, 166, 180, 202, 209, 217-20, 261
Hawking, Stephen, 46, 66, 79, 81, 82, 85, 86, 87
Helm, Paul, 5, 159, 175-79, 181, 184, 240, 247, 273
Hilbert, David, 124-5, 271
Hirsch, Eli, 17, 23
Horwich, Paul, 33
Hudson, Hud, 16, 39, 215
Hunt, David P., 275

Immutability. *See* God, immutability
Incarnation, 1, 232-5, 255
indexicals, 21, 25-29, 30, 32, 34, 98, 99, 173, 180, 189, 190, 198, 210, 221, 230, 238, 240, 243, 254

kairos (καιρός), 94, 103, 104
Kalām Cosmological Argument, 176, 214, 224, 231, 266, 268, 269-72
Kant, Immanuel, 49, 50, 95, 112, 120-6, 133, 245

Kaplan, David, 27
Kenny, Anthony, 153, 161, 164, 186
Kim, Jaegwon, 17, 163
Klein, G. L., 24, 101
Kneale, William, 2, 138, 246
Knight, George A. F., 100
Kretzmann, Norman, 130, 135-9 passim, 142, 151, 159-66, 168, 170, 175, 179, 181, 184, 186, 195, 213, 230, 233, 254, 258, 259

Language and Time, 20-29
Le Poidevin, Robin, 16, 42-3, 58, 72
Leftow, Brian, 58, 60, 114, 127, 128, 129, 137-9, 143, 145, 150, 154, 155, 159, 166, 167-75, 177, 178, 179, 181, 183, 184, 228, 230, 234, 246, 254, 256, 259, 267
Leibniz, Gottfried Willhelm, 122, 124, 234, 272
Lewis, C. S., 32
Lowe, E. J., 24, 234
Lucas, J. R., 58, 62, 120, 240, 267

MacBeath, Murray, 16, 58
McCall, Storrs, 61, 62, 83, 121, 157, 251
McTaggart, J. M. E., 16, 30, 226
McTaggart's Paradox, 4, 30
Mellor, D. H., 20, 24, 25, 27-34, 40, 41, 49, 227
Merricks, Trenton, 16, 137, 216
Middle Knowledge, 3, 185, 202, 204, 206-7, 218, 219, 264, 274
Miller, Ed. L., ix, 5
Minkowski manifold, ix, 35, 60, 71, 75, 76, 80, 121, 244, 246
mô'ēd (מאֵד), 94, 99, 100
Molina, Luis de, 3, 8, 185, 200, 202-8, 218, 264
Moreland, J. P., ix, 269, 272
Morriston, Wes, ix, 214, 271-2

Neoplatonism, 2, 3, 8, 103, 104, 111, 114, 127-30, 132, 133, 134, 139-42, 146, 158, 167, 185, 219, 221, 235, 258, 259, 260
Newman, Andrew, 23, 40
Newton, Sir Isaac, 39, 51, 56, 124, 197

Newtonian space and time, 65, 71, 81, 141, 231,
nhm (נָחַם), 106, 107, 110
Novikov, Igor D., 37
nētsakh (נֵצַח), 94, 97

Oaklander, L. Nathan, 21, 25
Ockham, William of, 3, 8, 185, 194-200, 201, 202, 203, 208, 235, 263, 264
'ôlām (עוֹלָם), 94, 96, 97, 105
omnitemporality, 8, 229, 232, 237–55, 257-76
Orelli, Conrad von, 95
Origen, 2, 131

Padgett, Alan, viii, 176, 209, 217, 225-9, 265
Parmenides, 131, 139, 140, 141, 142, 145, 187-9
Parunak, H. Van Dyke, 106
Penrose, Roger, 46, 59, 66, 79, 81, 86
Perry, John, 27
Philo of Alexandria, 2, 101, 103, 131
Physics. *See* Relativity, Special and General Theories; and Quantum Mechanics
Pike, Nelson, 233, 247
Plantinga, Alvin, 194, 202, 234, 264
Plato, 2, 3, 101, 102, 103, 127, 129, 131, 136, 139, 140, 141, 142
Plotinus, 2, 3, 112, 127-31 passim, 139, 140-42 passim.
prayer, 1, 8, 162, 255, 265, 275-6
Prior, Arthur N., 26, 32, 179
Proclus, 141

Quantum Mechanics, 4, 7, 15, 46, 59, 65, 72, 81, 84-88, 125, 147
Quine, Willard van Orman, 21, 23

Reichenbach, Hans, 37, 44, 45
Relativity
   Einsteinian interpretations, 10, 20, 66, 70-75, 174, 184, 246, 252
   General Theory of, 4, 7, 9, 10, 15, 65, 66, 75-83, 124, 174, 238
   Lorentzian interpretations, 68, 71-3, 257

   Special Theory of, 4, 7, 15, 48, 60, 65-75, 76, 83, 84, 160, 164, 168, 174, 184, 222, 226, 228, 230, 246, 252
Rudavsky, Tamar, 5
Russell, Bertrand, 22, 122

Schlesinger, George N., 32, 38-9
Scotus. *See* Duns Scotus, John
Searle, John, 23
sempiternity, 8, 96, 136, 239, 243, 250-52
Senor, Thomas D., 232-35
Sider, Theodore, 16
simultaneity, 10, 48, 52, 59, 66-73, 83, 85, 150, 170, 252
   relativity of, 10, 18, 48, 66-9, 160, 164, 174, 184, 227, 250
Sklar, Lawrence, 45, 46, 69, 252
Smart, J. J. C., 29, 33, 46, 68, 227
Smith, Quentin, 18, 21, 25, 26, 49, 59, 73, 85, 176, 246
Sorabji, Richard, 2, 55, 102, 120, 122, 127, 131, 139, 140, 141, 142, 145, 179
STR. *See* Relativity, Special Theory of
Stump, Eleonore, 130, 135-8, 150, 160-67, 168, 170, 175, 179, 181, 184, 230, 254, 258, 259
Swinburne, Richard, 2, 20, 25, 37, 51, 58, 60, 130, 139, 209, 217, 220-24, 226, 228, 237, 239, 250, 251, 252, 253, 258, 260, 265, 267, 269

temporal becoming, 4, 15, 34-6, 37, 42, 43, 112, 120, 121, 133, 180, 191, 194, 196, 197, 200, 235
temporality, 3, 8, 17, 22, 58, 102, 107, 150, 159, 177-8, 185-208, 209-35, 239, 243-47, 250, 253, 255, 257-76
Tertullian, viii, 6
time
   causal theory of, 7, 18, 20, 32, 36-63, 65, 68, 75, 80, 81, 82, 83, 84, 85, 86, 87, 88, 133, 162, 164, 173, 197, 220, 221, 222, 232, 247, 270, 271
   cosmic, 10, 68, 73, 74, 231
   direction of, 7, 37, 43-52, 63, 77, 88

dynamic, 3, 4, 7, 15-63, 65-89 passim, 112, 145, 151, 179, 182, 184, 185, 186, 191, 194, 203-8 passim, 210-235 passim, 240, 241, 245, 250, 253, 255, 256, 259, 260, 261, 262, 263, 268, 269, 271, 272, 273, 274
experience of, 29-35
flow of, 37-43
metaphysical, 10-11, 45, 60, 68, 80, 206, 228, 230, 242, 243, 252, 253, 254, 265-72 passim
metric of, 9, 10, 36, 52-3, 59, 63, 69, 71, 80, 220, 223, 224, 227, 228, 231, 243, 253, 267, 268
personal, 10
physical, 9, 10, 11, 53, 60, 68, 80, 151, 176, 206, 209, 223, 224, 227, 228, 230, 231, 232, 242, 243, 245, 248, 251, 252, 253,254, 262, 265, 266, 267, 271, 272, 273, 274
static, 2, 4, 7, 8, 15-36 passim, 37, 53, 66, 84, 88, 94, 111, 116, 121, 126, 134, 138, 139, 140-45, 156, 157, 159, 164-5, 179-84, 185, 190, 194, 215, 216, 218, 226, 230, 265, 274, 275
tensed theory of, 16-18, 20-31, 137, 167, 180, 183, 184, 193, 206, 207, 210, 212, 262, 275
tenseless theory of, 8, 16, 20-31, 33, 38, 42, 137, 138, 167, 179, 180, 181, 183, 184, 210, 212, 215, 216, 219, 261
topology of, 53-62
Tooley, Michael, ix, 17, 18, 19, 20, 21, 26, 28, 30, 33, 36, 37, 40, 41, 42, 44, 47, 51, 52, 55, 56, 72, 73, 173, 245, 246, 254, 264

Van Inwagen, Peter, 16, 41, 58, 138

Watts, Isaac, ix
Whitrow, G. J., 9, 37
Wieringa, Edward, 130, 259
Williams, Ruth M., 66, 75, 78
Wolterstorff, Nicholas, 93, 179, 210-13, 214, 215, 234, 258, 261, 267
Wood, David, 22, 120
wormholes, 76, 82

yôm (יוֹם), 94, 98, 99

Zagzebski, Linda Trinkaus, 19, 132, 144, 166, 191, 194, 197, 200, 202, 264